Scorpionates

The Coordination Chemistry of
Polypyrazolylborate Ligands

Scorpionates

The Coordination Chemistry of Polypyrazolylborate Ligands

Swiatoslaw Trofimenko
University of Delaware, Newark, DE, USA

Imperial College Press

Published by

Imperial College Press
57 Shelton Street
Covent Garden
London WC2H 9HE

Distributed by

World Scientific Publishing Co. Pte. Ltd.
5 Toh Tuck Link, Singapore 596224
USA office: 27 Warren Street, Suite 401-402, Hackensack, NJ 07601
UK office: 57 Shelton Street, Covent Garden, London WC2H 9HE

British Library Cataloguing-in-Publication Data
A catalogue record for this book is available from the British Library.

First published 1999
Reprinted 2005

SCORPIONATES — THE COORDINATION CHEMISTRY OF POLYPYRAZOLYLBORATE LIGANDS

Copyright © 1999 by Imperial College Press

All rights reserved. This book, or parts thereof, may not be reproduced in any form or by any means, electronic or mechanical, including photocopying, recording or any information storage and retrieval system now known or to be invented, without written permission from the Publisher.

For photocopying of material in this volume, please pay a copying fee through the Copyright Clearance Center, Inc., 222 Rosewood Drive, Danvers, MA 01923, USA. In this case permission to photocopy is not required from the publisher.

ISBN 1-86094-172-9

Printed in Singapore by World Scientific Printers (S) Pte Ltd

Preface

Whenever I think of poly(pyrazolyl)borates I think of Swiatoslaw, better known to his close friends as Jerry Trofimenko. Whenever one makes a new member of this group of compounds, or does some chemistry with their complexes, one can be sure of hearing from him: a request for a reprint or some other expression of real interest. This is one of the joys of working in an active and growing field which has such an energetic and imaginative creator.

This group of remarkable compounds is, in a sense, Jerry's scientific "baby". It is more than 30 years since he synthesized and described them in a series of landmark papers in the Journal of the American Chemical Society, and it still seems something of a surprise to me that they took quite a long time to catch on, to "grow up". Maybe the reason had something to do with the transition metal chemists' obsession with, and the dominance of, organometallic chemistry in the 60s and 70s. Perhaps there was a subconscious feeling that "hard" N-donor atoms could not do much for metal-carbon bonds. How times have changed! Now, one can hardly pick up a major journal dealing with the chemistry of metals without finding at least one paper dealing with the consequences of using these ligands and their many derivatives. So there is no doubt that poly(pyrazolyl)borates have finally come to maturity!

The early coordination chemistry of these ligands was certainly influenced by organometallic thinking. The analogy between the cyclopentadienide ion and tris(pyrazolyl)borates and between β-diketonates and bis(pyrazolyl)borates was particularly useful. However, Jerry himself has commented many times that while this view undoubtedly had its bvenefits, and this was what attracted me to use the ligands in the first place, it is rather naive. Jerry's own introduction of the term *scorpionate* which he explains and develops in this book, conveys a more meaningful three-dimensional picture, although the notion of a "sting" is perhaps not quite what was in his mind.

Poly(pyrazolyl)borates now find uses in an extraordinarily wide range of chemistry, from modelling the active site of metallo-enzymes, through analytical chemistry and organic synthesis, to catalysis and materials science. Furthermore, the versatility of these ligands, their relative ease of synthesis, and the attendant important stereochemical consequences on coordination has stimulated other chemists to devise similar types of ligands but with different donor atoms, different charges and, ultimately different shapes. There can be no more powerful testimony to the significance and importance of Jerry's inventions.

This book represents the culmination of a professional career devoted, almost by accident, to synthetic chemistry. This makes Jerry's development of the field all the more remarkable since his employers at the time, the du Pont Company, did not formally contract him to prepare novel ligand systems! The book encapsulates all that is presently known about the poly(pyrazolyl)borate family of ligands and most certainly does not represent the decline of the field into old age. To paraphrase Winston Churchill, it represents not the beginning of the end of poly(pyrazolyl)borate chemistry, but rather the end of the beginning.

Jon McCleverty
Bristol, April 1999

Contents

1	**INTRODUCTION**		**1**
	1.1	General Considerations	1
	1.2	Polypyrazolylborates — Scorpionates	3
	1.3	The Abbreviation System for Scorpionate Ligands	5
	1.4	Comparison of the Tp and Cp Ligands	9
	1.5	Historical Development	10
	1.6	Reviews	12
	1.7	Synthesis of Scorpionate Ligands	13
	1.8	List of Known Scorpionate Ligands	18
	1.9	Analogs of Scorpionate Ligands	23
2	**HOMOSCORPIONATES — FIRST GENERATION**		**27**
	2.1	General Considerations	27
	2.2	Group 1: H, Li, Na, K, Rb and Cs	28
	2.3	Group 2: Be, Mg, Ca, Sr, Ba	30
	2.4	Group 3: Sc, Y	31
	2.5	Group 4: Ti, Zr, Hf	32
	2.6	Group 5: V, Nb, Ta	33

	2.7	Group 6: Cr, Mo, W	36
	2.8	Group 7: Mn, Tc, Re	61
	2.9	Group 8: Fe, R, Os	66
	2.10	Group 9: Co, Rh, Ir	74
	2.11	Group 10: Ni, Pd, Pt	82
	2.12	Group 11: Cu, Ag, Au	85
	2.13	Group 12: Zn, Cd, Hg	87
	2.14	Group 13: B, Al, Ga, In, Tl	89
	2.15	Group 14: C, Si, Ge, Sn, Pb	91
	2.16	Group 15: N, P, As, Sb, Bi	93
	2.17	Lanthanides	93
	2.18	Actinides	95

3 HOMOSCORPIONATES — SECOND GENERATION 99

    3.1   General Considerations 99

          3.1.1.   Regiochemistry in Ligand Synthesis 99

          3.1.2.   Quantification of Steric Effects in Tp^x Ligands 102

          3.1.3.   Ligand Rearrangments 107

    3.2   Individual Ligands 108

          3.2.1.   B-substituted Ligands, RTp^x ($R \neq pz^x$) 109

          3.2.2.   3-Monosubstituted Ligands, Tp^R 113

          3.2.3.   4-Monosubstituted Ligands, Tp^{4R} 131

CONTENTS

    3.2.4. 5-Monosubstituted Ligands, Tp^{5R} 132

    3.2.5. C-Disubstituted Ligands 133

        3.2.5.1   3,4-Disubstituted Ligands 134

        3.2.5.2a   3,5-Disubstituted Ligands of Type Tp^{R2} 137

        3.2.5.2b   3,5-Disubstituted Ligands of Type $Tp^{R,R'}$ 142

        3.2.5.3   4,5-Disubstituted Ligands 148

    3.2.6. Trisubstituted Ligands 150

4   HETEROSCORPIONATES, RR'Bp^x 155

  4.1 General Considerations 155

  4.2 Specific Ligands 157

    4.2.1. Bp^x Ligands 157

    4.2.2. R$_2$Bp^x Ligands 167

  4.3 Ligands R(R'Z)Bp^x 173

    4.3.1. General Considerations 173

    4.3.2. Specific Ligands 174

  4.4 Pyrazaboles $(= R_2(\mu\text{-}pz^x)_2BR'_2)$ 178

    4.4.1. Symmetrical pyrazaboles 179

    4.4.2 Unsymmetrical pyrazaboles 182

5	APPLICATIONS OF SCORPIONATE LIGANDS		183
	5.1	General Considerations	183
	5.2	Catalysis	183
		5.2.1 Polymerization and Oligomerization	183
		5.2.2 Carbene/Nitrene Transfer	187
		5.2.3 Oxidation	188
		5.2.4 Miscellaneous Catalysis	188
	5.3	Enzyme Modelling	188
		5.3.1 Vanadium	189
		5.3.2 Molybdenum	189
		5.3.3 Tungsten	190
		5.3.4 Manganese	191
		5.3.5 Iron	191
		5.3.6 Nickel	194
		5.3.7 Copper	194
		5.3.8 Zinc	196
	5.4	C—H Bond Activation	198
	5.5	Metal Deposition	204
	5.6	Metal Extraction	205
	5.7	Miscellaneous Studies	205
	5.8	Concluding Remarks	208
REFERENCES			**209**
INDEX			**275**

Chapter 1

Introduction

1.1 General Considerations

Polypyrazolylborates are today a well-established ligand system, known in the literature for over 32 years. They combine the features of a tetrasubstituted boron anion, with the donor atoms of two or more pyrazol-1-yl substituents (denoted as "pz") attached to the boron, and can be represented by the general formulas **1** for the parent system, and **2** for its substituted variants. In structure **2**, R is a non-pyrazolyl substituent (H, D, alkyl, aryl, substituted aryl, and F have been reported) and n can be 0, 1 or 2. Furthermore, the pyrazolyl rings may contain R^1, R^2, and R^3 substituents (alkyl, aryl, heteroaryl, halo, cyano, fused benzo- and naphtho-rings, etc.) in the 3-, 4-, and 5-positions, and these substituents may themselves contain additional donor sites. This gives rise to an enormous number of possible structural variations for polypyrazolylborate ligands, permitting their design with specific steric and electronic features, which define and control their coordination chemistry.

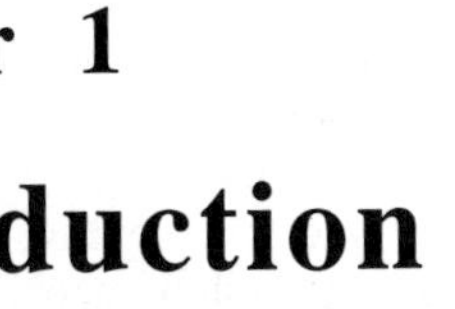

Polypyrazolylborate ligands have been generally used as molecular vises to keep the metal ion in a firm tridentate (C_{3v} symmetry) grip, so that chemical operations could be performed at the remaining coordination sites; as components of models for various enzymes, mimicking an array of three histidine N-donors; as components of catalyst systems for polymerization of alkenes and alkynes, and for other organic reactions; and as stabilizing groups to cap the corners of diverse metal clusters.

CHAPTER 1. INTRODUCTION

Since the first communication on polypyrazolylborates in 1966,[1] and especially since the introduction of their "second generation",[2,3] which provided steric coordination control via 3-substituents, a very substantial body of literature (over 1400 publications) has appeared, describing the synthesis and the application of polypyrazolylborate ligands in various areas of chemistry. Many ligand modifications have been reported, and by now over 170 different polypyrazolylborate ligands are known. Their complexes have been described for all elements of groups 1 through 13 (excluding Fr and Ra), for phosphorus, for all lanthanides (except for Pm), as well as for the actinides U, Th, Np and Pu. The popularity of polypyrazolylborate ligands seems to be growing. And yet, there has been no comprehensive source of information tying together the currently existing, but scattered, body of knowledge. The reviews that have appeared so far, covered only certain time periods, or were converned with specialized sub-areas, while the inorganic chemistry textbooks dealt with this subject very cursorily, if at all.

This book attempts to remedy the currently existing situation, by providing a comprehensive coverage of all aspects of polypyrazolylborate chemistry up to the end of 1998, also including a number of 1999 papers, so that it can be used as a convenient reference source, permitting the researcher to establish what has already been done and, just as importantly, which sub-areas are still relatively unexplored.

In view of the enormous number of complexes synthesized from polypyrazolylborate ligands, and the very diverse research goals in the context of which these complexes were prepared, organizing the material of this book necessitated extremely concise coverage of the various ligands and of their reactions. While the scheme that I have adopted may not be perfect, it seemed to be the most suitable one to achieve complete coverage of the information on hand.

The first chapter covers the general features of this ligand class, explains why the trivial name "scorpionates" is suitable for polypyrazolylborates, and includes an updated system for abbreviating these ligands, taking into account the most recent developments. It also offers a detailed comparison of the trispyrazolylborate and cyclopentadienide ligands, since they are generally regarded as "comparable" in their coordination chemistry. A fairly compact history of the development of polypyrazolylborate ligands, and of their coordination chemistry, is followed by a brief presentation of all the reviews in this area. Several general procedures for the synthesis of most types of polypyrazolylborate ligands are provided next, followed by a list of all the ppublished, and of some unpublished, ligands. Finally, a brief overview of related ligands, derived by replacing either pyrazole or boron with other moieties, is presented.

The second chapter is devoted to the coordination chemistry of the "first generation" ligands, $[HB(pz)_3]^-$, $[B(pz)_4]^-$ and $[HB(3,5Me_2pz)_3]^-$, which were the mainstay of the first 20 years of research in this area, and are still very extensively used even today. The simplest parent ligand, $[H_2B(pz)_2]^-$, is covered along with other

1.1 GENERAL CONSIDERATIONS

dihydrobis(pyrazol-1-yl)borates in Chapter four. The material is organized according to the groups of the periodic table. It includes non-metals, as well as the free acids $[H_nB(pz)_{4-n}]H$. Chapter three covers the tris- and tetrakispyrazolylborates containing substituents on the pyrazole ring, but excluding $[HB(3,5Me_2pz)_3]^-$. Included in it are the "second generation" ligands with bulky 3-substituents, and all other substituted ligands, even those reported in the early papers. The material is organized according to the ligand structure, that is, the type of substitution on the boron and on the pyrazolyl carbons. Ligand rearrangements of the type $[HB(3Rpz)_3]^- \rightarrow [HB(3Rpz)_2(5Rpz)]^-$ are also discussed.

Heteroscoropionates, defined as ligands having the structure $[RR'B(pz^x)_2]^-$, are dealt with in Chapter four. They can act in either bidentate or tridentate fashion, depending on the nature of the R and R' substituents on boron.

The fifth Chapter deals with specific applications of polypyrazolylborate ligands in areas such as modelling of enzymes in bioinorganic chemistry, their use as possible analytical reagents, their use in C—H bond activation (photolytic and thermal), and in catalysis. It also includes some physical studies to the extent that they go beyond simple techniques for structure determination, and other miscellaneous information which did not fit into any of the previous categories. The subjects are arranged thematically, and all ligand types are covered. The conclusion section summarizes the current status of this area, along with some tentative prognosis of future developments.

1.2 Polypyrazolylborates — Scorpionates

The defining feature of polypyrazolylborate complexes which are almost always at least bidentate, is the six-membered ring $B(\mu-pz)_2M$ within a more general structure $RR'B(\mu-pz)_2ML_n$ containing unspecified additional boron, pyrazolyl, and metal substituents. The $B(\mu-pz)_2M$ ring is in most instances in the boat form of varying depth, and this makes R and R' unequal: the pseudoequatorial R is pointing away from the metal roughly along the B-M axis, but the pseudoaxial R' is directed towards the metal, as shown in **3**:

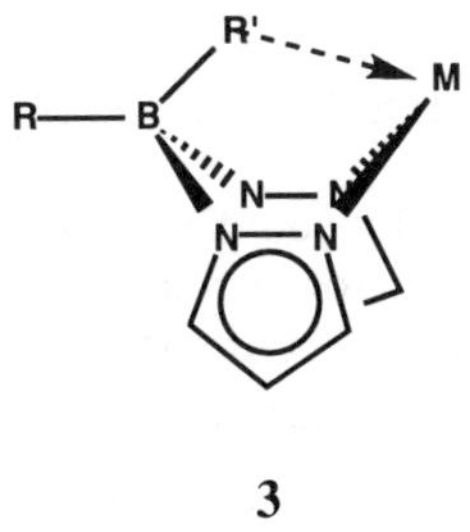

3

As a result of this reaching over the $B(\mu-pz)_2M$ ring, the R' substituent may form a full or partial bond to the metal, engage in agostic interaction, or simply screen the access to the metal for other potential ligands. This R' bridge may be either monoatomic (-H, -OR, -SR, -NMe$_2$) or diatomic (-pz, agostic C—H). It was this feature, unique at the time of their introduction, that prompted me to coin the term "scorpionates" for polypyrazolylborates, as the coordination behavior of the $[RR'B(\mu-pz)_2]^-$ ligands closely resembles the hunting habits of a scorpion: this creature grabs its prey with two claws (coordination of M through the two 2-N atoms of the $B(\mu-pz)_2$ groups), and then may, or may not, proceed to sting it with its overarching tail (the R' group). Thus, many aspects of scorpionate chemistry may be viewed as variations on the sting theme.

The nature of the boron substituents R and R' defines the ligand type further. For instance, R' may be another pyrazolyl group pz^x (pz^x = a pyrazol-1-yl group with unspecified substituents), identical to the two bridging pz^x groups, in which case the ligand is tridentate of C_{3v} symmetry, $[RB(pz^x)_3]^-$, and the "sting" becomes another "claw". Such ligands will be referred to as "homoscorpionates", and they are the most frequently used polypyrazolylborates. If neither R nor R' is pz^x, then the ligand is a heteroscorpionate. This category includes ligands where R and/or R' are H, alkyl, aryl, F, OR, SR, SAr, NMe$_2$, yielding complexes of type **4**, which may also involve additional bonding to M by the R' substituent. Heteroscorpionates also include ligands where R' is another pyrazolyl group, pz^y, different from pz^x either in the types of substituents, or in their regiochemistry (for instance, pz^x may be 3-R-pyrazol-1-yl, while pz^y would be 5-R-pyrazol-1-yl).

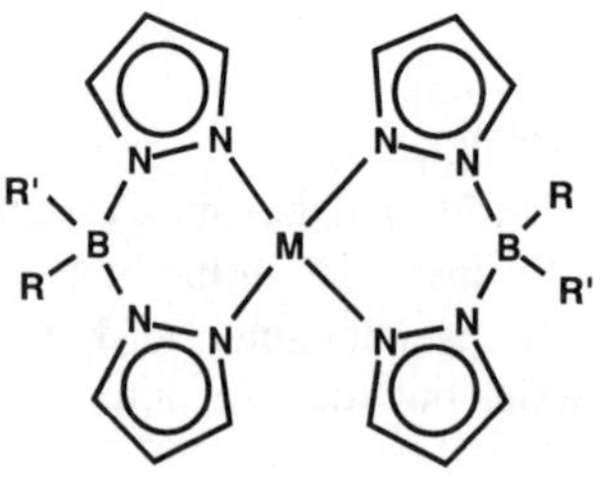

4 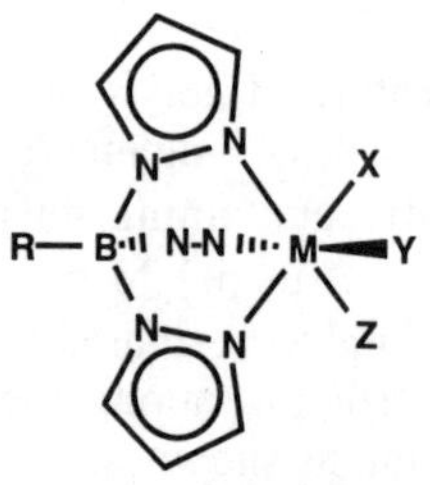   **5**
(N–N represents the third, hidden, μ-pz ring)

Homoscorpionates typically coordinate to the metal in tridentate fashion, as in the half-sandwich complex **5**, or in the octahedral homoleptic complex **6**. Nevertheless, there are instances where the $[RB(pz^x)_3]^-$ ligand coordinates only in bidentate fashion. This is often accompanied by rapid exchange of the coordinated and uncoordinated pz^x groups, observable on the NMR time scale. Conversely, a heteroscorpionate ligand may coordinate in tridentate fashion, not only in cases where R' is pz^y or a heteroatom, but even in cases where R' is a hydrogen or an alkyl group (agostic bonding).

6

When both boron substituents R and R' are pz^x groups, the $[B(pz^x)_4]^-$ ligand is still a homoscorpionate, as it usually coordinates through the three pz^x groups, with the fourth pz^x group remaining unattached to the metal. This happens with metals preferring octahedral coordination, e.g. cobalt(II). With zinc, the coordination may be octahedral in the crystal, but in solution all pz^x groups are identical by NMR, implying a tetrahedral structure with rapid exchange of the four pyrazolyl groups on the NMR time scale. The same thing happens in compounds such as $[B(pz)_4]Pd(\eta^3$-allyl).[4] Steric effects do play a role, however. Thus, the ligand $[B(3-Pr^ipz)_4]^-$ coordinates with Fe^{II}, Co^{II}, Ni^{II}, Cu^{II}, and Zn^{II} only in bidentate fashion, forming tetrahedral complexes, and without exchange of the coordinated and uncoordinated pz groups.[5]

1.3 The Abbreviation System for Scorpionate Ligands

Writing the systematic names for polypyrazolylborate ligands, such as "hydrotris(pyrazol-1-yl)borate" is cumbersome. While denoting them as $[HB(pz)_3]^-$ is more streamlined, a still better method of representing the trispyrazolylborate ligand system, proposed by Curtis,[6,7] is to use "Tp" for the $[HB(pz)_3]^-$ ligand, and "Tp*" for $[HB(3,5$-dimethylpyrazol-1-yl$)_3]^-$, as these two ligands — easiest to make among all the homoscorpionates — have been used most frequently. After the introduction in 1986 of the "second generation" ligands, which contained bulky substituents in the 3-position, this abbreviation system was expanded in order to accomodate most of the many new trispyrazolylborates that were being reported. The rules of creating these abbreviations, slightly modified from the original version,[8] are as follows:

1. The basic [HB(pz)$_3$]$^-$ structure is denoted by Tp, and any non-hydrogen substituent in the 3-position is denoted by a superscript. Thus, [HB(3-methylpz)$_3$]$^-$ is denoted as Tp^{Me}, **7**, [HB(3-phenylpz)$_3$]$^-$ is written as Tp^{Ph}, [HB(3-isopropylpz)$_3$]$^-$ as Tp^{iPr}, and so forth. The reason for giving such priority to the 3-substituent is that in the reaction of KBH$_4$ with 3(5)-monosubstituted pyrazoles the asymmetric R-substituent ends up in the 3-position of the ligand. In the few cases where rearrangement of the ligand has taken place, and Tp^R has become [HB(3-Rpz)$_2$(5-Rpz)]$^-$, the unsymmetrical ligand will be written as Tp^{R*}, **8**. When there are four identical pyrazolyl groups bound to boron, as in [B(3R-pz)$_4$]$^-$, the ligand will be denoted as pz^{o}Tp^R, **9**, while the simple parent ligand [B(pz)$_4$]$^-$ is abbreviated as pzTp. Boron substituents are written preceding "Tp": for instance, butylhydrotris(pyrazol-1-yl)borate is BuTp, **10**.

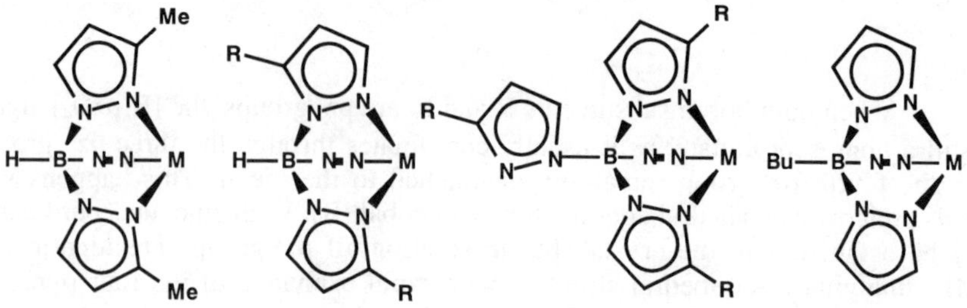

Tp^{Me}	Tp^{R*}	pz^{o}Tp^R	BuTp
7	**8**	**9**	**10**

(—N=N— is a pyrazolyl group, matching the one at the bottom of the drawing)

2. The 5-substituent follows the 3-substituent as a superscript, separated by a comma. Thus, [HB(3-isopropyl-5-methylpz)$_3$]$^-$ is denoted as Tp^{iPr,Me}, **11**. When both, the 3 and the 5 substituents are identical, the superscript R-substituent is followed by a 2: thus [HB(3,5-diphenylpz)$_3$]$^-$ is Tp^{Ph2}, **12**. In the case of the most commonly used ligand, [HB(3,5-dimethylpz)$_3$]$^-$ the systematic abbreviation would be Tp^{Me2} although, considering the long historical use of Tp*, **13**, that abbreviation can also be used, and will be adhered to in this book. Since Tp* defines uniquely the position of the two methyl substituents, a substituent in the 4-position follows the asterisk: thus, [HB(3,5-dimethyl-4-chloropz)$_3$]$^-$ is simply Tp*^{Cl}, **14**.

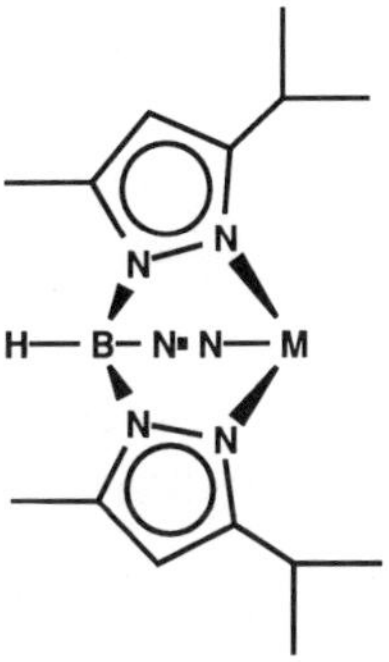

1.3 ABBREVIATION SYSTEM

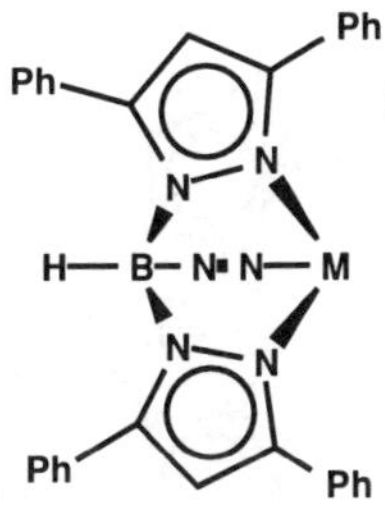

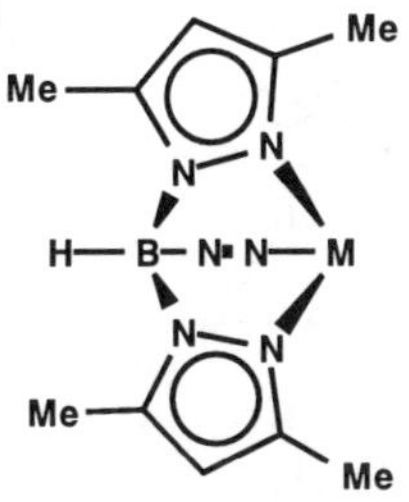

Tp^{iPr,Me}	Tp^{Ph2}	Tp^{Me2} = Tp*	Tp*^{Cl}
11	**12**	**13**	**14**

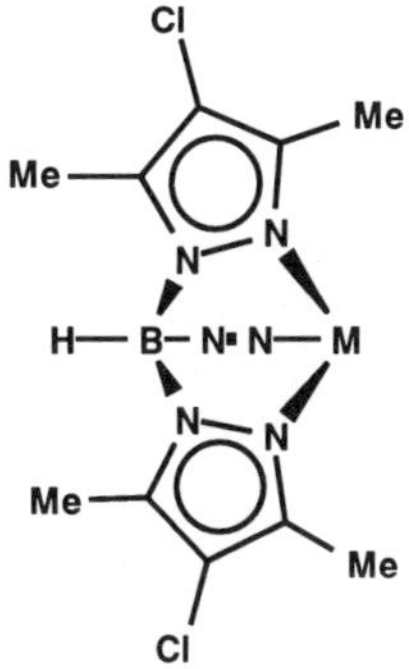

3. A substituent in the 4-position is denoted as a 4R superscript. Thus, [HB(3-isopropyl-4-bromopz)$_3$]$^-$ is Tp^{iPr,4Br}, **15**, and [HB(4-chloropz)$_3$]$^-$ is Tp^{4Cl}, **16**.

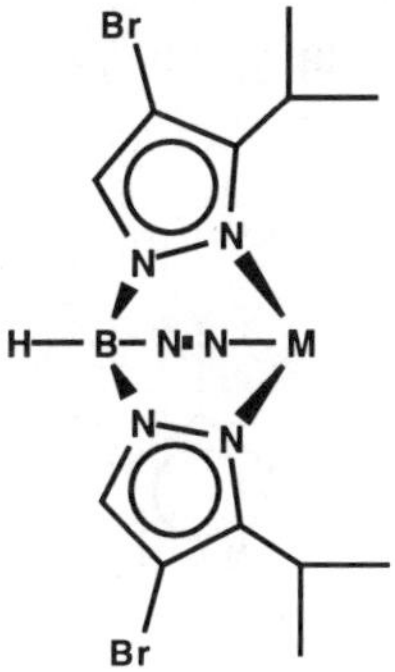

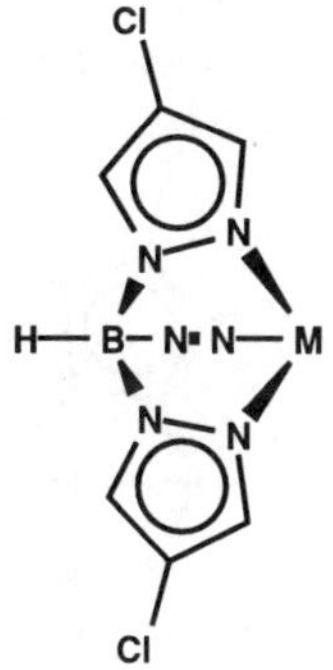

Tp^{iPr,4Br}	Tp^{4Cl}
15	**16**

4. In the less common case of polyindazolylborates, they will be represented as benzopyrazolylborates, Tp^{Bo}, with the mode of fusion of the benzo ring to pz indicated by the superscript of 3 or 4 preceding "Bo" (i.e., Tp^{3Bo} (**17**) and Tp^{4Bo} (**18**)) to indicate a 3,4- or 4-5 fusion of the benzo ring, and with the position numbering following the indazole numbering system.

CHAPTER 1. INTRODUCTION

Tp³ᴮᵒ (**17**)   Tp⁴ᴮᵒ (**18**)

5. In the few instances of very complicated substitution on the pyrazolyl ring, each ligand structure will have to be denoted individually as Tpᵃ, Tpᵇ, etc., or else written up in a fully systematic way.

6. A general homoscorpionate ligand with unspecified substituents will be denoted as Tpˣ, **19**, and a general pyrazolyl group will be pzˣ, **20**.

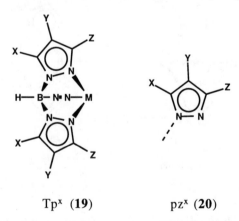

Tpˣ (**19**)   pzˣ (**20**)

7. Heteroscorpionate ligands will be abbreviated as "Bp", with the C-substituents denoted as defined above for Tp, and with the non-hydrogen substituents on the boron written before the abbreviation. For instance diethylbis(pyrazol-1-yl)borate will be denoted as Et₂Bp, **21**, and H₂B(3-tBupz)₂ as Bpᵗᴮᵘ, **22**.

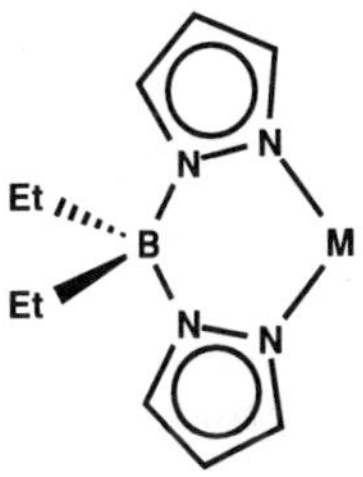

1.4 COMPARISON OF Tp AND Cp LIGANDS

Et$_2$Bp (**21**) 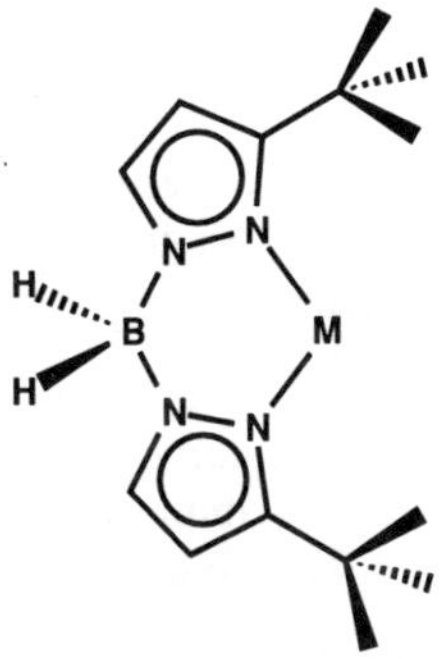 Bp^{tBu} (**22**)

1.4 Comparison of the Tp and Cp Ligands

Over the years it has been customary to compare the Tp ligand system to the cyclopentadienide ions, Cp or Cp*, inasmuch as they formed many similar metal complexes, such as L$_2$M, [LM(CO)$_3$]$^-$, [LM(CO)$_2$NO], etc. Both ligands are uninegative, each is donating six electrons and can occupy three coordination sites. However the similarity ends here, as the Tp ligand exhibits many distinguishing and unique features:

1. The symmetry of a TpM fragment is C$_{3v}$ as opposed to C$_{5v}$ for CpM.

2. There are ten substitutable positions on the Tp ligand: one on the boron, and a total of three on each of the 3-, 4-, and 5-positions of the pyrazole ring. This alone permits a wider range of modifications for the Tp ligand.

3. If we consider the number of possible R-substituted ligands with retention of the original symmetry (isosymmetric), then there is only one such possibility with Cp (C$_5$R$_5$), whereas in the case of Tp there are 15 such possibilities, including the different ways to place a given number of R substituents on a Tp ligand: four monosubstituted, six disubstituted, four trisubstituted, and one tetrasubstituted. If the R-substituents are non-identical, the number of these possibilities is greatly increased. It should be noted, that the symmetry of the Tp ligand does not change if different R-substitutents are used, provided they are of the same regiochemistry. For instance, the ligand [Tp^{tBu,Me,Br}]$^-$ has the same symmetry as the parent ligand Tp.

4. In terms of the capability to produce monomeric LMX species (X = halide), Cp can do it with only with M = Be, while stable Tp^{R}MX species are readily

obtained for numerous main group and transition metals, using sterically hindered Tp^R ligands.

5. The alkali metal salts of Tp ligands are air-stable solids, and require no unusual precautions for storage. By contrast, such Cp salts are air-sensitive, and require special handling.

6. The "free acids", TpH, are also stable compounds; moreover, they can act as ligands, so that Tp derivatives may be obtained even in acid media. By contrast, cyclopentadiene itself is a diene ligand, rather than ligating as the Cp anion, and it also tends to dimerize on storage.

7. Another advantage of Tp over Cp is the existence of a carbon-based, neutral, isosteric, and isoelectronic analog: trispyrazolylmethane, HC(pz)$_3$. This ligand, and its substituted variants, coordinate in essentially the same way as Tp, but the charge of the resulting complex is more positive by one unit. No neutral, isosteric and isoelectronic analogs of Cp exist.

8. One can also envisage an isosymmetric and isoelectronic analog of Tp based on beryllium: [RBe(pz)$_3$]$^{2-}$ which could lead to unique neutral octahedral chelates with tetravalent metals, [M{RBe(pz)$_3$}$_2$] and to the appropriate anionic species with M^{II} and M^{III} ions. None of that is, of course, possible in the Cp system.

As can be seen from the above, the Tp ligand system offers an astounding degree of versatility for the construction of ligands with judiciously chosen and appropriately located substituents. The well-developed chemistry of the pyrazole heterocycle offers a veritable cornucopia of substituted pyrazoles, capable of imparting the desired steric and electronic features to the derived Tp^x ligand.

1.5 Historical Development

In the early publications, the synthesis of the parent ligands, Bp, Tp and pzTp, of their transition metal complexes,[9] and of pyrazaboles [R$_2$B(μ-pz)$_2$BR$_2$][10] was described, as well as that of their B- and C-substituted analogs.[11,12] The magnetic moments and UV and visible spectra of Tp complexes were determined, NMR studies of the paramagnetic complexes of Co^{II}, and Mössbauer studies of various Fe^{II} complexes were conducted.[13,14] Churchill determined the first structure of an octahedral homoscorpionate complex, Tp$_2$Co, by X-ray crystallography.[15] Half-sandwich complexes TpM(X)(Y)(Z), where X, Y, and Z were diverse other ligands, were prepared, many of them similar to the Cp analogs, others unique.[16,4] Stereochemical nonrigidity of Tp^{x}Mo(CO)$_2$(η^3-allyl) complexes was studied by

1.5 HISTORICAL DEVELOPMENT

NMR.[17] Complexes of the type Tp^{x}Mo(CO)$_2$NO and Tp^{x}Mo(CO)$_2$N=NAr, were also synthesized.[18] Alkylation of [TpMo(CO)$_3$]$^-$ yielded products,[16] later shown to be the η^2-acyls TpMo(CO)$_2$(η^2-COR), and not Mo-alkyls TpMo(CO)$_3$R.[6,7]

Even at that early stage, it was noted that some reactions with the more hindered Tp* ligand proceeded anomalously. For instance, the reaction of diazonium salts with [Tp*Mo(CO)$_3$]$^-$ gave, instead of the expected diazo compounds Tp*Mo(CO)$_2$N=NAr, as was the case with Tp,[19] the η^2-acyls, Tp*Mo(CO)$_2$(η^2-COAr).[20]

Heteroscorpionates of the types R$_2$Bp and Bp^{R2} were found to form nominally 16-electron complexes of the type LMo(CO)$_2$(η^3-allyl),[21,22] and the first instance of an agostic B—H—M bond was structurally established for Bp*Mo(CO)$_2$(η^3-allyl).[23] In the case of Et$_2$BpMo(CO)$_2$(η^3-allyl), it was proposed on the basis of IR and NMR spectra that an interaction between the methylene hydrogens of the pseudoaxial ethyl group and the metal is taking place.[21,22] This was later confirmed by Cotton,[24] who established in detail the dynamic processes for the R$_2$BpMo(CO)$_2$(η^3-allyl) systems,[25] and who also found that the only truly 16-e complexes of this type were those of Ph$_2$Bp.[26]

By far, the largest amount of work with the Tp and Tp* ligands was done with Mo and W, starting with their easily prepared, and versatile intermediates [Tp^{x}M(CO)$_3$]$^-$. McCleverty investigated mono- and polynuclear complexes of the type Tp*Mo(NO)(X)(Y) establishing a vast sub-area of homoscorpionate chemistry.[27] Angelici explored the related Tp*W(CO)$_2$-based thiocarbenes and thiocarbynes,[28,29] and Lalor discovered an easy route to the carbyne Tp*Mo(CO)$_2$≡CCl, a convenient starting material for many derivatives.[30,31]

While work with the Tp and Tp* ligands continued, the second generation ligands, Tp^R, where R was a bulky substituent (Bu^t, aryl), were introduced in 1986.[2,3] They were capable of exerting considerable steric control over the immediate surroundings of the coordinated metal. These features initiated the second growth phase of scorpionate coordination chemistry, and led to many unique and unusual complexes. Among the numerous interesting contributions during this phase were those of Parkin, who prepared stable monomeric Tp^{tBu}MR complexes for M = Be, Mg and Zn,[32-34] Theopold, who investigated oxygen activation in cobalt complexes of Tp^{tBu,Me} and Tp^{iPr,Me},[35,36] Tolman, who prepared Tp^{tBu}CuNO (the first mononuclear CuNO complex) and Tp^{tBu}CuNO$_2$,[37,38] Vahrenkamp, who used Tp^{tBu,Me} and Tp^{Cum,Me} ligands to explore biorelevant zinc chemistry,[39,40] Takats, who synthesized homoleptic and heteroleptic lanthanide complexes of Tp^{tBu,Me},[41,42] Kitajima, who studied mono- and dinuclear complexes (mostly of copper) based on the Tp^{iPr2} ligand,[43-45] and Reger, who prepared numerous interesting main-group complexes, including Tp^{tBu}CdH.[46] Extensive work employing the "first generation" Tp and Tp* ligands was done by Etienne, who investigated mainly niobium chemistry,[47] and Templeton, who made important contributions to the area of tungsten carbenes and carbynes.[48-51]

At present, novel scorpionate ligands keep appearing with ever greater frequency, their number as of the latest count surpassing 170, and this trend continues.

1.6   Reviews

Twenty three reviews and chapters dealing with scorpionates were published up to the end of 1998. Some were devoted to this subject in their entirety, others included a significant component of Tp-based complexes within unrelated specialized subjects, and still others were devoted to specialized sub-areas of Tp chemistry.

The first review by Trofimenko in 1971 covered the earliest developments,[52] while the next one was on pyrazole-derived ligands, and contained polypyrazolylborates as a separate category.[53] Tp ligands were also included by Niedenzu and Trofimenko in 1975 among the boron-pyrazole compounds described in Gmelin's handbook,[54] and in a 1986 review.[55] Shaver contributed a chapter on Tp chemistry in 1977.[56] McCleverty presented an overview of his work on alkoxy, amido, hydrazido and related derivatives of the structure Tp*M(NO)(A)(B), for M = Mo and W.[57] Trofimenko wrote reviews on Tp ligands in 1986,[58] and in 1993,[8] the latter covering the 1984-1993 period. Niedenzu reviewed the pyrazaboles ($R_2B(\mu$-pz$)_2BR_2$ systems) in 1988,[59] while in 1992 Canty wrote, in part, about the Tp chemistry of Pd and Pt.[60] In 1993 Parkin discussed Tp-derived models for carbonic anhydrase.[61] Kitajima and Moro-oka included many Tp-based compounds in their 1994 review on copper-dioxygen complexes.[62]

Three reviews appeared in 1995: Santos and Marques covered the Tp-chemistry of lanthanides and actinides,[63] Parkin wrote about metal alkyls, hydrides, and hydroxides, derived from hindered Tp^x ligands,[64] while Kitajima and Tolman reviewed the organometallic and bioinorganic chemistry of hindered Tp^x ligands.[65] In 1996, Etienne dealt with the Tp chemistry of V, Nb and Ta,[47] while the review of Reger was devoted to the Tp^x complexes of Ga and In,[66] and Parkin discussed the effect of Tp^x-ligation on Grignard reagents.[67] In 1997, Young and Wedd mentioned a number of Tp*-based Mo and W pterin enzyme models,[68] Theopold and coworkers reviewed dioxygen activation with sterically hindered Tp^x cobalt complexes,[69] and Janiak included many data on TlTp^x complexes in a review of Tl^I and Tl^{II} chemistry.[70] He followed it up with a 1998 review of Tp^{x}Tl compounds, including their synthesis, structures, and applications.[71] The use of the scorpionate moiety [Tp*Mo(NO)(Cl)]$^-$ to form a variety of bridged structures linked by aromatic and conjugated systems, as a way of controlling the electronic and magnetic properties of the resulting polynuclear complexes, has been reviewed by McCleverty and Ward.[72] The latest review by Kirchner was concerned with the coordination chemistry and some applications of ruthenium homoscorpionate complexes, mostly with the parent homoscorpionate ligands, but also including a few of the more substituted ones.[73] No reviews have been published in early 1999.

1.7 Synthesis of Scorpionate Ligands

Scorpionate ligands may be prepared from a variety of boron sources, including boron trihalides, boron hydrides, alkyl- or arylboronic acids, their esters, halides, tosylates, etc.

$$R_nBX_{3-n} + [pz^x]^- + (2-n)[Hpz^x] \rightarrow [R_nB(pz^x)_{4-n}]^- + (2-n)HX + X^- \quad (1.1)$$

It is important when generating the $R_2B(pz^x)$ species, that there be sufficient pyrazolate ion, $(pz)^-$, present to convert it quickly to $[R_2B(pz^x)_2]^-$, otherwise the 1,3-dipole $R_2B(pz^x)$ will dimerize to the stable pyrazabole $R_2B(\mu-pz^x)_2BR_2$ and scorpionate ligands will not be obtained.

$$2\ R_2B(pz^x) \rightarrow R_2B(\mu-pz^x)_2BR_2 \quad (1.2)$$

$$R_2B(pz^x) + (pz^x)^- \rightarrow [R_2B(pz^x)_2]^- \quad (1.3)$$

The most general and convenient route to scorpionate ligands is through the borohydride ion, $[BH_4]^-$. This reaction can be controlled to yield bis-, tris-, and in the case of 5-unsubstituted pyrazoles, tetrakis(pyrazolyl)borates:

$$[BH_4]^- + \text{excess } Hpz^x \rightarrow [H_nB(pz^x)_{4-n}]^- + (4-n)H_2\uparrow \quad (1.4)$$

An almost unlimited variety of 1-H pyrazoles may be employed to synthesize the ligands by this route, with the exception of those containing functionalities incompatible with the borohydride ion, such as pyrazole-carboxylic or -sulfonic acids, and certain nitropyrazoles.

Synthesis of the parent ligands Bp, Tp, and pzTp has been described in detail.[74] Procedures for synthesizing specific, diversely substituted scorpionates, are given in the references from the table of known scorpionate ligands. In all syntheses it is recommended to measure the amount of hydrogen evolved, preferably with a wet-test-meter, in order to follow the reaction rate and the extent of its completion. While for some specific syntheses, the reader is referred to the original literature, a general set of conditions can be applied to prepare ligands from hitherto untried pyrazoles.

1.7.1 Bp^x Ligands

In general, the reaction components, substituted pyrazole and KBH_4 ($NaBH_4$ is cheaper, and can also be used, although the K salts are easier to crystallize) in a 2.3:1.0 mole ratio, are refluxed in anhydrous DMF (about 300 ml per 0.1 mol KBH_4), with the emanating hydrogen being measured by a wet-test-meter (protected by a -80 °C trap), until the theoretical amount has been evolved. When the reaction is complete, DMF is distilled out at reduced pressure. The residue is boiled with toluene, which should dissolve most of the unreacted pyrazole, and the mixture is filtered. The solid, which is crude $K[H_2B(pz^x)_2]$ is contaminated by only small amounts of Hpz^x and is usually suitable for complex formation. Additional purification can be achieved by converting the crude K salt to the Tl salt (which shows no tendency to retain Hpz^x) by dissolving the K salt in a minimum amount of THF (or DMF, if the solubility in THF is too low), and mixing this solution with a saturated aqueous solution of a soluble Tl^I salt (nitrate, sulfate, or acetate, taken in small excess). The $Tl[H_2B(pz^x)_2]$ salt precipitates immediately and can be isolated by filtration, or by extraction with methylene chloride, followed by filtration through a bed of alumina, and evaporation. The final purification is achieved through recrystallization from an aromatic (or in some cases aliphatic) high-boiling solvent.

1.7.2 Tp^x Ligands

Here, the procedure depends in part on whether the pyrazole is a liquid, or a low-melting solid, and also on whether it is 3-monosubstituted, 3,5-disubstituted, or with higher degree of substitution.

a. *The pyrazole is a liquid or a low-melting solid.* The method of choice is a neat reaction of the pyrazole with KBH_4 in a 4:1 ratio, the extent of the reaction being monitored by hydrogen evolution. When the theoretical amount of hydrogen has been evolved, the excess pyrazole is distilled out at reduced pressure, keeping the temperature as low as possible, in order to prevent the formation of the tetrasubstituted $pz°Tp^x$ ligand. The residual $K[Tp^x]$ salt is usually suitable for

1.7 SYNTHESIS OF SCORPIONATE LIGANDS

complex formation, or it can be converted to the Tl(I) salt, and purified further by recrystallization.

b. The pyrazole is 3-substituted and high-melting. The preferred method here is to reflux with rapid stirring the ingredients (in a 3.5:1 ratio of the pyrazole to KBH_4) in 3- or 4-methylanisole, using 400 ml of solvent per 0.1 mol of KBH_4. Anisole can also be used, but methylanisole is higher boiling, and the reaction proceeds faster. When the theoretical amount of hydrogen has evolved, which in some instances may take up to several days, the product may be in solution, or it may have precipitated.

If it has precipitated, the mixture is filtered hot, and the solid is washed with a small amount of hot methylanisole. After drying, it usually is very pure KTp^x. The filtrate is distilled at reduced pressure to recover the solvent, which may be reused. If the yield of KTp^x is high, the residue can be hydrolyzed to recover Hpz^x. If the yield of KTp^x is low, and much of the ligand still remains in solution, the filtrate is stripped, and the residue is dissolved in a minimum amount of THF or DMF, and converted to the Tl salt as described above and extracted with methylene chloride. The extracts are filtered through a short layer of alumina, stripped, and stirred with methanol in which Hpz^x (but not Tp^xTl) is usually soluble. After filtration of the mixture, and thorough washing with methanol, Tp^xTl of good purity is obtained.

Occasionally, after completion of the reaction, there are still a few particles of KBH_4 present, and they can be removed by decantation of the solution, or by filtration. Also, the Tl salt is sometimes not totally extracted with methylene chloride, and is present as a white solid in the organic layer. The organic layer is then filtered, to recover the Tl salt of high purity, and the filtrate is processed, as indicated above.

If the theoretical amount of hydrogen has evolved, and the solution remains clear, then the solvent is distilled out under vacuum, at oil-bath temerature not above 160-170 °C, to prevent formation of the $[pz^oTp^x]^-$ ligand, and the residue is converted to the Tl salt, as described above.

c. The pyrazole is 3,5-disubstituted. In this case the melt method is preferred, since tetrasubstitution does not take place, and higher temperatures can be safely employed. However, one has to be careful, since at times the solubility of the K salt in the melt is low, and it may precipitate partway through the reaction, with the possibility of hot-spotting, and decomposition. To prevent this, one uses a large excess of Hpz^x (6-8 :1 mole ratio of Hpz^x to KBH_4). More Hpz^x can actually be added when KTp^x starts precipitating midway through the reaction, until again a clear melt is obtained. After completion of the reaction, the solidified residue is broken up, and excess Hpz^x is sublimed out. The unsublimed residue is crushed and resublimed, until no further sublimation of Hpz^x is noted. The crude KTp^x is then converted to $TlTp^x$ in the usual way.

1.7.3 pz^oTp^x Ligands

These ligands, which are limited to 5-unsubstituted pyrazoles, are prepared by the reaction of Hpz^x with KBH_4 in a 5-6:1 mol ratio. No solvent is used, and the reaction can proceed at high temperatures, adjusted for the controlled evolution of hydrogen in the early stages (which may be rather vigorous), and raised as necessary, since evolution of the fourth equivalent of hydrogen is quite slow. After completion of the reaction, excess Hpz^x is either distilled off, or sublimed in vacuo, and the residue can be used for complex formation directly, or it can be converted to the Tl salt first. Presence of the fourth pz^x group makes $Tl[pz^oTp^x]$ salts more soluble in methanol than $TlTp^x$, and care should be exercised to avoid excessive solubility losses at this stage.

Caution: An attempt to prepare $pz^oTp^{iPr,4Br}$ by this method resulted in violent decomposition, accompanied by HBr evolution (even though $Tp^{iPr,4Br}$ was prepared without problems),[5] and the melt method should not be used for the preparation of $pz^oTp^{R,4Br}$ ligands.

1.7.4 Ligands Containing C—B Bonds

1.7.4.1 R_2Bp^x ligands

In general, such ligands are prepared from precursors already containing the C—B bond, such as trialkylboranes, triarylboranes, or the tetraphenylborate ion. A typical reaction of an R_3B or Ar_3B species with pyrazole has to be preceded by the formation of an anionic species $[R_3Bpz^x]^-$, through the reaction of R_3B with a pyrazolate ion, $[pz^x]^-$. The R groups in $[R_3Bpz^x]^-$ can be replaced by pz^x groups upon reaction with excess pyrazole. In the absence of pyrazolate ion, pyrazole reacts with the R_3B species yielding exclusively the pyrazaboles, $R_2B(\mu-pz^x)_2BR_2$, which are not readily convertible to scorpionate ligands.

$$R_3B + [pz^x]^- \rightarrow [R_3B(pz^x)]^- \tag{1.5}$$

$$[R_3B(pz^x)]^- + Hpz^x \rightarrow [R_2B(pz^x)_2]^- + RH \uparrow \tag{1.6}$$

With pyrazole itself, this reaction stops at the disubstitution stage. With high-boiling substituted pyrazoles, it can be driven one step further:

$$[R_2B(pz^x)_2]^- + Hpz^x \rightarrow [RB(pz^x)_3]^- + RH \uparrow \tag{1.7}$$

1.7 SYNTHESIS OF SCORPIONATE LIGANDS

As with the KBH_4 reaction, the pyrazole 3(5)-substituent ends up in the 3-position. One can also start with a tetraalkyl- or tetraarylborate salt, such as the commercially available $Na[Ph_4B]$, in which case only pyrazole needs to be used.

$$[R_4B]^- + 2\ Hpz^x \rightarrow [R_2B(pz^x)_2]^- + 2\ RH \uparrow \quad (1.8)$$

A rather convenient route to $[Ph_2B(pz^x)_2]^-$ ligands starts with the very stable Ph_3BNH_3 complex, which upon treatment with one equivalent of a pyrazolate salt and excess pyrazole eliminates ammonia and one equivalent of benzene to yield $[Ph_2B(pz)_2]^-$.

$$Ph_3BNH_3 + [pz^x]^- + Hpz^x \rightarrow [Ph_2B(pz^x)_2]^- + NH_3 \uparrow + PhH \uparrow \quad (1.9)$$

1.7.4.2 $RB(pz^x)_3$ ligands

The synthesis of $[RB(pz^x)_3]^-$ ligands, where R is alkyl or aryl, is possible by a number of routes. One of them involves the reaction of RBX_2 or $ArBX_2$ (where X is a halogen, or a leaving group such as tosylate) with the pyrazolate ion plus excess pyrazole:

$$RBX_2 + 3\ [pz^x]^- \rightarrow [RB(pz^x)_3]^- + 2X^- \quad (1.10)$$

The necessary aromatic boron precursors can be obtained by the borylation of aromatics (including ferrocene) with BX_3.[11,75]

Another route utilizes the reaction of alkyl- or arylboronic acids with the pyrazolate ion and excess pyrazole.[11,76]

$$RB(OH)_2 + [pz^x]^- + 2\ Hpz^x \rightarrow [RB(pz^x)_3]^- + 2\ H_2O \quad (1.11)$$

The various boronic acids can be prepared through hydrolysis of their esters, $RB(OR)_2$, obtainable from boric esters, $B(OR)_3$, via their reaction with alkyl- or aryllithium. A variant of the above method consists of direct reaction of the $RB(OR)_2$ ester with pyrazolide ion and pyrazole.[77]

$$RB(OR)_2 + [pz^x]^- + 2\ Hpz \rightarrow [RB(pz^x)_3]^- + 2\ ROH \quad (1.12)$$

A still different route employs RBH_2 (such as the commercially available $MeS(CH_2)_3BH_2$) in the reaction with pyrazolate ion plus excess pyrazole.[78]

$$RBH_2 + [pz^x]^- + 2\,Hpz^x \rightarrow [RB(pz^x)_3]^- + 2\,H_2 \uparrow \quad (1.13)$$

In each instance one can either convert the crude ligand directly to metal complexes, or isolate and purify it first as the Tl^I salt, $[RB(pz^x)_3]Tl$.

1.8 List of Known Scorpionate Ligands

The known Tp^x, pz^oTp^x and Bp^x ligands, and references to their synthesis, are listed in Table 1, below. If more than one synthesis has been reported, the two that seem the most convenient are cited. In some instances the details of ligand synthesis have not been fully disclosed.

Table 1

	Scorpionate Ligand	Comments	References
Unsubstituted			
1	Tp		9, 74
2	pzTp		9, 74
3-Monosubstituted			
3	Tp^{Me}		79
4	pz^oTp^{Me}		79
5	Tp^{iPr}		5
6	pz^oTp^{iPr}		5
7	Tp^{tBu}		3
8	pz^oTp^{tBu}		3
9	Tp^{Np} and Tp^{Np*}	Np = neopentyl	80
10	Tp^{Cpr}	Cpr = cyclopropyl	81
11	pz^oTp^{Cpr}		82
12	Tp^{Cbu}	Cbu = cyclobutyl	83
13	pz^oTp^{Cbu}		83
14	Tp^{Cpe}	Cpe = cyclopentyl	83
15	pz^oTp^{Cpe}		83
16	Tp^{Cy}	Cy = cyclohexyl	84
17	pz^oTp^{Cy}		1254
18	Tp^{Ph}	Ph = phenyl	3
19	Tp^{Ph*}	obtained via Tp^{Ph} rearrangement	85
20	Tp^{Tol}	Tol = *p*-tolyl	86, 470
21	pz^oTp^{Tol}		470
22	Tp^{An}	An = *p*-anisyl	86

1.8 LIST OF KNOWN SCORPIONATE LIGANDS

Table 1 (Continued)

	Scorpionate Ligand	Comments	References
23	Tp^{oAn}		87, 88
24	Tp$^{Ph(oSMe)}$		1560
25	Tp^{Ant}	Ant = 9-anthryl	89
26	Tp^{Fn}	Fn = 2-furyl	90
27	Tp^{Tn}	Tn = 2-thienyl	91
28	Tp^{2Py}	Py = pyridyl	92
29	Tp^{2Py6Me}		1433
30	Tp^{PhF}	PhF = 4-fluorophenyl	90
31	Tp$^{\alpha Nt}$	Nt = naphthyl	92
32	Tp$^{\beta Nt}$		92
33	Tp^{Ms}	Ms = mesityl	93
34	Tp^{Ms*}		93
35	Tp^{CF_3}		94, 95
36	Tp^{C_2F_5}		96
37	Tp^{C_3F_7}		96
38	Tp^{Trip}	Trip = triptycyl	97
39	Tp^{Menth} = HB(7(R)iPr-4(R)Me-4,5,6,7-tetrahydroindazolyl)$_3$		98
40	Tp^{Menth*}		98
41	Tp^{Mementh} = HB(7(S)tBu-4(R)Me-4,5,6,7-tetrahydroindazolyl)$_3$		98
42	Tp^{CHPh_2}		83
43	Tp$^{CON(CH_2)_4}$		83
44	Tp$^{2,4(OMe)_2Ph}$		83

4-Monosubstituted

45	Tp^{4Me}	99
46	Tp^{4iPr}	11
47	Tp^{4tBu}	83
48	Tp^{4Cl}	11
49	Tp^{4Br}	100
50	Tp^{CN}	83

5-Monosubstituted

No examples of Tp^{5R} ligands are known.

3,4-Disubstituted

51	Tp^{iPr,4Br}	5
52	Tp^{tBu,4Br}	83

Table 1 (Continued)

	Scorpionate Ligand	Comments	References
53	Tp^{Cy,4Br}		83
54	Tp^{Bn,4Ph}	Bn = benzyl	83
55	Tp^{3Bo,7Me}		101
56	Tp$^{(3Bo,7tBu)}$		102
57	Tp$^{(3Bo,7tBu)*}$		102
58	Tp^{a*}	= [HB(2H-benz[g]indazol-2-yl)$_3$]	103
59	Tp^{iPr,4tBu}		83
60	Tp^{Pr,4Et}		83
61	Tp^a	Tp^{Ph}, -CH$_2$CH$_2$-linking 4 and phenyl ortho	104
62	Tp^b	Tp^{Ph} with -CH$_2$-linking 4 and phenyl ortho	104
63	Tp$^{(3,4-(CH_2)_n)}$ and Tp$^{(3,4-(CH_2)_n)*}$ where n = 3, 4, 6 or 10		83

3,5-Disubstituted

	Scorpionate Ligand	Comments	References
64	Tp^{Me2} = Tp*		11
65	pz^{o}Tp*	Known only as the free acid, [pz^{o}Tp*]H	105
66	Tp^{Et2}		11
67	Tp^{iPr2}		107
68	Tp^{tBu2}		108
69	Tp$^{(CF_3)2}$		116, 117
70	Tp^{Ph2}		107
71	Tp^{Bn,Me}	Bn = benzyl	83
72	Tp^{iPr,Me}		106
73	Tp^{tBu,Me}		86
74	Tp^{tBu,iPr}		92
75	Tp^{tBu,Tn}	Tn = 2-thienyl	83
76	Tp^{Ph,Me}		104
77	Tp^{Tol,Me}		109
78	Tp^{Cum,Me}	Cum = cumyl (4-isopropylphenyl)	110
79	Tp$^{p-tBuPh,Me}$		111, 1538
80	Tp^{3Py,Me}		112
81	Tp^{3Pic,Me}	Pic = picolyl	112
82	Tp$^{(p-tBuPh)2}$		113
83	Tp^{CF_3,Me}		114
84	Tp^{CF_3,Tn}		115
85	Tp^{Ph,Tn}	ligand synthesis not reported	1564
86	Tp^{Tn,Me}	ligand synthesis not reported	1564

1.8 LIST OF KNOWN SCORPIONATE LIGANDS

Table 1 (Continued)

	Scorpionate Ligand	Comments	References

4,5-Disubstituted (all indazolylborates: numbering follows indazole system)

87	Tp^{4Bo}		118
88	pz^oTp^{4Bo}		119
89	$Tp^{4Bo,5Me}$		101
90	$Tp^{4Bo,5Et}$		101
91	$Tp^{4Bo,5tBu}$		101
92	$Tp^{4Bo,5Ph}$		101
93	$Tp^{4Bo,2,6Me_2}$		101
94	$Tp^{4Bo,5NH_2}$		1382
95	$Tp^{4Bo,5NO_2}$, $pz^oTp^{4Bo,5NO_2}$ and 6-NO_2 analogs		1380, 1563

Trisubstituted

96	$Tp^{*Me} = Tp^{Me3}$		11
97	Tp^{*Et}		1384
98	Tp^{*Bu}		11
99	Tp^{*Am}	Am = amyl	1384
100	Tp^{*Cl}		114, 120
101	Tp^{*Br}		979
102	Tp^{Br3}		83
103	Tp^{*Bn}	Bn = benzyl	121
104	$Tp^{iPr2,Br}$		122
105	$Tp^{Ph,Me,Ph}$		123
106	$Tp^{4Bo,3Me}$		101
107	$Tp^{a*,3Me}$ = hydrotris(3-methyl-benz[g]indazol-2-yl)borate		101
108	$Tp^{(a*,3Me)*}$		101
109	$Tp^{a,Me`}$ $Tp^{Ph,Me}$ -CH_2CH_2-linking 4 and phenyl ortho		104

B-Substituted

110	EtTp	Known only as the [(EtTp)BEt]$^+$ cation	124
111	iPrTp		76
112	BuTp		11
113	$MeS(CH_2)_3Tp$		78
114	$MeS(CH_2)_3Tp^R$		83
115	PhTp		11
116	$PhTp^{tBu}$		1215
117	C_6D_5Tp		125
118	TolTp	mixture of *m*- and *p*-isomers	126

Table 1 (Continued)

	Scorpionate Ligand	Comments	References
119	p-BrPhTp		127
120	FcTp	(Fc = ferrocenyl)	128
121	FcTp^{Me}		1217
122	FcTp^{Ph}		1217
123	Fc(Tp^{Me})$_2$		1217
124	Fc(Tp^{Ph})$_2$		1217
125	Me$_2$NTp		129, 130
126	MeTp^{Me}		77
127	Tp-Tp	= [(pz)$_3$B-B(pz)$_3$]$^{2-}$	131
128	Tp^{Me}-Tp^{Me}	= [(pz^{Me})$_3$B-B(pz^{Me})$_3$]$^{2-}$	1219
129	Tp^{Py}-Tp^{Py}	= [(pz^{Py})$_3$B-B(pz^{Py})$_3$]$^{2-}$	1219

Heteroscorpionates

130	Bp		9
131	Bp^{Me2} = Bp*		11
132	Bp^{Me}		132
133	Bp^{tBu}		3
134	Bp^{Trip}		97
135	Bp^{Fc}		133
136	Bp$^{(CF_3)_2}$		134
137	Bp^{Br_3}		83
138	Bp^{4Bo}		141
139	Bp^{4Bo,5NO_2}		142
140	Bp^{4Bo,5NH_2}		1382
141	Me$_2$Bp		135
142	Et$_2$Bp		11
143	Et$_2$Bp^{Fc}		133
144	Pr$_2$Bp		1456
145	Bu$_2$Bp		11
146	Ph$_2$Bp		136
147	(Me)(Ph)Bp^{Me}		77
148	(BBN)Bp	BBN = (cyclooctane-1,5-diyl)	137
149	(BBN)Bp^{Me}		138
150	(BBN)Bp^{Ph}	known only as the free acid	1463
151	(BBN)Bp^{Ph,Me}	known only as the free acid	1463
152	(BBN)Bp*	known only as the free acid	1463
153	(BBN)Bp^{Fc}		133
154	Cl$_2$Bp		1422

Table 1 (Continued)

#	Scorpionate Ligand	Comments	References
155	Br$_2$Bp		1422
156	I$_2$Bp		1422
157	F$_2$Bp*		11
158	(p-TolS)Bp*		139
159	(MeBnS)Bp	MeBn = p-MeC$_6$H$_4$CH$_2$-	140
160	Bp$^{2,4(OMe)_2Ph}$		83
161	[H$_2$B(pz)(pz*)]		143
162	H$_2$B(pz)(pz^{tBu2})		144
163	H$_2$B(pz*)(pz^{tBu2})		144
164	H$_2$B(pz^{Trip})(pz^{tBu2})	Trip = triptycyl	144
165	(MeO)Bp^{tBu}		145, 1088
166	(EtO)Bp^{tBu}		145, 1088
167	(Pr^{i}O)Bp^{tBu}		145, 1088
168	Bp^{tBu,iPr}		146
169	(MeO)Bp^{tBu,iPr}		146
170	CpBp*	only in complex (CpBp*)SmTp*	1470
171	Bp^{Py}		147
172	(pz*)Bp		825
173	(pz)Bp*		825
174	(pz^{4CN})Bp		52
175	(MeO)Bp*	ligand itself not isolated	1565
176	(EtO)Bp*	ligand itself not isolated	1565
177	(MeO)$_2$Bp	ligand itself not isolated	1565
178	(EtO)$_2$Bp	ligand itself not isolated	1565

1.9 Analogs of Scorpionate Ligands

In addition to placing regiospecifically various substituents on carbon and/or boron in scorpionate ligands, one can also modify them in two more fundamental ways:

    a.   through replacement of pyrazole with another heterocycle or donor system, which would retain the uninegativity of the ligand, and form chelate rings of comparable size.

    b.   through replacement of boron with another element. This could either retain, or change the charge of the ligand.

1.9.1 Replacement of Pyrazole

The first example of forming a Tp analog based on 1,2,4-triazole, and its conversion to the water soluble [HB(1,2,4-triazol-1-yl)$_3$]$_2$Co, was reported in 1967,[9] and little happened thereafter,[148] until in 1993 Janiak embarked upon a systematic study of polytriazolyl- and polytetrazolylborates, and found them leading to many extended structures, due to the additional coordinating nitrogen atoms.[149-158] Other reports have also appeared.[159] Several poly(benzotriazolyl)borates were also described,[160-163] as was the [HB(3,5-Me$_2$-1,2,4-triazolyl)]$^-$ ligand,[164,165] and the ligand [Me$_2$B(2-pyridyl)$_2$]$^-$.[166]

Hydrotris(methimazolyl)borate was also a related S-bonding ligand and, since the boron-to-metal bridge was triatomic, it formed eight-membered B(N-C-S)$_2$ rings.[167] On the other hand, six-membered B(C-S)$_2$M rings were formed by ligands of the type [R$_n$B(CH$_2$SR')$_{4-n}$]$^-$, where n was 0, 1 or 2.[168,169]

1.9.2 Replacement of Boron

1.9.2.1 Aluminum and Indium

The ligands [Me$_2$Al(pz)$_2$]$^-$, [Al(pz)$_4$]$^-$ and [Me$_2$In(pz)$_2$]$^-$ were prepared, but no metal complexes could be obtained from them.[135]

1.9.2.2 Gallium

Storr studied polypyrazolylgallates, which were homo- and heteroscorpionate analogs containing gallium instead of boron. Ligands R$_2$Ga(pz^x)$_2$ were prepared with R = Me, and pz^x = pz or pz*.[170-178] The higher reactivity of the Ga—H bond compared with B—H, made it impossible to obtain complexes with Ga—H bonds, and the greater Ga—N bond length (1.99 Å vs. 1.56 Å for B-N) altered the ligand bite, and limited the number of scorpionate analogs that could be synthesized. [RGa(pz^x)$_3$]$^-$ ligands were also prepared, and converted to a number of complexes, resembling those of Tp ligands (e.g. LMn(CO)$_3$, LNi(NO), LMo(CO)$_2$(η^3-allyl), and others) but they were somewhat less stable.[179-186] Over-all, replacement of B with Ga involved more difficult syntheses, and reduced the versatility of the resulting ligand system.

1.9.2.3 Carbon

It was demonstrated in 1970 that geminal polypyrazolylalkanes, R$_2$C(pz^x)$_2$ and HC(pz^x)$_3$ yielded the same types of complexes as R$_2$Bp^x and Tp^x ligands, differing

1.9 ANALOGS OF SCORPIONATE LIGANDS

only by having a charge higher by one unit per ligand.[187] Some possible benefits of these ligands : one can construct complexes of easily reducible metal ions such as silver(I) or palladium(II), unobtainable with Bp ligands. Also, sometimes it may be beneficial to have a cationic [L$_2$M]$^{2+}$ complex (e.g. for solubility in water). Moreover, unlike in the boron-based ligand system, N,N,O ligands such as RC(pz^x)$_2$(OR) can be easily prepared, although they have not yet been exploited in coordination chemistry.[188]

Most of the work in this area was done with the R$_2$C(pz^x)$_2$ ligands, [185-258] to a lesser extent with HC(pz)$_3$ or HC(pz*)$_3$.[259-297, 1549-1552] Only quite recently has the synthesis of HC(pz^x)$_3$ ligands with bulky 3-substituents, and their coordination chemistry, been reported.[298-301]

1.9.2.4 Silicon

The ligand MeSi(pz*)$_3$ was synthesized and its structure was determined by X-ray crystallography, but no complexes based on it have been prepared.[302] However, an attempt to use this ligand for the preparation of copper complexes in DMF resulted in an unexpected isolation of a copper compound containing the new ligand, HC(pz*)$_2$NMe$_2$, arising from a reaction of MeSi(pz*)$_3$ with DMF.[1548]

1.9.2.5 Other elements

In the above scorpionate analogs the core atom was in a tetrahedral environment. In principle, any geminal polypyrazolyl species could be considered as a scorpionate analog, if its geometry permitted chelation of metal ions. Simple examples would be [E(pz)$_3$]$^-$ species with E = Ge and Sn, which are tridentate, but of variable coordination, geometry.[303,304] More complicated ones are exemplified by [(ArH)M(pz)$_3$]$^-$ and [(ArH)M(Cl)(pz)$_2$]$^-$, both of which formed bimetallic heteroleptic complexes exemplified by LCoTp^{iPr,4Br}.[305] Related to them was the N,N,O-bonding heteroscorpionate ligand [η^6-C$_6$Me$_6$)Ru(pz)$_2$(PO(OMe)$_2$)]$^-$.[306] The somewhat simpler [O=P(pz*)$_2$(O)]$^-$ ligand was also N,N,O-bonding.[307]

Chapter 2

Homoscorpionates – First Generation

2.1 General Considerations

This chapter deals with "first generation" homoscorpionate ligands and, more precisely, with the three ligands Tp, pzTp and Tp* (although in the first papers other ligands were also reported), while the ligand Bp will be covered, along with other heteroscorpionates, in Chapter 4. These three ligands have been used most extensively over the years, and despite the introduction of their more sophisticated "second generation" analogs, they are still widely employed. The reason for this popularity is the ease of their synthesis from readily available and inexpensive starting materials: pyrazole, or 3,5-dimethylpyrazole, and a borohydride salt. In the case of pyrazole, the reaction can be controlled to produce either Tp or pzTp, by controlling the reaction temperature, and using excess pyrazole.

$$3 \text{ Hpz } + [BH_4]^- \rightarrow [HB(pz)_3]^- + 3 H_2 \qquad (2.1)$$

$$\text{Hpz} + [HB(pz)_3]^- \rightarrow [B(pz)_4]^- + H_2 \uparrow \qquad (2.2)$$

The ligand Tp* is prepared similarly:

$$3 \text{ H}(3,5Me_2pz) + [BH_4]^- \rightarrow [HB(3,5Me_2pz)_3]^- + 3 H_2 \uparrow \qquad (2.3)$$

In contrast to pyrazole, borohydride reaction with 3,5-dimethylpyrazole stops at the Tp* stage, and does not proceed to pz^{o}Tp*. The ligand pz^{o}Tp* has been obtained only once, accidentally, as the free acid H[pz°Tp*] from the reaction of KTp* with Cp_2TaCl_2.[105]

Complexes of the ligands Tp, pzTp and Tp* will be discussed in this chapter according to the coordinated cationic species (metal or non-metal), following the Periodic Table. The ligands Tp and Tp* coordinate usually in tridentate fashion, forming octahedral full sandwich complexes $[Tp^x]_2M$, while their tetrahedral complexes, $[Tp^x]MX$ are labile and self-convert to the full sandwiches. The only Tp^xMX species isolated, were the unstable Cl-bridged dimers, $[TpMCl]_2$ (M = Cu and Co), of which the Cu complex was structurally characterized.[308]

While being generally tridentate, the Tp, pzTp and Tp* ligands can also act in bidentate fashion. This happens in complexes such as $TpPd(\eta^3$-allyl),[309] or Tp_2Pd,[310] where the ligand is bidentate in the crystal, but all the pyrazolyl groups exchange rapidly in solution, so that only one type of pyrazolyl group is observed in the NMR spectrum. The ligand pzTp can also act in bis-bidentate fashion: the isolated complex pzTpPd(η^3-allyl) is still a neutral bidentate ligand,[4] and can form the binuclear cation $[(\eta^3$-allyl)Pd(μ-pz)$_2$B(μ-pz)$_2$Pd(η^3-allyl)]$^+$.[311]

2.2 Group 1: H, Li, Na, K, Rb and Cs

2.2.1 H

Although the scorpionate ligands can be hydrolyzed under acid conditions to the corresponding pyrazoles and boric acid, their complexes with a proton, the "free acids" $H[Tp^x]$, can be obtained by careful acidification of the anions $[Tp^x]^-$. They are reasonably stable solids, containing a chelated proton and they also may contain a variable number of associated water molecules. Upon heating, they lose pyrazole, and are converted to the corresponding pyrazaboles. For instance,

$$2 \text{ H[HB(pz)}_3] \rightarrow (pz)HB(\mu\text{-pz})_2BH(pz) + 2 \text{ Hpz} \uparrow \qquad (2.4)$$

The structures of the free acids resemble those of their alkali metal salts,[312] and they are capable of forming directly complexes derived from their anions. For instance, $TpTcCl_2O$ and $TpReO_3$ have been synthesized under strongly acid conditions without destruction of the ligand.[313,314] In addition, a triprotonated cation of a second generation ligand $[Tp^{tBu}H_3Cl]^+$, containing a chloride ion coordinated to three NH protons, has been isolated and structurally characterized.[315] There are also a few examples of the free acids themselves coordinating to a metal, with retention of the NH bond, as in $\{[TpH]AuMe_2\}^+$, **23**, and $\{[pzTpH]AuMe_2\}^+$, **24**, in both of which the ligands were coordinating in κ^2 fashion. Their NMR spectra indicated fluxional behaviour involving five-coordinate intermediates.[316]

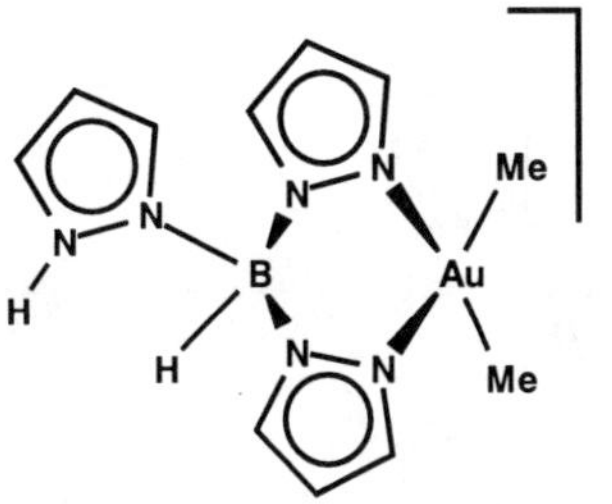

2.2 GROUP 1 COMPLEXES

23 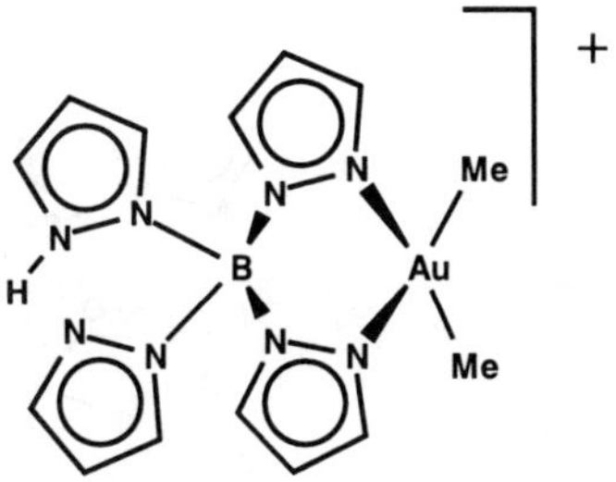 **24**

The free acids were useful in preparing salts of cations which were not readily available as borohydride salts (e.g. Cs), or which were unstable under the synthesis conditions (e.g. R_4N^+) through their reaction with the corresponding hydroxide.[9]

$$H[Tp^x] + (cation)(OH) \rightarrow (cation)[Tp^x] + H_2O \tag{2.5}$$

In a recent study of thermolysis of the free acid TpH it was found that three different products can be formed, depending on the decomposition conditions. Simple pyrolysis yielded the *trans*-4,8-dipyrazolylpyrazabole, **25**, while thermolysis in the presence of $ZrCl_4$ produced the *cis* isomer, *cis*-4,8-dipyrazolylpyrazabole, **26**. Finally, heating in the presence of moisture and HCl gave rise to a molecule, **27**, in which two 4,8-dipyrazolylpyrazaboles were linked through two oxo bridges, which have replaced the original B—H bonds.[1539]

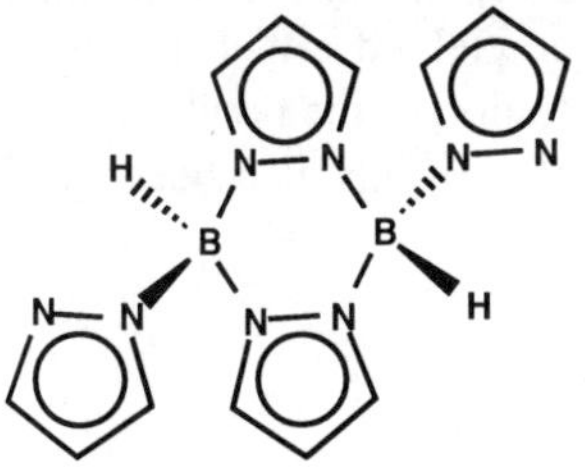

25 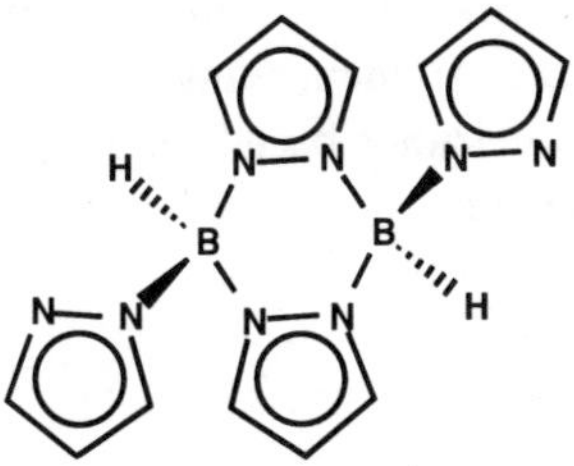 **26**

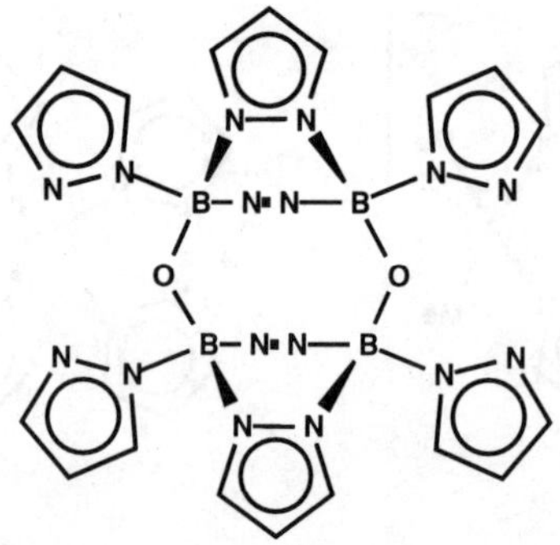

27

2.2.2 Li, Na, K, Rb and Cs

Homoscorpionate salts of these cations have been prepared by direct synthesis from the corresponding borohydrides, or from the free acids H[Tp^x] and a metal hydroxide, or [Me$_4$N][OH] as in Eq. 2.5.[9] The isomorphous salts NaTp and KTp were structurally characterized.[318] Evidence for the existence of the [KTp$_2$]$^-$ species was obtained from negative ion electrospray mass spectrometry,[1513] while the sodium anion, [NaTp$_2$]$^-$ was structurally characterized by X-ray crystallography.[319]

2.3 Group 2: Be, Mg, Ca, Sr, Ba

2.3.1 Be

Complexes of beryllium(II) were prepared from Tp and pzTp. While the latter formed a tetrahedral [pzTp]$_2$Be complex, the former yielded either Tp$_2$Be, or a trimeric species [TpBeOH]$_3$, depending on the concentration of the beryllium(II) cation.[320,321] The analogous trimer based on Tp*, [Tp*BeOH]$_3$, has also been synthesized, and structurally characterized.[322]

2.3.2 Mg

In a study of extracting alkaline earth cations as their known[9] scorpionate complexes, it was found that Tp is an effective extractant for Be^{2+}, Mg^{2+} and Ca^{2+}, while pzTp is effective only for Be^{2+} and Mg^{2+}.[320] The structures of Tp$_2$Mg and [pzTp]$_2$Mg were found to be octahedral. On the basis of molecular mechanics calculations, X-ray crystallography, and multinuclear NMR it was concluded that the extraction selectivity is governed by the stability of $Tp^x{}_2M$ complexes, which is controlled by steric effects.[322] The complex Tp*MgR was found to undergo ligand redistribution, forming the structurally characterized octahedral $Tp^*{}_2Mg$.[323-325]

2.3.3 Ca, Sr, Ba

Structures of Tp_2Ca, $Tp^*{}_2Ca$ and $Tp^*{}_2Sr$ have been determined, as were those of $Tp^*{}_2Ba$. The complex Tp*BaI was synthesized, and converted to $Tp^*BaI(HMPA)_2$, the structure of which was determined by X-ray crystallography, and also to the heteroleptic complex Tp*Ba[Bp](THF).[326-328]

2.4 Group 3: Sc, Y (lanthanides and actinides are listed separately)

2.4.1 Sc

The only scandium complex reported was the eight-coordinate $ScTp_3$.[329]

2.4.2 Y

Reported yttrium complexes included the homoleptic, eight-coordinate YTp_3,[329] and the ionic, six-coordinate $[Tp^*{}_2Y][OTf]$,[330] as well as a large number of heteroleptic complexes including those of various aliphatic and aromatic carboxylic mono-, and diacids. These complexes were exemplified by: $Tp_2Y(OOCPh)$,[331] and $Tp_2Y(OOCC_6H_4\text{-p-}Bu^t)$,[332] by the tropolone derivative $Tp_2Y(\text{tropolonato})$,[333,334] by the oxalato complex $[(Tp_2Y)_2(C_2O_4)]$,[335] and also by beta-diketonato species, such as $Tp_2Y(AcAc)$,[336,337] $Tp_2Y(Ph(O)CHC(O)Me)$ and $Tp_2Y(Ph(O)CHC(O)Ph)$.[338] The salicylaldehyde derivative $Tp_2Y(\text{salicylaldehydato})$ has also been reported.[339] The structures of complexes $Tp_2YCl(H_2O)$,[340] $Tp_2YCl(Hpz)$,[341] and of the dinuclear, seven-coordinate $[TpY(\mu\text{-OAc})_4YTp]$ were established by X-ray crystallography.[341] In the structurally characterized $[TpY(NCS)(\mu\text{-OH})]_2$ the thiocyanato ligand was found to be non-linear.[342] The complex $Tp^*YCl_2(THF)$ was converted to $Tp^*YR_2(THF)$ derivatives (R = Me, Bu^t, Ph, CH_2SiMe_3), by treatment with the appropriate carbanions. Such complexes were found to be active catalysts in ethylene polymerization.[343]

2.5 Group 4: Ti, Zr, Hf

2.5.1 Ti

The first Ti scorpionate complexes reported were Tp*TiCp$_2$, TpTiCl$_2$Cp, TpTiCp$_2$, TpTiCl$_2$(THF), and the cation [Cp$_2$Ti(μ-pz)$_2$B(μ-pz)$_2$TiCp$_2$]$^+$, containing one of the few examples of a bis-bidentate pzTp ligand.[344] Other titanium(IV) complexes, such as TpTiCl$_3$ and Tp*TiCl$_3$, proved to be convenient starting materials for a variety of titanium(IV) derivatives through the displacement of chloride ion with various nucleophiles.[345,346, 1559] The complex TpTi(OMe)Me$_2$ was prepared in this way,[347] as were Tp*TiCl$_2$(LL) species.[1559] The reaction of TpTiCl$_3$ with hydrazines or with organohydrazides yielded hydrazido complexes, which contained a side-on bonded hydrazido group.[348] On the other hand, with Tp*TiCl$_3$ one obtained titanium(III) complexes, exemplified by Tp*TiCl$_2$(Hpz*), the structure of which was determined by X-ray crystallography.[349] Other, somewhat more complicated species, such as Tp*Ti(NBu^t)Cl(4-Bu^{t}py), **28**,[350,351] the two cyclooctatetraenyl complexes TpTi(COT) and Tp*Ti(COT),[352] and the tetracarbonyl anion [TpTi(CO)$_4$]$^-$, **29,** were also reported.[353, 354]

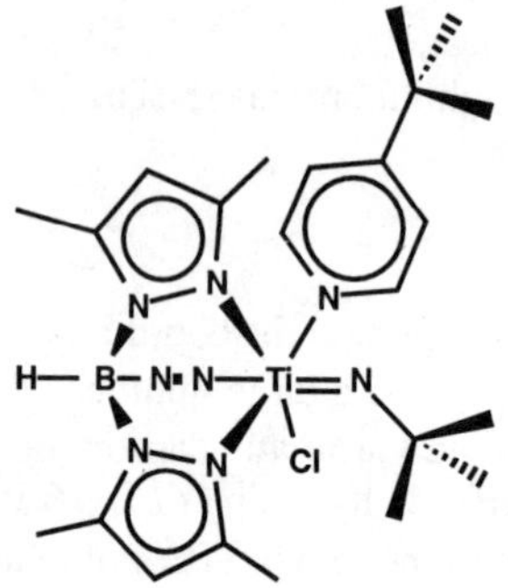

28

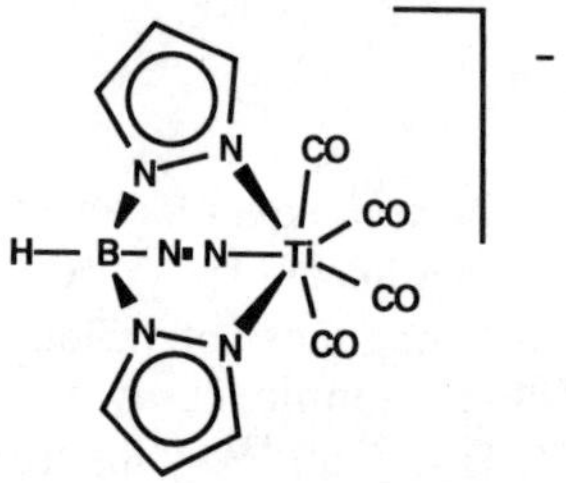

29

2.5.2 Zr

Like TiCl$_4$, ZrCl$_4$ was readily converted by KTp^x ligands to TpZrCl$_3$ and Tp*ZrCl$_3$,[355] although better yields were claimed using Tp*SnClBu$_2$ as the Tp* transfer agent, instead of KTp*.[356] From these intermediates other complexes, such as Tp*Zr(OAr)$_3$,[357] Tp*ZrX$_n$(OR)$_{3-n}$,[358] Tp*ZrCl$_2$(LL),[1559] and Tp*Zr(OBu^t)R$_2$ were synthesized.[359,360] Complexes including both, Tp^x and Cp ligands, such as TpZrCpCl$_2$, Tp*ZrCpCl$_2$,[361] and Tp*ZrCp(OAr)$_2$,[362,363] were also described.

2.5.3 Hf

The only reported hafnium scorpionate complex was TpHf(Cp)Cl$_2$.[363]

2.6 Group 5: V, Nb, Ta

The coordination chemistry of Group 5 scorpionates has been covered in considerable detail up to late 1995 in an excellent review by Etienne.[47]

2.6.1 V

The first reported examples of scorpionate vanadium species were TpVCl$_3$, TpVCp and TpVCl$_2$(THF),[344] as well as the homoleptic compounds Tp$_2$V and [pzTp]$_2$V.[364] The structure of TpVCl$_2$(THF) was determined by X-ray crystallography.[365] A useful entry into vanadium chemistry was provided by the complexes Tp^{x}VCl$_2$(DMF).[366] The chlorides in them were replaceable by nucleophiles, such as alkoxy groups, yielding a number of structurally characterized derivatives: TpVCl$_2$(OPr^i), TpVCl$_2$(OBu^t) and Tp*VCl$_2$(OBu^t).[367] The exposure to air of these compounds produced the analogous vanadyl species. The complex Tp*VO(AcAc) was obtained directly from VO(AcAc)$_2$ and KTp*, and its structure was determined by X-ray crystallography,[368] as was the structure of Tp*VO(PhCOCHCOPh),[369,370] of the cationic complex [Tp*$_2$V][BPh$_4$],[371] of the diphenolate Tp*VO(OC$_6$H$_4$-p-Br)$_2$,[372] and of the dithiocarbamate Tp*VO(S$_2$CNPr$_2$).[373] EPR studies were done on the latter compound,[374] and also on [Tp*VO(Hpz*)]$_2$(O$_2$C-R-CO$_2$).[375] The electronic structure of [TpVO] β-diketonates and dithiocarbamates was investigated.[376,377] Imido derivatives of the Tp*V core were synthesized,[378] and the t-butylimido complex Tp*V(=NBu^t)Cl$_2$ was readily converted to the mercapto species Tp*V(=NBu^t)(SR)$_2$, **30**[379] Related complexes were found to polymerize ethylene and propylene in the presence of aluminoxane.[380]

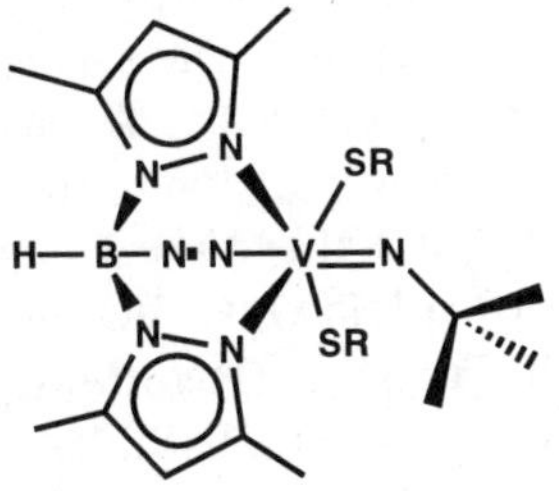

30

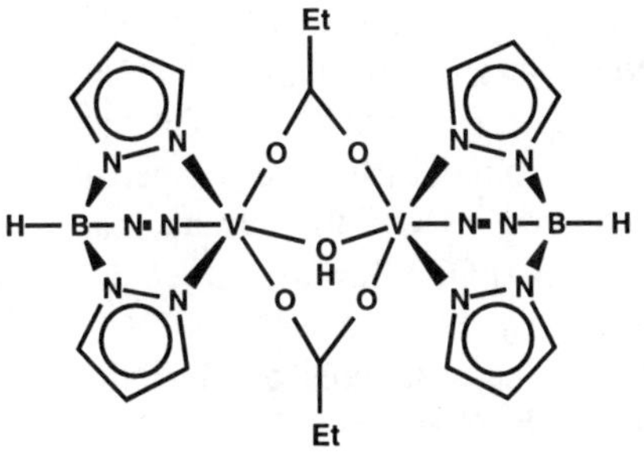

31

In addition to the above mononuclear vanadium complexes, many polynuclear species were investigated, such as the binuclear complex [(TpV)$_2$(μ-O)(μ-OAc)$_2$],[381] and the structures of its protonated and unprotonated analogs, of the antiferromagnetic [(TpV)$_2$(μ-OH)(μ-O$_2$CEt)$_2$], **31**, and also of the ferromagnetic [(TpV)$_2$(μ-O)(μ-OAc)$_2$] were determined.[382] The complex [(Tp*VO)$_2$(malonate)], which was structurally characterized, along with Tp*VOCl(Hpz*) and Tp*VO(O$_2$CPh)(Hpz*), contained a six-membered chelate ring between malonate ion to one vanadium, and a single oxygen bond to the other vanadium.[383] The ferrocenyl derivative Tp*VO(Hpz*)(O$_2$CFc) was synthesized, studied by cyclic voltammetry, and its structure was determined by X-ray crystallography.[384]

Numerous mononuclear and dinuclear vanadium(IV) complexes bridged with various phosphinate, phosphite, and similar anions were synthesized, and structurally characterized. These complexes are exemplified by [TpVO(μ-(PhO)$_2$PO$_2$)]$_2$, [Tp*VO(μ-(PhO)$_2$PO$_2$)]$_2$,[385] [(TpVO)$_2$(μ-(Ph)$_2$PO$_2$)], [(TpVO)$_2$(μ-(Ph)HPO$_2$)], [(TpVO)$_2$(μ-OH$_2$)], [(TpVO)$_2$(μ-OAc)(μ-OH)],[386] the tetranuclear Tp$_4$V$_4$(μ-ArOPO$_3$)$_4$,[387] [TpVCl(μ-(PhO)$_2$PO$_2$)]$_2$, [TpVCl(μ-(Ph)$_2$PO$_2$)]$_2$, [TpVCl(μ-(Ph)HPO$_2$)]$_2$, [TpV[(PhO)$_2$POS](DMF), [TpV[(PhO)$_2$PO$_2$](H$_2$O),[388] the cyclic tetramer [TpVO$_2$]$_4$,[367] and two complex trinuclear species, capped at each end with Tp ligands, and bridged to the central vanadium(III) by diphenyl phosphate anions.[389] Several dinuclear complexes containing [TpV] moieties bridged by one oxo and one carboxylato or phosphonato bridge were synthesized and structurally characterized. Their UV-visible, resonance Raman, paramagnetic NMR, and magnetic susceptibility properties were found to be dependent on the V—O—V angle.[390]

2.6.2 Nb

The first niobium scorpionate complexes reported were TpNbO(OMe)$_2$, TpNbCl$_4$, the dimer [TpNbCl$_3$]$_2$, and the salts K[TpNbCl$_5$] and K[TpNbCl$_4$].[391,392] Tp*NbCl$_4$ was obtained from NbCl$_5$ and KTp*, as was Tp*NbCl(pz*)$_3$,[105] and Tp*NbCl$_3$ was obtained from Tp*SnClBu$_2$ and NbCl$_4$(THF)$_2$.[356] The synthesis of Tp*NbOCl$_2$, Tp*NbSCl$_2$ and Tp*Nb(=NBu^t)Cl$_2$ was also reported.[378] A cluster of the composition Nb$_3$BO$_7$ was capped at each Nb with a Tp ligand,[393] and structurally characterized, as was the tetranuclear complex [TpNb(=O)O]$_4$.[394] The complexes Tp*Nb(O)(Cl)(OR) (R = Me, Et) could be hydrolyzed to [Tp*Nb(O)(Cl)]$_2$O, and were converted by PCl$_3$ to Tp*Nb(O)Cl$_2$, from which Tp*Nb(O)(HNSiMe$_3$)$_2$ was prepared. The structures of [Tp*Nb(O)(Cl)]$_2$O and Tp*Nb(O)(HNSiMe$_3$)$_2$ were determined by X-ray crystallography.[395]

2.6 GROUP 5 COMPLEXES

Much interesting chemistry evolved from niobium alkyne complexes exemplified by Tp*NbCl$_2$(PhC≡CR) and Tp*NbCl$_2$(MeC≡CMe),[396] and their analogs derived from Tp, which included also TpNbCl$_2$(Me$_3$SiC≡CSiMe$_3$).[397] The structures of TpNbCl$_2$(PhC≡CMe) and of TpCpNbCl(PhC≡CMe) were determined by X-ray crystallography, and their ground state geometries were probed by extended Hückel calculations. Alkyl groups on the coordinated acetylenes could be deprotonated next to the acetylenic bond, and alkylated. In this fashion TpNbCl$_2$(PhC≡CCR'HMe) was obtained from the precursor TpNbCl$_2$(PhC≡CEt).[398] The ethyl derivatives Tp*NbCl(Et)(PhC≡CR) have shown α-agostic, rather than β-agostic C—H—Nb interaction.[399] Complexes of the type Tp*NbCl(R')(PhC≡CR) underwent thermal metathesis of the Nb-bonded and alkyne-bound alkyl groups.[400] A high-yield one-pot synthesis of Tp*NbCl$_2$(alkyne) complexes, and their conversion to dimethyl or dibenzyl derivatives, Tp*NbR'$_2$(alkyne), has been described. However, the reaction with two equivalents of EtMgCl led to a niobacycle, characterized by NMR.[401] The propensity for niobacycle formation was a recurring theme in this chemistry. For instance, the reaction of Tp*NbMe(OMe)(PhC≡CR), which was obtained from Tp*NbCl(OMe)(PhC≡CR) and MeLi, with carbon monoxide produced the niobacycle **32**.[402] The structurally characterized metallacycle **33** was obtained from the reaction of Tp*NbCl$_2$(PhC≡CR) with one equivalent of allyl Grignard reagent.[403] Still another niobacycle, **34** was readily formed upon the addition of HBF$_4$ to Tp*Nb(CO)(RCN)(PhC≡CMe), which was produced by CO displacement from Tp*Nb(CO)$_2$(PhC≡CMe) by RCN.[404]

32   **33**   **34**

Dicarbonyl complexes, Tp*Nb(CO)$_2$(alkyne), were obtained via sodium amalgam reduction of Tp*NbCl$_2$(alkyne). They reacted with other alkynes undergoing replacement of one CO and formation of Tp*Nb(CO)(alkyne')(alkyne).[405] A detailed kinetic study of the reversible migratory insertion/β-alkyl elimination in α-agostic alkylniobium alkyne complexes was carried out, showing that alkyl migration is the key step of the rearrangement. Reaction of Tp*Nb(Et)Cl(PhC≡CEt) with N$_3$P(N-Pr$^i_2$)$_2$ produced the structurally characterized Tp*NbCl[=NP(N-Pr$^i_2$)$_2$](CPh=CEt$_2$).[406] Novel catalytic systems for ethylene polymerization were discovered, based on

$Tp^xNbMe_2(PhC\equiv CMe)/B(C_6F_5)_3$, where Tp^x was Tp or Tp*.[407] An unusual equilibrium between α- and β-agostic interactions was observed in the structurally characterized $Tp^*Nb(Pr^i)Cl(PhC\equiv CMe)$, while only α-agostic interaction was found in the related complex $Tp^*Nb(Et)Cl(PhC\equiv CMe)$.[408]

2.6.3 Ta

The first homoscorpionate complexes of tantalum were prepared by the reaction of $TaMe_3Cl_2$ with Tp, Tp* and pzTp salts, forming the respective Tp^xTaMe_3Cl species. The structure of $TpTaMe_3Cl$ showed it to be a capped octahedron.[409] Tp*K reacted readily with $TaCl_3(=CHBu^t)(THF)_2$ yielding $Tp^*TaCl_2(=CHBu^t)$, **35**, in which one Cl could be replaced by alkoxy groups, by the (=NR) moiety, and also by a side-on bonded $PhN=CH_2$.[410] The complex $[Tp^*TaCl_3][TaCl_6]$ was prepared from $TaCl_5$ and $Tp^*SnClBu_2$.[356]

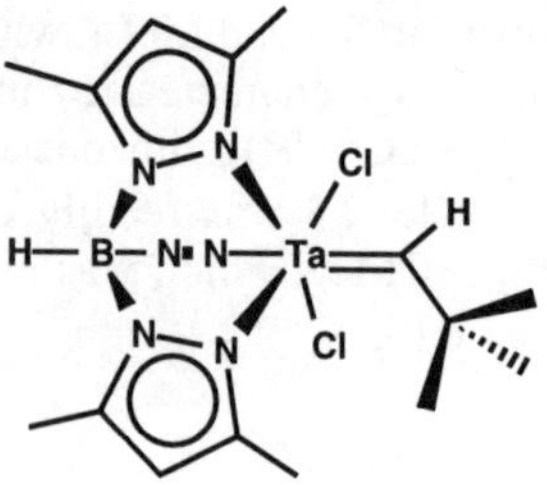

35

2.7 Group 6: Cr, Mo, W

2.7.1 Cr

The first chromium scorpionate was the unstable $pzTpCr(CO)_2(\eta^3$-allyl), obtained from $[pzTpCr(CO)_3]^-$ and allyl bromide. It underwent a complicated reaction with halocarbon solvents, producing ultimately the $[pzTp_2Cr]^+$ cation, isolated as the hexafluorophosphate salt.[4] The anion $[TpCr(CO)_3]^-$ was studied electrochemically.[411] Oxidation of the related Tp^x anions, produced 17-electron complexes $TpCr(CO)_3$, $pzTpCr(CO)_3$, and $Tp^*Cr(CO)_3$, which were studied by EPR. $TpCr(CO)_2PMe_3$ was obtained from $TpCr(CO)_3$ on treatment with PMe_3.[412] The structure of $pzTpCr(CO)_3$ was established by X-ray crystallography.[413] Also reported was the complex

2.7 GROUP 6 COMPLEXES

Tp*Cr(=NBu^t)$_2$Cl,[378] as were [TpCrCl$_3$][AsPh$_4$], [Tp$_2$Cr][PF$_6$], and the structurally characterized TpCrCl$_2$(py).[414] Spectroscopic properties and ligand field parameters of the cation [Tp$_2$Cr]$^+$ were determined.[415] During an unsuccessful attempt to synthesize the complex [TpCrCl]$_2$, several other Cr^{III} complexes were isolated and structurally chartacterized. They included, among others, [Tp$_2$Cr][TpCrCl$_3$], [HPMe$_3$][TpCrCl$_3$], and TpCrCl$_2$(THF).[416] Carbyne complexes such as Tp*Cr(≡CNPr$^i_2$)(CO)$_2$, **36**, and Tp*Cr(≡CNPr$^i_2$)(CO)(CNBu^t), **37**, were prepared and characterized.[417] Two chromium(III) dialkyls, Tp*CrMe$_2$(DMAP) and Tp*Cr(CH$_2$Ph)$_2$(DMAP), where DMAP = 4-dimethylaminopyridine, were synthesized, and structurally characterized. Analogous complexes with donor ligands other than DMAP were less stable.[418]

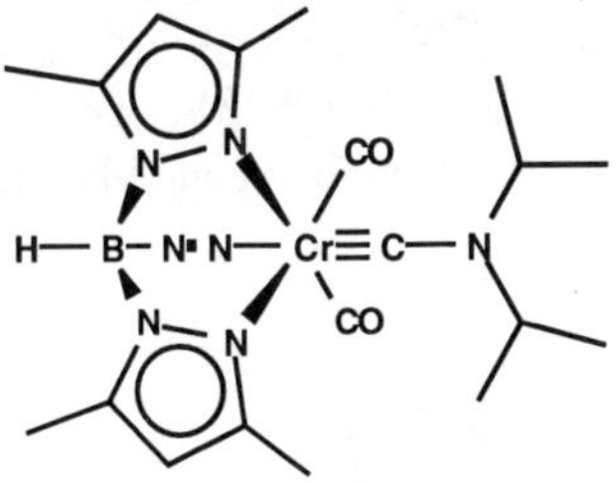

36 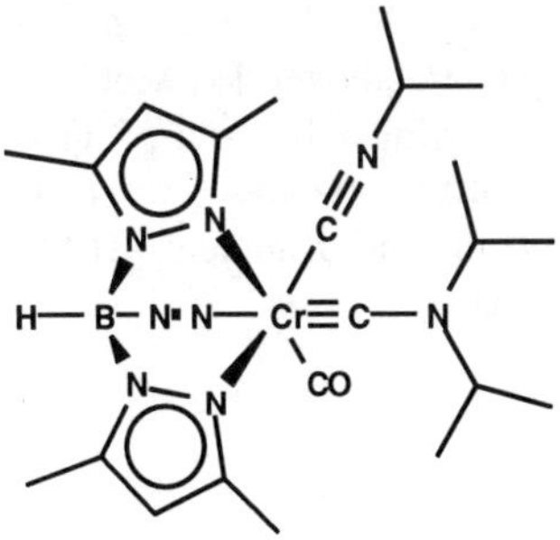   **37**

2.7.2 Mo

Among the "first generation" homoscorpionate compounds, those of molybdenum were the most numerous, comprising about 30 % of all such complexes, and together with tungsten, which displays comparable chemistry, they accounted for almost 40 % of them. In a fair number of references identical, or almost identical complexes of both, molybdenum and tungsten, have been reported. The reason for this popularity of molybdenum was the ease with which molybdenum complexes of various oxidation states could be synthesized, ranging from Mo^O in [TpMo(CO)$_3$]$^-$ to Mo^{VI} in TpMoO$_2$X. The reaction of Tp^x ligands with Mo(CO)$_6$ was a particularly convenient entry into the low-valent molybdenum area:

$$[Tp^x]^- + Mo(CO)_6 \longrightarrow [Tp^xMo(CO)_3]^- \longrightarrow [Et_4N][Tp^xMo(CO)_3] \quad (2.6)$$

Such tricarbonyl anions could be isolated as quaternary amonium salts,[16,4] of which [Et$_4$N][TpMo(CO)$_3$],[419] and [Et$_4$N][Tp*Mo(CO)$_3$],[420] were structurally characterized, and the relative stability in solution of [TpMo(CO)$_3$]$^-$ was determined by calorimetry.[421] An unusual complex, alleged to be Tp*Mo(CO)$_3$TiCp$_2$(β-diketonate), has been reported, but this structure has not been adequately established.[422]

From the above tricarbonyl anions a veritable cornucopia of other complexes could be obtained. Protonation yielded the metal hydrides, Tp^{x}Mo(CO)$_3$H,[16] and the structures of Tp*Mo(CO)$_3$H and of the 17-electron radical, Tp*Mo(CO)$_3$, were determined by X-ray crystallography. The rates of degenerate transfer of electrons, protons and hydrogen atoms between these species were determined and compared.[423] Gentle oxidation of [TpMo(CO)$_3$]$^-$ yielded the structurally characterized 17-electron radical TpMo(CO)$_3$, which on heating formed the dimer [TpMo(CO)$_3$]$_2$, containing a Mo≡Mo bond.[413,414,424,425] In the complex TpMo(μ-OAc)$_2$MoTp, one Tp was tridentate and the other bidentate.[426] The structurally characterized [TpMoCl]$_2$ dimer contained a ClMo≡MoCl unit, with each Tp ligand bonding with two pz groups to one Mo atom, and with one pz to the other.[427] Complexes such as TpMo(CO)$_3$—Rh(PPh$_3$),[428] **38**, TpMo(CO)$_3$—SnR$_3$,[429] **39**, and the trinuclear [TpMo(CO)$_3$]$_2$Cu,[430] containing molybdenum-metal bonds, were also reported.

38   **39**

The fairly stable tricarbonyl radical Tp*Mo(CO)$_3$ was converted to [Tp*Mo(CO)$_3$]$_2$S,[431] and from [Tp*Mo(CO)$_3$]$^-$ and N$_3$S$_3$Cl$_2$ the complexes Tp*Mo(CO)$_2$(NS) and [Tp*Mo(CO)$_2$]$_2$S were obtained,[432] while in the reaction with tetraalkylthiuram disulfide, the product was Tp*Mo(S$_2$CNR$_2$)$_2$, containing one bidentate, and one monodentate R$_2$NCS$_2$ ligand.[433,434] The reaction of KTp* with Mo(I)$_2$(CO)$_3$(MeCN)$_2$ yielded Tp*Mo(CO)$_2$I, along with some ligand degradation products.[435] Oxidation of [Tp*Mo(CO)$_3$]$^-$ with iodine gave Tp*Mo(CO)$_2$I, which added CNBu^t, forming Tp*Mo(CO)$_2$I(CNBu^t), which was reduced to the anion [Tp*Mo(CO)$_2$I(CNBu^t)]$^-$. Upon reaction with methyl iodide, it produced a mixture of three products, the aminocarbyne, η^2-iminoacyl, and η^2- acyl derivatives, **40, 41**, and **42**, respectively. When CNMe or CNPh were used instead of CNBu^t, only the

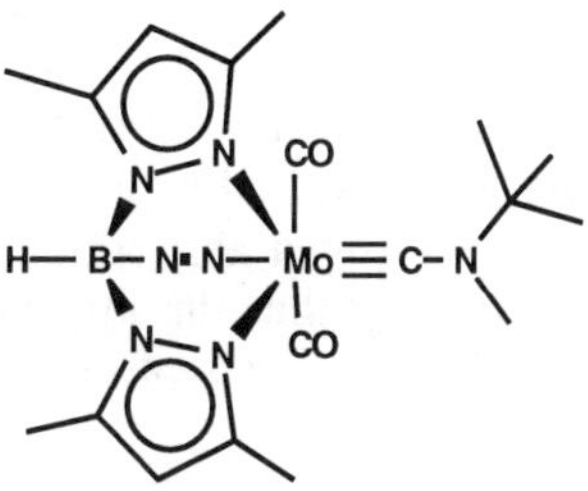

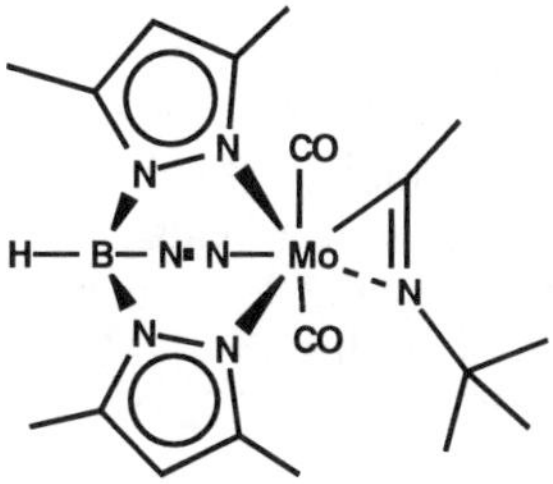

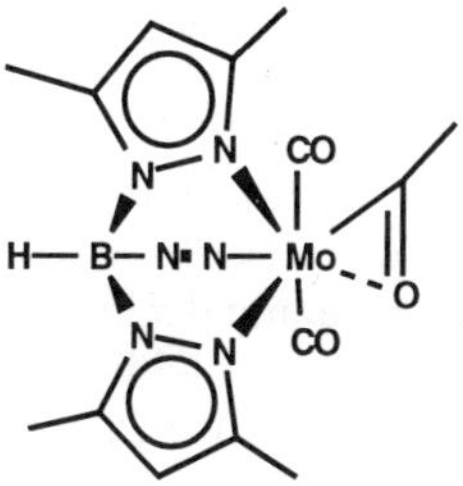

2.7 GROUP 6 COMPLEXES

40   **41**   **42**

aminocarbyne was formed.[436] Treatment of [Tp*Mo(CO)$_3$]$^-$ with sulfur or selenium produced the binuclear species [Tp*Mo(CO)$_2$]$_2$S and [Tp*Mo(CO)$_2$]$_2$Se, respectively.[437] Compounds Tp*(CO)$_2$EAr (E = S, Se) were obtained from [Tp*Mo(CO)$_3$]$^-$ and ArSCl (ArSO$_2$Cl could also be used) or ArSeX,[19] and Tp*Mo(CO)$_2$(SAr), where Ar was *p*-chlorophenyl, was structurally characterized.[438]

The reaction of [TpMo(CO)$_3$]$^-$ or [pzTpMo(CO)$_3$]$^-$ with allyl halides led directly to the η^3-allyl derivatives, [Tp^{x}Mo(CO)$_2$(η^3-allyl)], Tp^x including pzTp, which gave rise to **43**.[16,4] No reaction took place with the Tp* analog, but the complex [Tp*Mo(CO)$_2$(η^3-allyl)] could be readily obtained by treating [Tp*]$^-$ with an independently prepared precursor of the cation [Mo(CO)$_2$(η^3-allyl)]$^+$.[17] Structures of [Tp*Mo(CO)$_2$(η^3-methallyl)],[439] [Tp*Mo(CO)$_2$(η^3-cinnamyl],[440] of eight other diversely 1-, 1,3-, 1,1,3- and 1,2,3-(η^3-allyl)-substituted [TpMo(CO)$_2$(η^3-allyl)] complexes,[441] of the cyclopentadienone complex TpMo(CO)$_2$(η^3-C$_5$H$_4$O),[442] and of TpMo(CO)$_2$(PEt$_2$Ph),[413] were determined by X-ray crystallography. In pzTpMo(CO)$_2$(η^5-Cp), **44**, the pzTp ligand was bidentate.[443] Various other Tp and

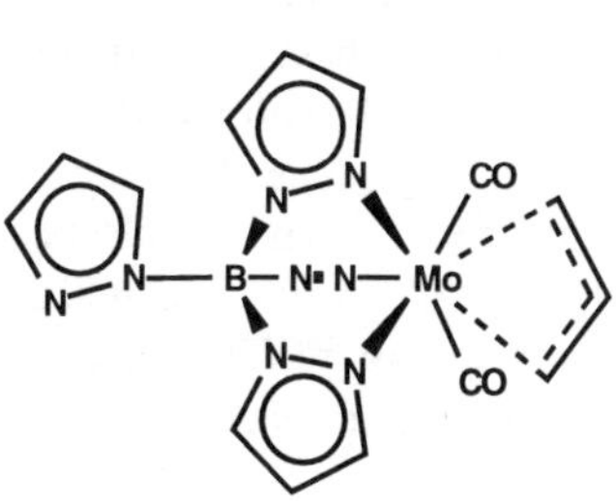

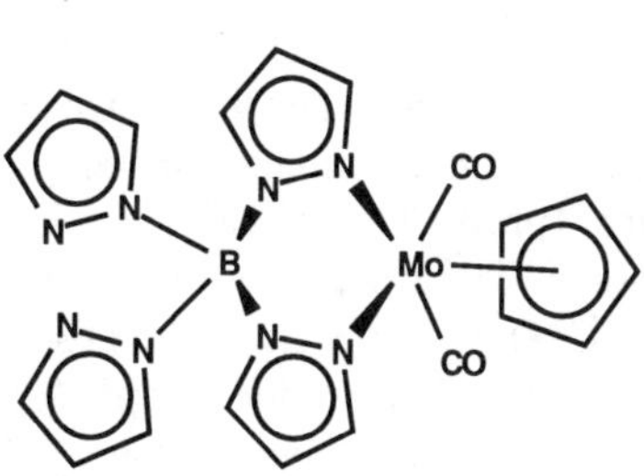

43   **44**

Tp* η^3-allyl species were synthesized,[444,445] and investigated as useful reagents or catalysts for organic synthesis.[446-450] The structurally characterized complex TpMo(CO)$_2$(η^3-C$_6$H$_7$O) resisted hydride abstraction with [Ph$_3$C]$^+$, in contrast to the η^3-cyclopentenone analog. A rationale for this behavior was provided by extended Hückel calculations combined with a Walsh analysis of hydrogen abstraction.[451] Stereochemical non-rigidity, related to a rotational process within the molecule around the B—Mo axis was studied by NMR.[17,452,453] Exhaustive treatment of TpMo(CO)$_2$(η^3-allyl) with NO produced TpMo(NO)$_2$NO$_2$.[454]

Alkylation of [TpMo(CO)$_3$]$^-$ with alkyl halides yielded not the expected seven-coordinate Tp^{x}Mo(CO)$_3$R species,[16] but rather the η^2-acyl complexes Tp^{x}Mo(CO)$_2$(η^2-COR)[6,7,455], which have been briefly reviewed.[456] The only authentic Tp^{x}Mo(CO)$_3$R complex reported was the structurally characterized, seven coordinate TpMo(CO)$_3$CH$_2$CN.[413] Related to the above η^2-acyl complexes were pzTpMo(CO)$_2$(η^2-OCNMe$_2$),[457] pzTpMo(CO)$_2$(η^2-SCNMe$_2$),[458] and thioacyl species.[1561] Bromination of [Tp^{x}Mo(CO)$_3$]$^-$ produced the seven-coordinate Tp^{x}Mo(CO)$_3$Br,[459] but, on the other hand, reaction of [Tp*Mo(CO)$_3$]$^-$ with iodine or with PhICl$_2$ yielded Tp*Mo(CO)$_2$I and Tp*Mo(CO)$_2$Cl, respectively, rare examples of 16-electron carbonyl derivatives with a spin triplet ground state. Oxidation of Tp*Mo(CO)$_2$I gave rise to the structurally characterized Tp*MoOI$_2$,[460] while oxidation of [Tp*Mo(CO)$_3$]$^-$ with dimethyldioxirane yielded [TpMoO$_3$]$^-$, along with the tetramer [TpMoO(μ-O)]$_4$.[461] The structure of TpMo(CO)$_2$(S$_2$CNR$_2$) was determined by X-ray crystallography.[462]

The above acyl complexes provided an entry into molybdenum carbene and carbyne derivatives through stereospecific alkylation of molybdenum(II) enolates.[463] For instance, the treatment of Tp*Mo(CO)$_2$(η^2-COMe) with base generated [Tp*Mo(CO)$_2$(C(O)=CH$_2$)]$^-$, which was converted to various products, including carbynes, vinylidene and ketenyl derivatives, among which the complex {Tp*Mo(CO)[P(OPh)$_3$](η^2-C(O)CHMeBn)} was structurally characterized.[464,465] Treatment of Tp*Mo(CO)$_2$(η^2-COR) complexes with excess base converted them to carbyne complexes, Tp*Mo(CO)$_2$($\equiv$CR). These could be deprotonated to anionic vinylidene species, capable of being alkylated at the vinylidene β-carbon.[466] The complex TpMo(CO)$_2$(η^2-OCNPr$^i_2$) has also been reported.[1545] The simplest carbyne derivative, Tp*Mo(CO)$_2$$\equiv$CH, dimerized slowly to **45**,[467] while the reaction of TpMo(CO)$_2$$\equiv$CAr with CS$_2$ produced the complex **46**.[468] The arylcarbyne Tp*Mo(CO)$_2$$\equiv$CAr produced Tp*Mo(CO)$_2$B(Et)CH$_2$Ar upon reacting with Et$_2$BH, and this complex contained an agostic bond.[469]

The chlorocarbyne Tp*Mo(CO)$_2$$\equiv$CCl,[30] obtained via radical reaction of the tricarbonyl anion with methylene chloride, turned out to be a very versatile starting

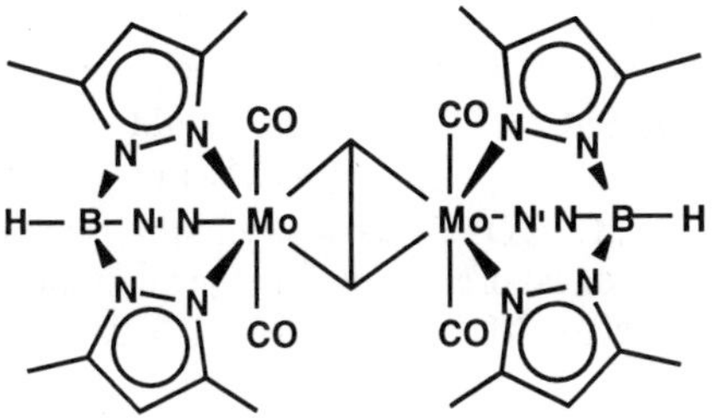

2.7 GROUP 6 COMPLEXES

45 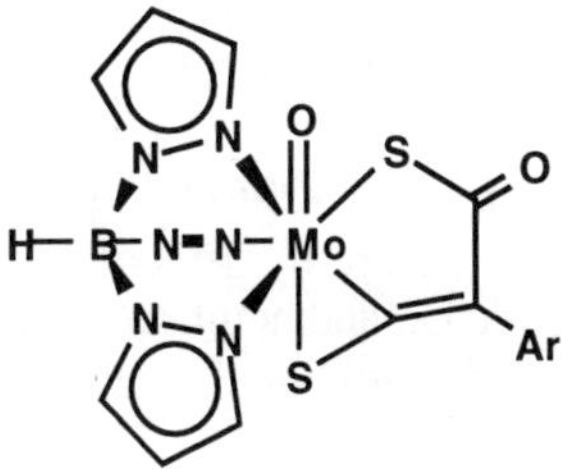   **46**

material for many complexes. The analogous bromocarbyne was prepared similarly. The formation of this chlorocarbyne was studied in considerable detail, and it was found to be strongly affected by the solvent, and also by steric effects.[470] Tp*Mo(CO)$_2$≡CCl reacted with [CpFe(CO)$_2$]$^-$ forming the structurally characterized dinuclear Tp*Mo(CO)$_2$≡C–Fe(CO)$_2$Cp,[471] and in a similar reaction Tp*Mo(CO)$_2$≡C–Mn(CO)$_2$Cp was obtained.[472] It also provided a convenient route to complexes [Tp*Mo(CO)$_2$≡C–P=C(NR$_2$)$_2$],[473,474] which could be readily oxidized with molecular oxygen to compounds exemplified by [Tp*Mo(CO)$_2$≡C-P(O)$_2$C(NR$_2$)$_2$],[475] or Tp*Mo(CO)$_2$≡COAr,[476] as well as to the carbynes Tp*Mo(CO)$_2$≡C(ER) (E = S, Se), to [Tp*Mo(CO)$_2$≡C-E]$^-$,[31] and to vinylidene metallacycles, such as **47**.[477]

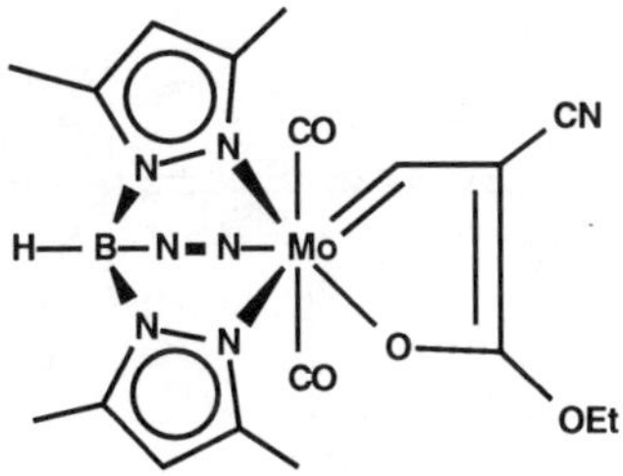

47

Photolysis of TpMo(CO)$_2$≡CR in the presence of various phosphines yielded species, such as TpMo(CO)(PR$_3$)(η^2-OCCR), which could be converted to the thio,[478] and to the seleno analogs.[479] LiMe$_2$Cu converted Tp*Mo(CO)$_2$≡CCl to Tp*Mo(CO)$_2$≡CCH$_3$ which, upon deprotonation, formed the reactive anion [Tp*Mo(CO)$_2$=C=CH$_2$]$^-$, exhibiting a rich chemistry as a nucleophile.[480] For instance, it reacted with Tp*Mo(CO)$_2$≡CCl yielding [Tp*Mo(CO)$_2$≡C]$_2$CH$_2$, which itself could be deprotonated to the dianion [Tp*Mo(CO)$_2$=C=C=C=Mo(CO)$_2$Tp*]$^{2-}$ and this could be readily dialkylated at the central carbon atom. Mixed analogs containing one Mo and one W moiety were also prepared.[481] By a similar sequence of deprotonations and alkylations, species such as **48, 49** and **50** were synthesized.[50] [Tp*Mo(CO)$_2$=C=CH$_2$]$^-$ reacted with TpW(CO)(I)(PhC≡CPh) producing the dinuclear complex, Tp*Mo(CO)$_2$≡CCH$_2$—W(CO)(PhC≡CPh)Tp, which could be converted by treatment with KOBu^t and I$_2$ to Tp*Mo(CO)$_2$(≡CC≡)W(CO)(PhC≡CPh)Tp.[482] Protonation of Tp or Tp* complexes Tp^{x}Mo(CO)(RC≡CR')Me, which resulted in loss of methane, and addition of various ketones, produced η^1-ketone complexes with different conformational preferences about the M-O bond. Their E/Z isomerization barriers were studied.[483]

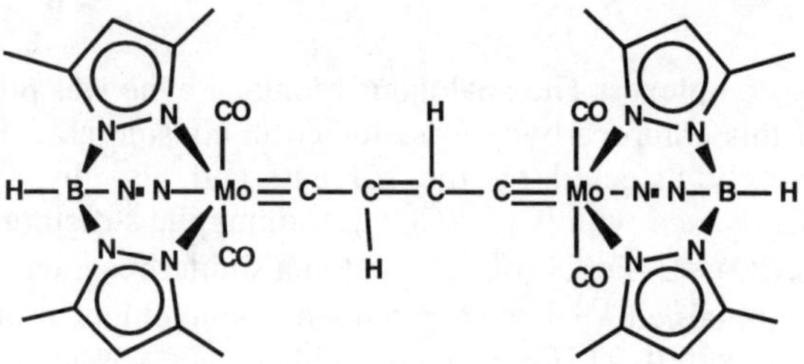

48

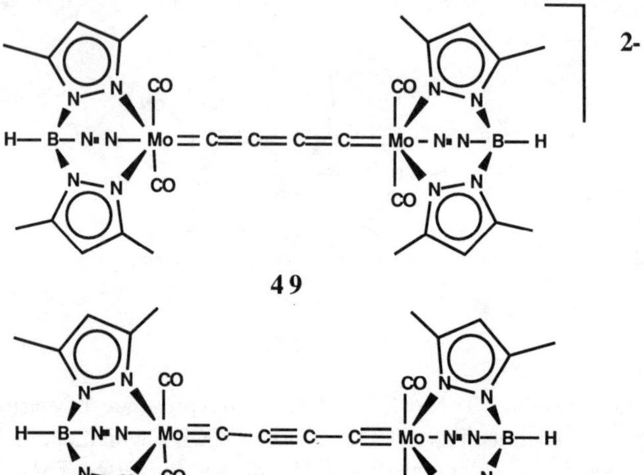

49

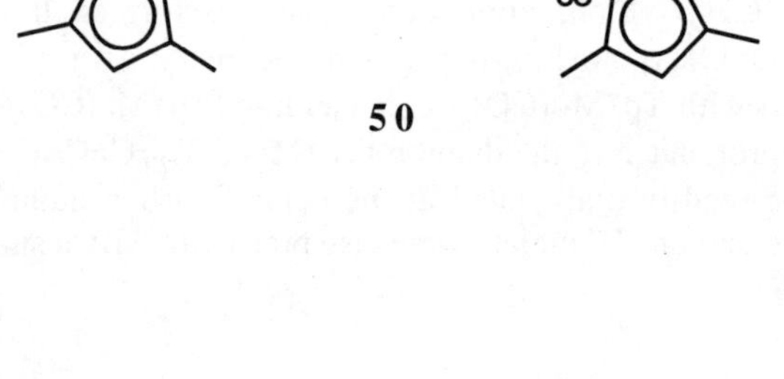

50

2.7 GROUP 6 COMPLEXES

While the reaction of [TpMo(CO)$_3$]$^-$ or [pzTpMo(CO)$_3$]$^-$ with [ArN$_2$]$^+$ ions produced the arylazo complexes Tp^{x}Mo(CO)$_2$(N=NAr),[18,484] such reaction with [Tp*Mo(CO)$_3$]$^-$ yielded products,[19] which were later identified as η^2-aroyl species, Tp*Mo(CO)$_2$(η^2-COAr).[20,30] Tp^{x}Mo(CO)$_2$(N=NAr) complexes were studied by Raman and IR spectroscopy,[485] and those with *m*- and *p*-fluoro substituents by ^{19}F and ^{15}N NMR.[486] Oxidative addition of halogen to TpMo(CO)$_2$(N=NAr) was claimed to produce dimeric species.[487] Reactions of TpMo(CO)$_2$(N=NAr) with PR$_3$ and with disulfides were studied, and the structure of TpMo(SR)$_2$(N=NTol) was established by X-ray crystallography,[488] while TpMo(NO)(Cl)(N=NAr) was produced by the reaction of TpMo(CO)$_2$(N=NAr) with nitrosyl chloride.[489,490] The arylazo cation [TpMo(CO)(N=NAr)(PPh$_3$)]$^+$ has also been reported,[491] but it was not fully characterized. Structures of TpMoF(N=NAr)$_2$,[492] and of Tp*Mo(NO)(SPh)$_2$,[493] were determined by X-ray crystallography.

The reaction of [Tp^{x}Mo(CO)$_3$]$^-$ with various sources of NO$^+$ yielded Tp^{x}Mo(CO)$_2$NO,[18,494] which served as starting material for a very large number of interesting derivatives, including Tp^{x}Mo(CO)NO(PPh$_3$),[495] Tp^{x}Mo(CO)NO(optically active phosphine),[496] and Tp^{x}Mo(CO)NO(NCR).[497] The ^{95}Mo-^{14}N spin coupling in Tp*Mo(CO)$_2$NO was studied,[498] and ^{95}Mo NMR showed diastereomer splitting in Tp*Mo(CO)NO(PR$_3$).[499] Tp*Mo(CO)$_2$(NS) was also reported.[500] In the reaction of ClNO with [Tp^{x}Mo(CO)$_3$]$^-$, one obtained, in addition to Tp^{x}Mo(CO)$_2$NO, also Tp^{x}Mo(NO)$_2$Cl and Tp^{x}MCl$_2$NO.[18] Tp*Mo(CO)$_2$NO was also converted to the Fischer carbene Tp*Mo(CO)(NO)=C(OMe)(Me) and the carbene methyl group was functionalized via deprotonation and alkylation.[501] The reaction of Tp*Mo(NO)I$_2$ with either [S$_{10}$]$^{2-}$ or with [Se$_6$]$^{2-}$ yielded Tp*Mo(NO)(E$_5$) species containing a MoE$_5$ ring, of which the one with E = Se was structurally characterized. Upon treatment with PBu$_3$, they were converted to the E-bridged complexes [Tp*Mo(NO)]$_2$(μ-E)$_2$.[502] Hydrogen bonding and polar group effects on the redox potentials of Tp*Mo(NO)(SR)$_2$ complexes, where the R groups were alkyls, or alkyls containing mono- or dialkylamido functions,[503] including the structurally characterized Tp*Mo(NO)[S(CH$_2$)CONH(CH$_2$)S] complex, were studied by electrochemistry.[504]

A very large sub-area of scorpionate chemistry was originated and developed by McCleverty. The core unit of this chemistry was the [Tp*MoNO] fragment, the other two coordination sites containing a vast variety of other substituents with diverse functionalities, as shown in **51**, permitting the construction of polynuclear species, where the electronic properties of Mo could interact with other metal sites. The synthesis of these compounds started with Tp*Mo(NO)X$_2$ (X being usually I, since bromination or chlorination halogenated the 4-position of the pyrazole rings),[505] obtained via halogenation of Tp*MoNO(CO)$_2$. Tp*Mo(NO)I$_2$ was reduced electrochemically to produce a paramagnetic species Tp*Mo(NO)I, thought to be the intermediate in substitution reactions at Mo.[506] Hydrolysis of Tp*Mo(NO)X$_2$

51 (Z = O, S, NH; R = alkyl, aryl, heterocycle)

produced first a monohydroxo species, and it reacted with another molecule of the starting material to yield the oxo-bridged [Tp*Mo(NO)X]$_2$(μ-O), which was structurally characterized along with the bis-hydroxo derivative Tp*Mo(NO)(OH)$_2$.[507] A related sulfur-bridged analog, [Tp*Mo(NO)Cl]$_2$(μ-S$_2$), was also reported.[508] Tp*Mo(NO)I$_2$ was readily converted to Tp*Mo(NO)(NCS)$_2$ and Tp*Mo(NO)(N$_3$)$_2$,[509] and the reaction of Tp*Mo(NO)I$_2$ with alcohols produced monoalkoxy species, Tp*Mo(NO)I(OR),[510,511] and with thiols the appropriate sulfur analogs.[512] These early results were briefly reviewed,[27] and the synthetic procedures for preparing some of the alkoxy- and alkylamide complexes were described in detail.[513] Bis-alkoxy species containing the same,[514] or different,[515] alkoxy groups, such as Tp*Mo(NO)(OR')(OR) could also be synthesized, as well as complexes Tp*Mo(NO)X(OR), in which the alkoxide was part of various cyclic systems,[516] of menthol,[517] of cholesterol,[518] of monosaccharides,[519] as well as of glucofuranoside or galactofuranoside.[520] Potentially chelating diols or diamines produced generally only monoalkoxy,[521] and monoamido derivatives,[522] respectively. In a rather unusual reaction of Tp*Mo(NO)I$_2$ with acetone the products obtained were Tp*Mo(NO)I(OEt) and the dinuclear [Tp*Mo(NO)I]$_2$O.[523,524] Bimetallic chelated complexes of the type [MoTp*(NO)(η^5-C$_5$H$_4$CH$_2$O)$_2$M] where the other metal was either Fe or Ru, and the non-chelated trimetallic [MoTp*(NO){Fe(η^5-C$_5$H$_5$)(η^5-C$_5$H$_4$CH$_2$O)}$_2$] were also prepared,[525] as were binuclear species containing cyclohexanediolate and cyclopentanediolate bridges.[526]

Some types of cyclic ethers reacted with Tp*Mo(NO)I$_2$, producing ω-iodoalkoxy derivatives, Tp*Mo(NO)I[O(CH$_2$)$_n$I].[527,528] The alkoxy complexes were converted by HX acids to Tp*Mo(NO)X$_2$. Mono- and bis-aryloxides containing the [Tp*Mo(NO)] core were also synthesized,[529] including those with long-chain alkyl substituents.[530] Detailed studies of electrochemical interaction and of the reduced species derived from [Tp*Mo(NO)Cl]$^-$ moieties linked by a variety of diphenol linkages,[531,532] and by four-carbon links with varying degrees of unsaturation, were carried out.[533] The reaction of Tp*Mo(NO)I$_2$ with 4,4-dihydroxybiphenyl led not only to the expected dinuclear species, but also to interesting triangular trinuclear and even

2.7 GROUP 6 COMPLEXES

square tetranuclear metallacyclophanes,[534] and a trimetallacyclophane based on hydroquinone was also obtained.[535] Unusual macrocycles of structure [{Tp*Mo(NO)}(diol)]$_2$ were prepared, where the diols were 4,4'-methylenebisphenol and 1,4-bis(hydroxymethyl)phenol, respectively.[536,1555] Reaction of Tp*Mo(NO)I$_2$ with 2,7-dihydroxynaphthalene produced the structurally characterized binuclear complex [Tp*Mo(NO)(2,7-O$_2$C$_{10}$H$_6$)]$_2$.[537] Complexes with bulky alkoxy, aryloxy and arylamido substituents were also reported,[538] as well as analogous thiophenolato and carboxylato derivatives.[539]

The alkyl- and arylamido derivatives, Tp*Mo(NO)I(NHR),[540] could be prepared, not only from simple aliphatic and aromatic amines, but also from aminopyridines,[541] from p-ferrocenylaniline,[542] from an arylamino derivative of retinal,[543] from stilbenamine containing long alkoxy substituents,[544] and from the optically acive (+)- and (-)-1-phenylethylamine, one of the pure diastereomers being structurally characterized.[545] Amido and hydrazido derivatives were also prepared,[546] and they were converted by acetone to -N=CMe$_2$ and -NHN=CMe$_2$ derivatives, respectively.[547] In the reaction of Tp*Mo(NO)I$_2$ with pyrrolidine or piperidine both, amino and amido complexes Tp*Mo(NO)I(NHC$_n$H$_{2n}$) and Tp*Mo(NO)I(NC$_n$H$_{2n}$) were obtained,[548] and structurally characterized,[549] as was the bis-amide complex Tp*Mo(NO)(NHBu)$_2$.[550] The reaction of Tp*Mo(NO)I$_2$ and o-phenylenediamine produced the bimetallic 52, instead of a chelated species.[551,552]

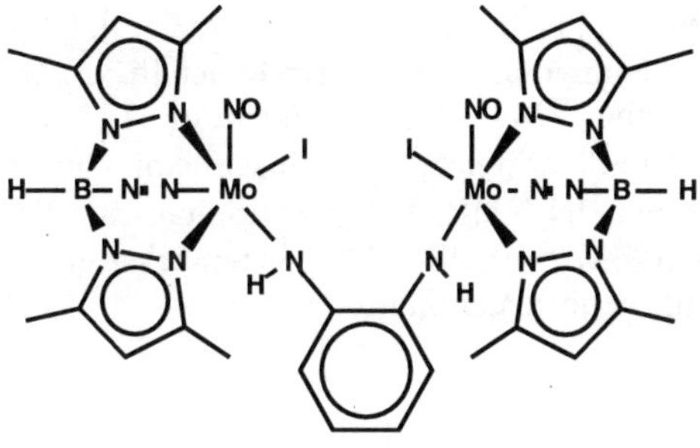

52

Other bimetallic complexes were also synthesized and studied by electrochemistry.[553,554] Valence-localized and valence-delocalized mixed valence states were obtained by electrochemical or chemical reduction of binuclear [Tp*Mo(NO)X] entities connected through 1,2-, 1,3-, and 1,4-phenylenediamine links.[555,556] Amido, bis-amido and dialkoxy complexes containing [OCH$_2$CH$_2$]$_n$ polyether loops, capable of cation binding, have been synthesized,[557,558] and the structure of one of them determined by X-ray crystallography.[559] An interesting S,S-chelated bimetallic complex, [TpMoNO(Fe(η^5-C$_5$H$_4$S)$_2$)], was obtained from ferrocene 1,1'-dithiol.[560]

The reaction of Tp*Mo(NO)I$_2$ with pyridine, pyrazole, imidazole and related bases yielded either neutral or cationic derivatives,[561] and the structure of [Tp*Mo(NO)(NCMe)$_2$]PF$_6$ was established by X-ray crystallography.[562]

Considerable attention was given to exploration of the paramagnetic complexes derived from Tp*Mo(NO)X$_2$, primarily by electrochemistry.[563] These studies included, among others, the complexes Tp*Mo(NO)I(NCR),[564] the anion [Tp*Mo(NO)Cl$_2$]$^-$ and the pyridine complex Tp*Mo(NO)Cl(py),[565] the cation [Tp*Mo(NO)(py)$_2$]$^+$,[566] the Hpz* complex Tp*Mo(NO)Cl(Hpz*), obtained from Tp*Mo(NO)Cl$_2$ through reduction with BuLi,[567] and complexes of the type Tp*Mo(NO)Cl(stilbazole) containing alkoxy chains of varying length in the 4'-position of stilbazole.[568] The effect of *meta*-substituents on the reduction potential of Tp*Mo(NO)Cl(NHC$_6$H$_4$-3-Z) was studied for a variety of electron-donating and electron-withdrawing substituents, Z.[569] Cyclic voltammetry was used to study Tp*Mo(NO)X complexes containing *meta*-substituted benzene- and naphthalene-diols or diamines,[570] arylmercapto derivatives,[571] *para*-substituted arylamides,[572] saturated heterocyclic amide ligands,[573] and *para*-substituted anilines and phenols,[574,575] Magnetic studies of a number of 17-electron complexes Tp*Mo(NO)I(L) and [Tp*Mo(NO)(L)$_2$]$^+$ (L = pyridine-derived bases) were conducted, and the structures of [Tp*Mo(NO)(4-Ph-py)$_2$]I and of [Tp*Mo(NO)(py)$_2$][BPh$_4$] were determined.[576] The pyrazine bridged bimetallic complex [{Tp*Mo(NO)Br}$_2$(μ-pyrazine)] was found to exist in a five-membered electron transfer chain, with z = -2, -1, 0, +1 and +2,[577] and the structure of the related [{Tp*Mo(NO)Cl}$_2$(μ-pyrazine)], **53**, was determined by X-ray crystallography.[561] Numerous redox-active bimetallic complexes were discussed in a mini-review,[578] the ^{95}Mo NMR spectra of Tp*Mo(NO)$_2$Cl,[579] and of Tp*MoO(S$_2$CNR$_2$) were determined,[580] and a series of Tp*Mo(NO)XY complexes was studied by ^{95}Mo and ^{14}N NMR.[581] The cation radical [Tp*Mo(NO)(MeCN)$_2$]$^+$ showed the ^{95}Mo and ^{97}Mo satellites at low temperatures, thus confirming the presence of the paramagnetic center on Mo.[582]

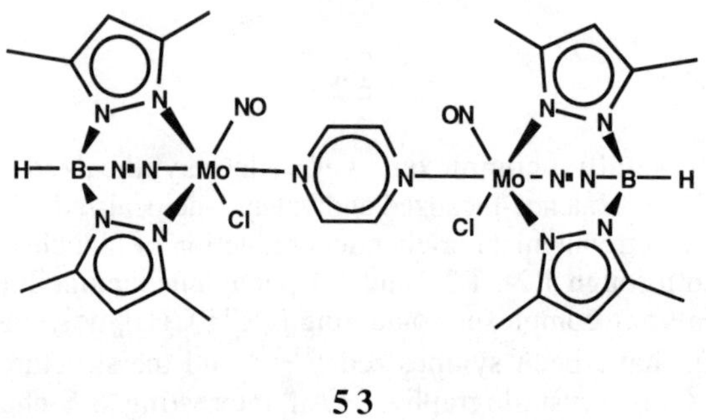

53

2.7 GROUP 6 COMPLEXES

EPR spectra demonstrated magnetic exchange between three metal centers in complexes containing [Tp*Mo(NO)Cl] moieties,[1553] which were linked to pyridyl or aryloxy termini of 1,3,5-trisubstituted benzenes containing either 4-pyridylvinyl or 4-hydroxyphenylvinyl substituents.[583] Dinuclear seventeen-electron [Tp*Mo(NO)Cl] species, linked through bipyridyl ligands,[584,585] through polyene-connected dipyridines with varying number of polyene links,[586-588] and also with oligothienyl spacers,[589] were studied by EPR and by electrochemistry. These studies also included a series of complexes where from two to four [Tp*Mo(NO)Cl] units were linked to benzene, tris- or tetrakis-substituted with ethenyl-4-pyridyl or ethynyl-4-pyridyl substituents,[590] and also a hexanuclear complex with [Tp*Mo(NO)Cl] units linked to the pyridine termini of 1,3,5-tris[3,5-bis(2-pyridyl)phenyl]benzene.[591] In all instances exchange between all Mo nuclei was found, with stronger exchange occurring between *ortho*- and *para*-linked Mo centers, than with those *meta*-linked.

A series of mixed-valence dinuclear [Tp*Mo(NO)X] complexes containing benzenediamido and dianilido bridges was synthesized, and compared with the related phenolato and dipyridyl species.[592] Dinuclear complexes containing Rh, Hg, Pd, Pt or Cd were synthesized from Tp*Mo(NO)X(NH(CH$_2$)$_3$PPh$_2$) and suitable derivatives of the other metal.[593] Complexes of the type Tp*Mo(NO)Cl(NHC$_6$H$_4$-4-N=N-4'-Fc, and related ones, were investigated for their non-linear optical properties,[594-598] and structures of the above azo compound, as well as of [Tp*Mo(NO)Cl(p-O-tetraphenylporphyrinyl] were determined by X-ray crystallography.[599] The effect of linkage position (*ortho*, *meta* or *para*) of the [Tp*Mo(NO)Cl(Cl)-O]- moiety to the tetraphenylporphyrin on the intramolecular electron transfer was studied, including the structure determination of the *ortho*-isomer,[600] as was the photochemistry of such systems containing ruthenium within the porphyrin ring.[601] Some of the macrocyclic polynuclear complexes containing the Tp*Mo(NO) unit were discussed in a general review on molecular containers.[1535]

Further studies of this type employed complexes where the [Tp*Mo(NO)(Cl)] fragments were attached to pyridyl rings, para-linked via diverse polyene units to ferrocene,[602] systems where, in addition to the above ferrocenyl moieties, there were OR groups instead of Cl, the R group being chiral alcohols of fairly complicated structure,[603] as well as complexes containing [Tp*Mo(NO)(X] linked either through O or NH bonds to stilbene derivatives, containing diverse substituents on the remote phenyl ring.[604] The role of bridging ligand topology and conformation in controlling exchange interactions between paramagnetic molybdenum units in dinuclear and trinuclear complexes was discussed in detail, and the structure of the trinuclear molybdenum(V) species derived from 1,3,5-trihydroxybenzene, [Tp*MoO(Cl)]$_3$(C$_6$H$_3$O$_3$) was determined by X-ray crystallography.[605] Very unusual host molecules were obtained from the reaction of Tp*Mo(NO)I$_2$ with tetrahexylcalix[4]resorcinarene. They contained two, three or four Tp*Mo(NO)I units incorporated into the rim of the *calix*. The structure of the tetrametallated derivative showed that one nitrosyl formed a hydrogen bond to a methylene chloride molecule included in the host site.[606] A variety of cyclic dimers, trimers and monometallomacrocycles was prepared starting with Tp*Mo(NO)I$_2$ and 1,3- and 1,4-bis(mercaptomethyl)benzenes, and the stuctures of some of these complexes were determined by X-ray crystallography.[607]

Another way of achieving electronic interaction between redox-active molybdenum entities and other metals was to connect the [Tp*Mo(NO)(X)] group to a metal complex derived from a Schiff base, and containing aromatic hydroxyl groups serving as links, as exemplified by structure **54**, where Z was -(CH$_2$)$_2$- or -(CH$_2$)$_3$-.

54

The first such system investigated contained a central copper, and two molybdenum centers.[608] Related complexes with nickel or palladium were also studied,[609] although it was noted that hydrolysis of the azomethine linkage took place, being accelerated by the presence of the electronegative [Tp*Mo(NO)(X)] group, and a monometallic species with free aldehyde groups was structurally characterized.[610] The reaction of [Tp*Mo(CO)$_3$]$^-$ with thionyl chloride produced Tp*MoCl$_3$,[19] also obtainable on oxidation of [Tp*MoCl$_3$]$^-$.[611] Upon treatment with alumina, it was converted to Tp*MoOCl$_2$, which was studied by ESR in the isostructural Tp*SnOCl$_2$ matrix.[612] A general synthesis of Tp*MoOX$_2$ complexes (X = F, Cl, Br) entailed the reaction of Tp*MoO(OMe)$_2$ with halo acids,[613] and the EPR parameters of these complexes were compared.[614] Metal-metal interaction in a paramagnetic complex containing two oxo-Mo^V centers, [Tp*MoO]$_2$(dianion of ellagic acid), was studied by EPR.[615]

Partial hydrolysis of TpMoOCl$_2$ produced the dinuclear [TpMoOCl]$_2$O, and both of these complexes were structurally characterized. On further hydrolysis in methanol a linear tetranuclear complex, capped at each end with the Tp ligand, and containing oxo- and methoxo bridges was isolated,[616,617] and studied by electronic and vibrational spectroscopy.[618] A related mixed-valence complex Tp*MoO(Cl)(μ-O)MoO$_2$Tp* has also been reported.[619] Structures of two dinuclear complexes [Tp*MoO]$_2$(μ-O)(μ-S$_2$) and Tp*MoS(μ-S$_2$)MoS(S$_2$CNEt$_2$), and of the tetranuclear [Tp*MoS(μ-S)$_2$MoO(μ-OH)$_2$]$_2$ were determined by X-ray crystallography.[620]

2.7 GROUP 6 COMPLEXES

A large area of molybdenum chemistry dealing with its higher oxidation states, mainly Mo^{IV} and Mo^{VI}, was developed, mostly by Enemark and by Young, in conjunction with efforts to synthesize models for the molybdenum cofactor, present in a number of enzymes such as sulfite oxidase, xanthine oxidase or nitrite reductase.[621,622] In this cofactor the molybdenum atom is coordinated to the two sulfur atoms from the *cis*-enedithiolate function of molybdopterin, to an oxo group, and to either an oxo or sulfido group, depending on the enzyme in question. In this context a variety of complexes containing the structural component Tp*MoO(X)(Y) were synthesized and studied.[623] In most cases X was Y, and the synthesis entailed nucleophilic displacement of chloride ions from Tp*MoOCl$_2$.[624] The numerous structurally characterized complexes were Tp*MoO(OPh)$_2$,[625] Tp*MoO(NCS)$_2$,[626] Tp*MoO(OC$_6$H$_4$-p-Cl)$_2$,[627] Tp*MoO(S$_2$py),[628] Tp*MoO(O$_2$-o-C$_6$Cl$_4$),[629] Tp*MoO(S$_2$-o-C$_6$H$_4$),[630] Tp*MoO(NCS)$_2$ and Tp*MoO(N$_3$)$_2$,[631] as well as Tp*MoO(S$_2$P(OR)$_2$),[632] **55**, Tp*MoO(S$_2$CNPr$_2$),[633] **56**, and Tp*MoS(S$_2$CNEt$_2$), **57**.[634] EPR spectra of Tp*MoO(X)$_2$ for X = Cl, OMe, SEt and NCS, have been

55 **56** **57**

simulated.[635] Treatment of Tp*MoOCl$_2$ with B$_2$S$_3$ converted it to Tp*MoSCl$_2$, the first mononuclear Mo^V complex with a terminal sulfido ligand.[636] Numerous *para*-substituted phenolato complexes, Tp*MoO(OAr)$_2$, were synthesized and characterized,[637] as well as many related complexes with aliphatic diolato, dithiolato and alkoxo ligands, with emphasis on the effect of ring size on the properties of the Mo center.[638] The He I valence photoelectron spectra for Tp*Mo(O)(OR)$_2$ complexes containing also chelating diolato substituents were reported.[639] Electrochemical studies of a series of Tp*MoO(E-Z-E) complexes, containing aromatic and aliphating 1,2-chelating substituents, with E = O, or S, have revealed that replacement of O by S results in a decrease of the electron transfer constant.[640] The first ionization energy in the gas-phase photoelectron spectra of Tp*Mo(E)(3,4-toluenedithiolate), was independent of the nature of E (O, S, NO).[641] EPR spectroscopy of a series of [Tp*MoO] complexes with chelating dithiolate ligands suggested that these systems may be considered as a minimal structural and spectroscopic benchmark for the oxomolybdenum(V) center in molybdopterin.[642] Reduction of Tp*MoO$_2$(SPh) by cobaltocene in pyridine yielded the stable, structurally characterized, Tp*MoO(SPh)(py).[643] In a comparison of the Tp*Mo(NO)(OR)$_2$ and Tp*MoO(OR)$_2$ pairs of complexes it was shown that the nitrosyl ones are 0.8 eV more difficult to ionize than the oxo analogs.[644] Spectra at the molybdenum L$_2$ and L$_3$ edges of a series of Tp*MoO(X)(Y) complexes were analyzed in terms of ligand field theory,[645] and optical spectra in high- and low-symmetry Tp*MoOX$_2$ complexes were assigned on the basis of magnetic circular dichroism spectroscopy.[646]

The ^{31}P NMR spectroscopy was used in complexes Tp*MoO(OAr)$_2$, also including catecholate derivatives, with -OPO(OPh)$_2$ substituents on the aromatic rings, as a probe of models for the molybdenum phosphate interaction in oxo-type molybdoenzymes.[647] The complex Tp*MoO(OC$_6$H$_4$-o-S) was reported,[648] and its solution was oxidized in air to the Mo^{VI} species, containing a disulfide bond, [Tp*MoO$_2$(O-C$_6$H$_4$-o-S]$_2$.[649] The compound Tp*MoO(S)(S$_2$PEt$_2$) was found to exhibit intramolecular sulfur-sulfur interaction, of relevance to the behavior of the active site of xanthine oxidase.[650] Unusual di- and trinuclear metallamacrocycles were prepared by the reaction of Tp*MoCl$_2$ with bifunctional ligands, such as 1,3-(HO)$_2$C$_4$H$_4$, 1,3-(HS)$_2$C$_4$H$_4$, 2,7-(HO)$_2$C$_{10}$H$_6$, and 4,4'-[(HO)C$_6$H$_4$]$_2$CH$_2$. These were studied by electrochemistry, and strong interaction was found in the resorcinol derivatives.[651] Trinuclear oxomolybdenum(V) complexes, such as the structurally characterized **58** and its *meta*-analog, were studied electrochemically, and in terms of their magnetic exchange interactions. Such interactions were observed between the terminal and the central metal fragments, but not between the two terminal metal fragments. Ferromagnetic and antiferromagnetic exchange interactions were observed between adjacent metal atoms.[652]

58

Among the Mo^{VI} homoscorpionate complexes the nitrides Tp*Mo≡N(N$_3$)$_2$ and Tp*Mo≡N(N$_3$)(Cl),[653,654] and Tp*Mo(=NNPh$_2$)$_2$Cl were reported,[655] as well as Tp*MoO(S)(OR) and Tp*MoS$_2$(OR).[656] Tp*MoO$_2$X compounds, which were typical starting materials for the synthesis of Mo^{VI} complexes, for instance

2.7 GROUP 6 COMPLEXES

Tp*MoO$_2$Cl,[657] were prepared by the reaction of [Tp*]$^-$ with MoO$_2$X$_2$, and the halogen in Tp*MoO$_2$X could be replaced with NCS, OPh and SPh,[658] and such complexes reacted with PPh$_3$ producing, depending on the solvent, Tp*Mo(O)X, [Tp*Mo(O)X]$_2$, or Tp*Mo(O)X$_2$. Related compounds with alkyl groups were also prepared,[659,660] and were studied by NMR.[661] The structure of [Tp*MoO$_2$]$_2$O was established by X-ray crystallography.[662] Reaction of Tp*MoO$_2$X with aqueous ammonia yielded oxo-bridged polynuclear species, and related thio-bridged complexes were also prepared.[663] Substitution rates of X in Tp*MoO$_2$X were slow.[664] Reduction of the structurally characterized Tp*MoO$_2$(SPh) with cobaltocene produced the radical anion complex [Tp*MoO$_2$(SPh)]$^-$ which could be converted to Tp*MoO(OSiMe$_3$)(ISPh).[665] The reaction of Tp*MoO$_2$X complexes with sulfiding agents yielded the tetrasulfidomolybdenum complexes Tp*Mo(X)(S$_4$).[666] Treatment of Tp*Mo(NCS)(S$_4$) with dicarbamethoxyacetylene resulted in extrusion of sulfur, and formation of the chelated complex Tp*Mo(NCS)[S$_2$C$_2$(COOMe)$_2$].[667] The crystallographically characterized complexes [Tp*MoO{S$_2$P(OR)$_2$}] and [Tp*MoO$_2${S$_2$P(OR)$_2$}] showed in the first case a bidentate [S$_2$P(OR)$_2$]$^-$ ligand, but a monodentate one in the case of the latter complex.[668] The compound TpMo(=NAr)(OTf)(=CHCMe$_2$Ph) had covalently bound OTf,[669] and on hydrolysis yielded TpMo(=NAr)(O)(-CH$_2$CMe$_2$Ph) and [TpMo(=NAr)(O)]$_2$O.[670] Addition of LiMe to TpMo(=NAr)(=CHCMe$_2$Ph)(OTf), where Ar was 2,6-bis(isopropyl)phenyl, produced TpMo(=NAr)(=CHCMe$_2$Ph)Me. Abstraction of the methyl ligand from this complex readily yielded solvated cationic species, exemplified by [TpMo(=NAr)(=CHCMe$_2$Ph)(Et$_2$O)]$^+$. Reaction with potassium methoxide converted

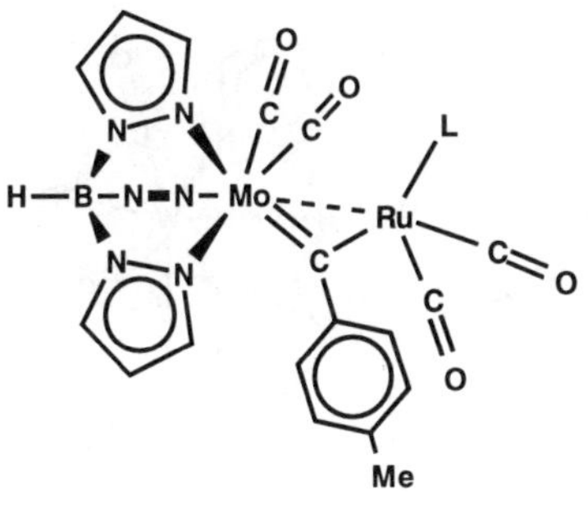

59 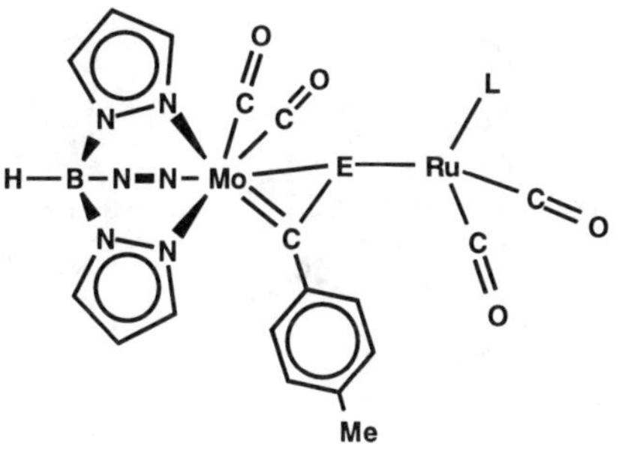   **60**

TpMo(=NAr)(=CHCMe$_2$Ph)(OTf) to the complex TpMo(≡CCMe$_2$Ph)(NHAr)(OMe), while under milder conditions the OTf ligand was replaced by methoxide ion, without any other changes, as was demonstrated by X-ray crystallography.[671] Unusual dinuclear complexes of structures **59** and **60** where the ligand L was (η^5-7,8-

$C_2B_9H_{11}$) and E was S or Se, were synthesized, and were found to be readily cleaved by donor solvents, but they were stable in dichloromethane or in toluene.[673] "First generation" homoscorpionates were also used as capping ligands in a number of clusters, including Mo_3S_4,[674] Mo_4S_4,[675] and various polymetallic ones.[676,677] It was found that p-anisaldehyde binds to [Tp*Mo(CO)(MeC≡CMe)]$^+$ in σ-fashion, but but in π-fashion to the analogous W complex.[678]

2.7.3 Tungsten

The reported homoscorpionate chemistry of tungsten is almost as rich as that of molybdenum, ranging in oxidation states from W^O to W^{VI}, with many complexes being identical to those of Mo, or very similar. For this reason, numerous papers dealt with analogous Mo and W complexes together, and so they were cited first in the molybdenum section, 2.7.2.

Thus, alkylation of Tp*WO$_2$Cl to Tp*WO$_2$Me was reported,[660] as was the redox chemistry of [TpWCl$_3$]$^-$,[611] and the synthesis of the cyclopentadienone complex TpW(CO)$_2$(η^3-C$_5$H$_4$O),[442] and of the bis-imide complex Tp*W(=NR)$_2$(Cl).[378] Photolytic reaction of TpW(CO)$_2$≡CR with phosphines yielded TpW(CO)(PR$_3$)(η^2-OCCR) complexes of structure **61** (E = O) which could be converted to the thio derivative (E = S).[478] The structures of the Se analog, **62**, and of **63** (R = p-tolyl) were determined by X-ray crystallography.[479] The carbyne Tp*W(X)$_2$≡CR (X = Cl, Br; R = Bu^t, Ph) was converted by alumina to Tp*W(O)(X)(=CHR), and by aniline to Tp*W(=NPh)(X)(=CHR). Both complexes were catalysts for acyclic diene metathesis

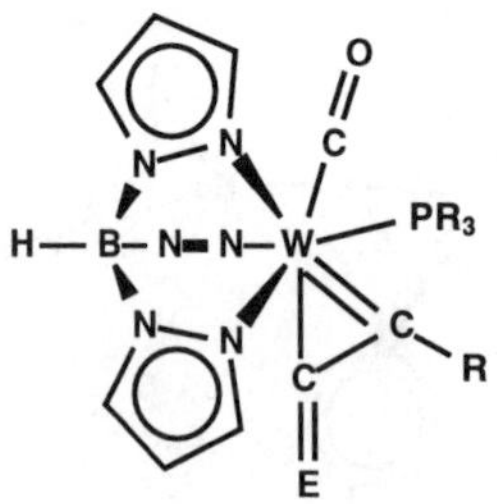

61 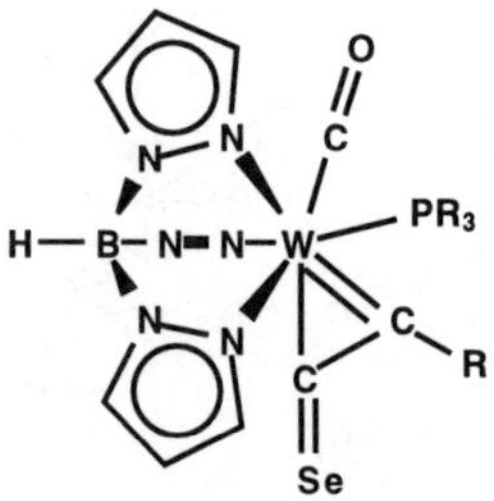   **62** 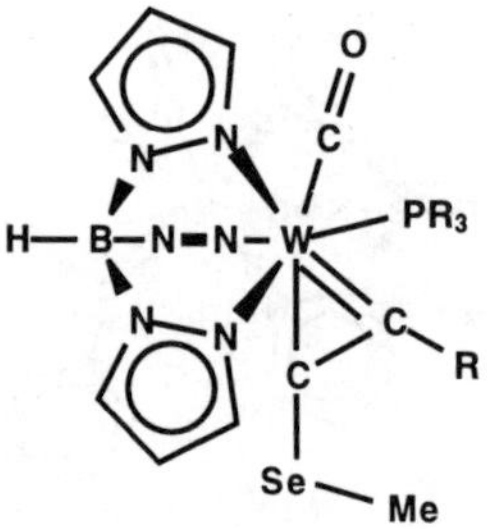   **63**

polymerization.[679] The structurally characterized complex **64** had a distorted tetrahedral geometry at the bridging methylene.[392] Complexes Tp*W(CO)$_2$≡COAr and [Tp*W(CO)$_2$≡CPR$_3$]$^+$ were prepared from Tp*W(CO)$_2$≡CCl.[476] Numerous compounds containing the [Tp*W(NO)] core were synthesized. Such complexes included Tp*W(NO)Cl(OR) and Tp*W(NO)(OR)$_2$,[515] similar complexes where OR also contained an cyclic ether structure,[516] where OR was a mentholate,[517] a

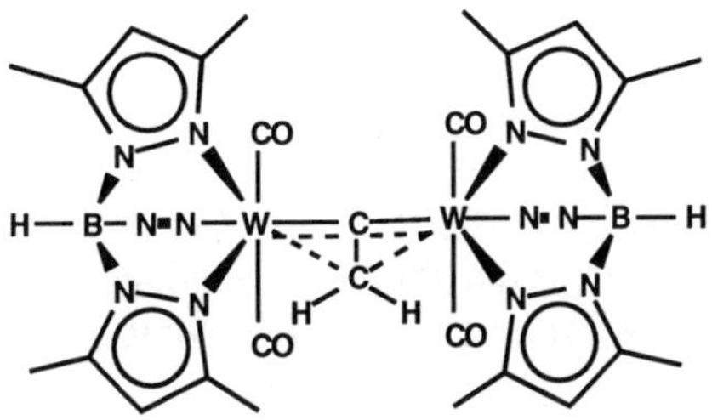

2.7 GROUP 6 COMPLEXES

64

saccharide,[519] a glucofuranoside or galactopyranoside,[520] as well as dinuclear structures involving cyclopentanediolate or cyclohexanediolate bridges,[526] those obtained via ring-opening of oxacyclobutane,[528] and hydroquinone-based tungstacyclophanes of structure [Tp*W(NO)(OC$_6$H$_4$O)]$_3$.[535] Also reported were Tp*W(NO)(NHAr)$_2$ species where Ar was 2-pyridyl,[541] as well as optically resolved 1-methylbenzylamido analogs, Tp*W(NO)I(NHMeCHPh),[545] and various Tp*W(NO)(NHR)$_2$ complexes,[550] including amido complexes of the NHAr type with a variety of *para*-substituents, the effect of which on the reduction potentials of these complexes, and of the OAr analogs,[574] was studied.[569,572] Electrochemical studies were also carried out on analogs with OPh, SPh and NC$_4$H$_4$ substituents,[571] on those with saturated heterocyclic amide,[573] and other ligands,[538,578] on complexes such as [Tp*W(NO)Cl]$_2$[(NHC$_6$H$_4$)$_2$O],[575] and on analogs linked with diphenolic bridges.[526] Related complexes with a phosphine moiety present on the amido substituent coordinated to other metals were also prepared.[587] Amido complexes linked to ferrocene through arylamido or arylazo bridges were investigated for non-linear optical properties,[594-598,604] The complex TpW(NO)(Cl)(N=NPh) was obtained from TpW(CO)$_2$(N=NPh) and NOCl.[491]

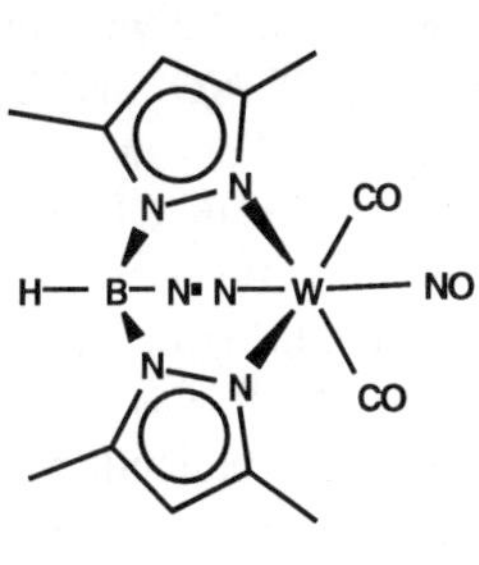

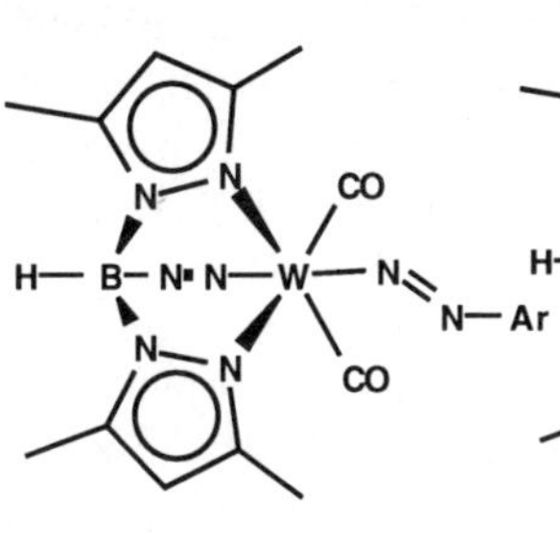

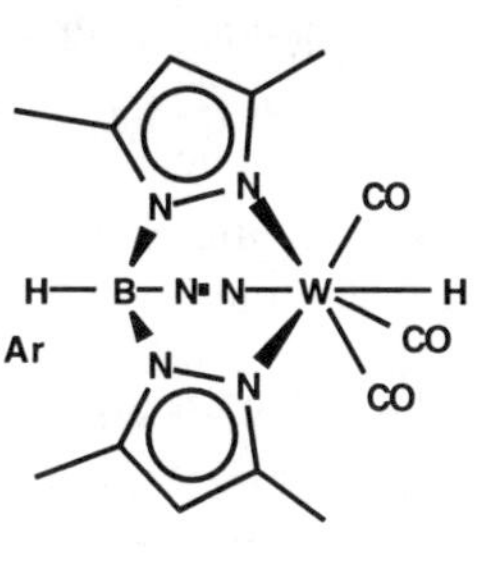

65   **66**   **67**

Generally, an easy entry into low-valent tungsten scorpionate chemistry was the reaction of [Tp^x]$^-$ with W(CO)$_6$, yielding the anion [Tp^{x}W(CO)$_3$]$^-$, which was converted to various η^3-allyl derivatives, $Tp^xW(CO)_2(\eta^3$-allyl),[4] to nitroso and arylazo derivatives $Tp^xW(CO)_2(NO)$, **65**, and $Tp^xW(CO)_2(N=NAr)$, **66**, respectively,[18,491] and was protonated to the 7-coordinate species $Tp^xW(CO)_3H$, **67**.[4] The complexes $Tp^*W(CO)_3H$ and $Tp^*W(CO)(PhC\equiv CMe)H$ were structurally characterized.[680] Seven-coordinate halides, $TpW(CO)_3X$, were prepared,[425,459] as was $Tp^*W(CO)_3I$.[681] The anion $[Tp^*W(CO)_3]^-$ was oxidized to the unreactive 17-electron radical,[341] and further to the 16-electron cation $[Tp^*W(CO)_3]^+$, which reacted with phosphines forming a variety of structurally characterized 7-coordinate complexes $[Tp^*W(CO)_3(PR_3)][PF_6]$.[682] A detailed synthesis of $Tp^*W(CO)_3X$ complexes (X = Cl, Br, I) and their conversion to $Tp^*W(CO)_2X$ species has been reported.[683] The dynamics of asymmetric η^3-allyl systems were studied by ^{183}W and 2D ^1H NOESY NMR.[684] $Tp^*W(CO)_2(\eta^3$-allyl) complexes could also be obtained by irradiation of $[Tp^*W(CO)_3H]$ in the presence of alkynes or unconjugated dienes.[685] Thiolate complexes $Tp^*W(CO)_2(SR)$ were obtained by the reaction of $[Tp^*W(CO)_3]^-$ with $ArSO_2Cl$,[19] and in better yield by the reaction of thiolates with $Tp^*W(CO)_2I$, which was itself obtained by thermolysis of $Tp^*W(CO)_3I$.[686] Air oxidation of $Tp^*W(CO)_3X$ yielded the structurally characterized complex $Tp^*WO(CO)X$.[687] Tetraalkylthiuram disulfides converted $[Tp^*W(CO)_3]^-$ to the structurally characterized $[Tp^*W(CO)_2(S_2CNR_2)]$, and on further reaction, to a variety of other thio derivatives.[688] In a similar reaction with propylene sulfide, $[Tp^*W(CO)_2]_2S$ was obtained.[689] The reaction of $Tp^*WI(CO)_n$ (n = 2,3) with $[S_2P(OR)_2]^-$ yielded dithio ligand complexes $Tp^*W(CO)_2[S_2P(OR)_2-S]$, fluxional by NMR, which were converted by various oxygen donors to $Tp^*W(O)(CO)[S_2P(OR)_2]$. The structure of $Tp^*W(O)(CO)[S_2P(OPr^i)_2]$ was established by X-ray crystallography.[690]

Tungsten chloro- and bromohalocarbynes, $Tp^*W(CO)_2\equiv CX$ were prepared just like their Mo analogs,[30,470] and were converted to metallacycles via anionic vinylidene species, such as $\{Tp^*W(CO)_2[=C=C(CN)_2]\}^-$.[477] Angelici synthesized the thiocarbonyl derivative, $[TpW(CO)_2(CS)]^-$ from $[IW(CO)_4(CS)]^-$, and it was converted by iodine to $trans$-$TpW(CO)_2(CS)I$,[28] by NO^+ to $TpW(CO)(CS)NO$, and by alkyl halides to the stable mercaptocarbynes, $TpW(CO)_2\equiv CSR$, the one with R = Me being also obtainable from MeLi and $TpW(CO)_2(CS)I$.[29] With $ClAuPR_3$, the bimetallic complex with W—Au bonds, $TpW(CO)_2(CS)$—$Au(PR_3)$, was obtained and structurally characterized. This complex contained one semibridging CO, and one semibridging CS.[691] When the alkylthiocarbyne $TpW(CO)_2\equiv CSR$ was treated with SMe^+, a η^2-dithiocarbene, $[TpW(CO)_2[\eta^2$-$C(SMe)SMe]^+$, was obtained, bonded to the metal through both the carbon and the sulfur atoms. This species reacted with various nucleophiles. For instance, with MeS^- the complex **68** was produced, while R_2NH yielded $TpW(CO)_2\equiv CNR_2$. The complex cation $[TpW(CO)_2[\eta^2$-$C(SMe)SMe]^+$ also reacted with PR_3 forming **69**.[692] Alkylidyne complexes $TpW(\equiv CAr)(CO)_2$ and their Tp^* analogs were also reported.[693] Related products were also obtained from

2.7 GROUP 6 COMPLEXES

[TpW(CO)$_2${η^2-CH(SMe)}]$^+$,[694] and several reactions of [TpW(CO)$_2$(η^2-C(SMe)SMe)] with various nucleophiles were studied.[695]

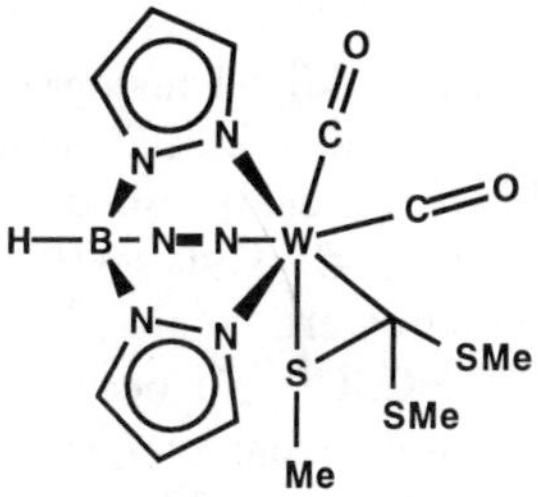

68 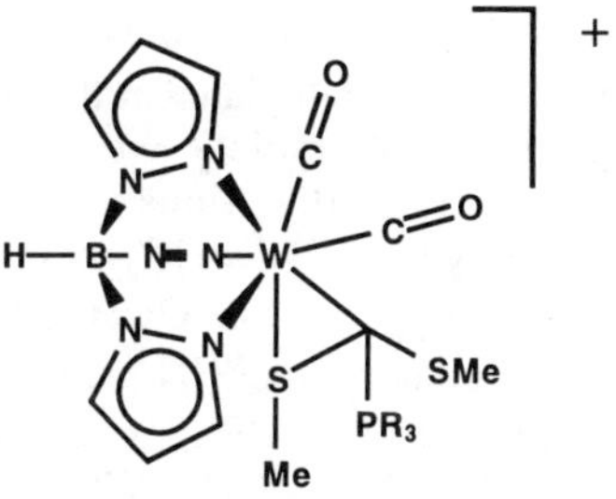 **69**

A side-on bridging thiocarbonyl ligand was found in the bimetallic complex TpW(CO)$_2$(CS)Mo(CO)$_2$(indenyl).[696] Among other reactions of TpW(CO)(CS)NO, that with MeNH$_2$ converted the thiocarbonyl ligand to an isonitrile, producing TpW(CO)(CNMe)NO.[29] Protonation of TpW(CO)$_2$≡CSMe formed TpW(CO)$_2$(η^2-CHSMe),[697] while the reaction with phosphines yielded a η^2-ketenyl derivative, TpW(CO)(PR$_3$)(η^2-MeSC=C=O), which upon methylation was transformed into the cation [TpW(CO)(PR$_3$)(MeSC≡COMe)]$^+$.[698] The structure of Tp*W(CO)$_2$≡C–P=C(NEt$_2$)$_2$, prepared from Tp*W(CO)$_2$≡CCl and Me$_3$SiP=C(NEt$_2$)$_2$, was established by X-ray crystallography.[473] This complex could be readily oxidized with molecular oxygen at the phosphorus to yield [Tp*W(CO)$_2$≡C-P(O)$_2$C(NR$_2$)$_2$].[475] Protonation of [Tp*W(CO)$_2$≡C–P=C(NR$_2$)$_2$] yielded [Tp*Mo(CO)$_2$≡C-P(H)-C(NR$_2$)$_2$]$^+$, which was rearranged to a metallaphosphirene.[474] The structurally characterized carbyne pzTpW(CO)$_2$≡C(p-C$_6$H$_4$Me) was a four-electron donor in the preparation of heteronuclear polymetallic clusters.[699] Other tungsten carbynes reported included Tp*WBr$_2$≡CPh,[700] Tp*W(CO)$_2$≡CN(R)Et, in which the carbyne link was elaborated from the EtNC ligand,[701] and the carbyne cation [Tp*W(CO)$_2$≡CPR$_3$]$^+$ which reacted with PR$_3$ forming the structurally characterized carbene cation [Tp*W(CO)$_2$=C(PR$_3$)$_2$]$^+$.[702] Treatment of TpW(CO)$_2$(η^2-OCNPr$^i_2$) with sodium ethoxide produced in low yield the complex TpW(CO)$_2$(≡CNPr$^i_2$).[1545] The carbyne Tp*W(CO)$_2$≡C(p-C$_6$H$_4$Me) was converted by Et$_2$BH to the structurally characterized Tp*W(CO)$_2$B(Et)CH$_2$(p-C$_6$H$_4$Me), containing an agostic bond to W from the benzylic methylene.[469]

Interesting tungsten chemistry resulted from the work of Templeton, much of which dealt with Tp*W complexes, entailing transformations of coordinated alkynes, or carbyne ligands. Isomeric, geometric and thermodynamic features for complexes such as Tp*W(CO)(NO)=C(OMe)(Me) were discussed.[467] The methylcarbyne Tp*W(CO)$_2$≡CCH$_3$ formed, upon deprotonation, the reactive anion [Tp*W(CO)$_2$=C=CH$_2$}$^-$, exhibiting a rich chemistry as a nucleophile.[405] For instance, it reacted with Tp*W(CO)$_2$≡CCl yielding [Tp*W(CO)$_2$=C]$_2$CH$_2$, which itself could be deprotonated to the dianion [Tp*W(CO)$_2$=C=C=C=W(CO)$_2$Tp*]$^{2-}$ and this could be readily dialkylated at the central carbon atom. Mixed analogs containing one Mo and one W moiety were also prepared.[481] By a similar sequence of deprotonations and alkylations, it was possible to synthesize species such as Tp*W(CO)$_2$≡C-CH=CH-C≡(CO)$_2$WTp*, [Tp*W(CO)$_2$=C=C=C=C=W(CO)$_2$Tp*]$^{2-}$, and Tp*W(CO)$_2$≡C-C≡C-C≡(CO)$_2$WTp*, which were analogous to their molybdenum counterparts.[50] The anion [Tp*W(CO)$_2$=C=CH$_2$]$^-$ underwent reaction with TpW(CO)(I)(PhC≡CPh) producing the structurally characterized dinuclear complex, Tp*W(CO)$_2$(≡CCH$_2$)-W(CO)(PhC≡CPh)Tp, which could be converted by treatment with KOBu^t and I$_2$ to Tp*W(CO)$_2$(≡CC≡)W(CO)(PhC≡CPh)Tp.[482]

The synthesis of a family of alkyne intermediates, valuable for further elaboration of the organic moieties bonded to tungsten, was based on the reaction of Tp*W(CO)$_3$I with a variety of acetylenes, leading to Tp*W(CO)(I)(RC≡CR') complexes, including the structurally characterized Tp*W(CO)(I)(PhC≡CMe). The iodide ion could be replaced by neutral ligands, such as acetonitrile, or phosphites, yielding cationic species.[681] It could also be replaced by cyanide ion, producing Tp*W(CO)(CN)(PhC≡CMe), in which the alkyne was aligned parallel to the M—CO axis. This species was converted to the dinuclear, CN-bridged, complex [Tp*W(CO)(PhC≡CMe)]-C≡N-[Tp*W(CO)(Bu^{n}C≡CMe)], and complexes of this type were subjected to extended Hückel MO calculations.[703] The cation [Tp*W(CO)$_2$(PhC≡CH)]$^+$, was obtained by the reaction of Tp*W(CO)$_2$I with AgBF$_4$ and phenylacetylene. It could be readily reduced to the structurally characterized β-agostic carbene derivative, Tp*W(CO)$_2$(=CPhMe),[704] and it reacted with LiHBEt$_3$ or with MeLi to produce neutral η^2-vinyl complexes of structures **70** and **71**, respectively. This reaction was studied in considerable detail, and a related complex, **72**, could be obtained from P(OMe)$_3$ and [Tp*W(CO)$_2$(PhC≡CH)]$^+$.[705]

70   **71**   **72**

2.7 GROUP 6 COMPLEXES

Deprotonation of the structurally characterized Tp*W(CO)$_2$(PhC≡CMe),[706] produced an anion which reacted with alkyl halides, R'X, to yield complexes of structure Tp*W(CO)$_2$(PhC≡CCH$_2$R'), and with aldehydes to produce the corresponding alcohols.[707] In an approach to non-linear optics, the complex Tp*W(CO)I(PhC≡CMe) was converted to the structurally characterized ferrocene derivative, Tp*W(CO)I(PhC≡C-CH=CMe(Fc).[708] Acetonitrile, coordinated in the cation [Tp*W(CO)(RC≡CR')(MeCN)]$^+$, was stepwise reduced via sequential addition of nucleophiles and electrophiles to its triple bond, and all the intermediate products could be readily isolated and identified.[709,48] Cationic bis-alkyne complexes, [Tp*W(CO)(PhC≡CH)(PhC≡CR)]$^+$ were synthesized by the reaction of [Tp*W(CO)$_2$(PhC≡CH)]$^+$ with PhC≡CR/AgBF$_4$, and the structure of the product, [Tp*W(CO)(PhC≡CH)$_2$], was determined by X-ray crystallography. This compound, upon monodeprotonation yielded readily the phenylacetylido species, Tp*W-(C≡CPh)(CO)(PhC≡CH). Carbenes and vinyl complexes could also be obtained from such precursors.[710] When [Tp*W(CO)$_2$(PhC≡CMe)]$^+$ was treated with LiCuR$_2$, neutral η^1-acyl complexes [Tp*W(CO)(PhC≡CMe)(η^1-C(O)R)] were obtained in which the alkyne ligand remained as a four-electron donor. The alkyne methyl group could be deprotonated and the anion could be alkylated with alkyl halides. Carbenes were obtained via protonation or alkylation of the η^1-acyl ligand, and the complex Tp*W(CO)(PhC≡CMe)=CMe(Bu) was structurally characterized.[711] The reaction of Tp*W(CO)(I)(PhC≡CMe) with LiCuMe$_2$ afforded Tp*W(CO)(Me)(PhC≡CMe) which could be protonated in the presence of ketones or aldehydes to yield the derived η^1-carbonyl complexes, convertible through hydride addition to the corresponding alkoxides, of which Tp*W(CO)(PhC≡CMe)(OCH$_2$CMe$_3$) was structurally characterized.[712] A fairly stable, structurally characterized ether complex, Tp*W(CO)(OEt$_2$)(PhC≡CMe), was obtained by treating Tp*W(CO)(Me)(PhC≡CMe) with HBAr$_4$. It reacted with methylene chloride to produce the metallacycle **73**.[713] The Tp or Tp* complexes Tp^{x}W(CO)(RC≡CR')Me were protonated, which resulted in loss of methane, followed by the addition of various ketones, producing η^1-ketone complexes with different conformational preferences about the M—O bond. The E/Z isomerization barriers were studied.[484]

Oxidation of the amido complex Tp*W(CO)(PhC≡CMe)(NHPh) yielded the structurally characterized nitrene, [Tp*W(CO)(PhC≡CMe)=NPh]$^+$, which reacted with LiBHEt$_3$ forming the hydride, Tp*W(H)(PhC≡CMe)(=NPh), and with MeMgBr forming the unusual metallacycle **74**.[714] Electrophilic tungsten(II) methylene carbene cation, [Tp*W(CO)(PhC≡CMe)=CH$_2$]$^+$ and its Tp analog, were prepared by the reaction of Tp*W(CO)(PhC≡CMe)Me with [Ph$_3$C][PF$_6$]. They formed adducts, such as the structurally characterized Tp*W(CO)(PhC≡CMe)(CH$_2$PPh$_3$), transferred a methylene group, and catalyzed aziridine formation.[715] Displacement of CO from [Tp*W(CO)$_2$(HC≡CH)]$^+$ by iodide ion produced the neutral [Tp*W(CO)(I)(HC≡CH)], which could be sequentially deprotonated and alkylated, giving rise to a variety of novel compounds, including the unusual, structurally characterized, cyclodecyne complex, **75**.[716] The cationic, seven-coordinate, complexes [Tp*W(CO)$_3$(PMe$_3$)]$^+$, and its PMe$_2$Ph analog, were structurally characterized,[717] as were Tp complexes TpW(CO)(PPh$_3$)(η^2-ArC=C=O) and TpW(CO)(Cl)(η^3-ArC≡CMe).[718]

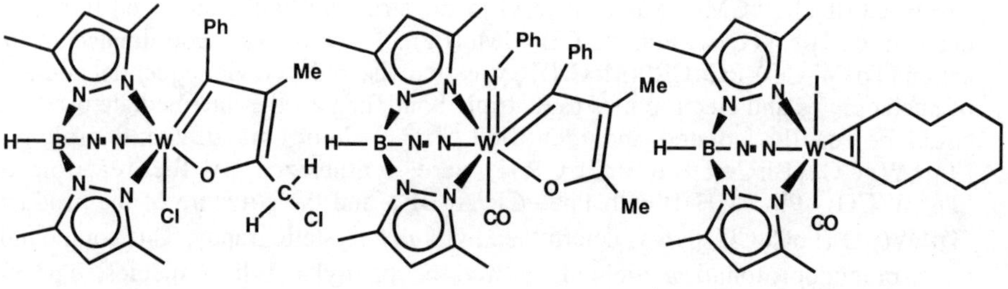

73 **74** **75**

Reduction of Tp*W(CO)$_2$(=NPh) with [BH$_4$]$^-$ generated the formyl complex Tp*W(CO)(CHO)(=NPh) which underwent hydride migration from carbon to nitrogen, yielding Tp*W(CO)$_2$(HNPh).[719] Tosyl azide reacted with Tp*W(CO)$_3$H forming Tp*W(CO)$_2$(NHTs), which could be oxidized to the imino complex Tp*W(CO)$_2$(=NTs), and its reaction with PMe$_3$ produced, among others, the cationic species [Tp*W(CO)$_2$(PMe$_3$)$_2$]$^+$ and [Tp*W(CO)(PMe$_3$)(=NTs)]$^+$.[720] Oxidation of the amido complexes [Tp*W(CO)(PhCCMe)(NHCHRR')] led to chiral η^1-imine complexes, [Tp*W(CO)(PhCCMe)(NH=CRR')], also available by other routes.[721] Details of the synthesis and reactivity of a variety of complexes such as the unstable nitrido derivative Tp*W(CO)$_2$≡N, Tp*W(CO)$_2$(NHR), [Tp*W(CO)$_2$(=NR)]$^+$, Tp*W(CO)$_2$(NR'R), Tp*W(CO)(CHO)(=NPh), and related structures, have been reported,[722] as also was the synthesis of the tosylimido derivative TpW(=NSO$_2$C$_6$H$_4$Me)$_2$Cl.[723] Refluxing Tp*W(CO)$_3$Br in acetonitrile along with Na(SPr^i) led to reductive coupling of MeCN, and isolation of the complex Tp*W(CO)$_2$(=NCMe=CMeN=)W(CO)$_2$Tp*.[724] A general route to tungsten cationic nitrene complexes, [Tp*W(CO)$_2$(=NR)]$^+$, of which two were structurally characterized, consisted of oxidizing Tp*W(CO)$_2$(NHR) with iodine.[725] The structurally characterized acetylimido complex Tp*W(CO)(SPh)(=NCOMe) was synthesized by the reaction of PhS$^-$ with Tp*W(CO)(I)(=NCOMe), itself prepared by the reaction of Tp*W(CO)(I)(NCMe) with pyridine N-oxide.[726] Cationic complexes of general structure [Tp*W(CO)$_2$(RC≡CH)]$^+$ reacted with R'NH$_2$, forming Tp*W(CO)$_2$[N(R')CH=CHR], and the structure of the complex with R' = benzyl was determined by X-ray crystallography.[727]

Remarkably stable tungsten(VI) imido alkylidene complexes were synthesized, and the structure of Tp*W(=NAr)(=CHCMe$_2$Ph)(Hpz*) was determined by X-ray crystallography.[728] Several related Tp-based analogs, having the general

2.7 GROUP 6 COMPLEXES

structure TpW(=CHCMe$_2$R)(=NAr)(CH$_2$CMe$_2$Ph) [R = Me, Ar = Ph; R = Ph, Ar = Ph; R = Ph, Ar = 2,6-bis(isopropyl)phenyl], has been synthesized, and some of them were converted to a variety of different cationic species, as exemplified by [TpW(=NPh)(CHCMe$_3$)(solvent)]$^+$. In the presence of AlCl$_3$, the above compounds catalyzed the ring-opening polymerization of cyclooctene.[729] Reactions of Tp*W(CO)$_2$(S$_2$PR$_2$) with pyridine N-oxide yielded carbonyl oxo-complexes Tp*WO(CO)(S$_2$PR$_2$), while thermal decarbonylation of Tp*W(CO)$_2$I in nitrile solvents produced Tp*W(CO)(η^2-RCN), convertible to Tp*W(CO)$_2$(S$_2$PR$_2$), as well as to Tp*WSX(CO) or to Tp*WS(CO)$_2$(S$_2$PR$_2$).[730] Refluxing [Tp*W(CO)$_3$X] in nitriles gave rise to [Tp*W(CO)X(RCN)], which reacted with [R$_2$PS$_2$]$^-$ forming dithio ligand derivatives, [Tp*W(S$_2$PR$_2$)(RCN)] in which the nitrile was κ^2, and [R$_2$PS$_2$] was monodentate.[731] Primary amines R'NH$_2$ were found to add across the nitrile triple bond of [Tp*W(CO)(PhC≡CR")(NCR)]$^+$, forming the well characterized cationic amidine species such as [Tp*W(CO)(PhC≡CR")(NH=C(R)NHR')]$^+$, **76**.[732]

76

Oxidative decarbonylation of [Tp*W(CO)$_3$]$^-$ with tetraethylthiuram disulfide led to a variety of products, including the structurally characterized Tp*WS(S$_2$CNEt$_2$).[733]

A convenient entry into homoscorpionate W^{VI} chemistry was provided by the reaction of Tp^x ligands with WO$_2$Cl$_2$, which produced Tp^{x}WO$_2$Cl complexes, and the related Tp*W(=NR)$_2$Cl species were prepared similarly.[378] The reaction of Tp*WO$_2$Cl with Grignard reagents led to alkyl or aryl complexes, of which Tp*WO$_2$Et and Tp*WO$_2$Ph were structurally characterized. Boron sulfide converted them to the oxothio and dithio analogs.[734,735] An overview of thio-tungsten chemistry, with emphasis on species such as Tp*WOSX, Tp*WS$_2$X, Tp*WSX$_2$ and Tp*WSX was published.[736] The structure of Tp*WOS[O-(2-iPr-5-Me-C$_6$H$_3$)] was determined by X-ray crystallography,[737] as was that of the mixed-valence complex Tp*W^{VI}O$_2$(μ-O)W^{IV}O(CO)Tp*.[738] Details of halogen replacement in Tp*WO$_2$Cl by diverse nucleophiles to yield Tp*WO$_2$X species (X = NCS, OMe, HCO$_2$, OPh, SPh, S$_2$PPh$_2$, SePh) have been reported, along with the structure of Tp*WO$_2$SePh.[739] The complexes Tp*WO$_2$(alkenyl) were oxidized with singlet oxygen, forming allylic hydroperoxides, which could be converted, in some cases stereoselectively, to a variety of oxygenated products.[740] The complex Tp*WO$_2$(OH) was obtained by heating Tp*W(CO)$_3$I in DMSO, and upon treatment with Et$_4$N(OMe) it was converted to the structurally characterized Et$_4$N[Tp*WO$_3$]. A dinuclear complex [Tp*WO$_2$]$_2$O was also reported.[741] The structurally characterized alkylidene complex Tp*WO(Cl)[=CHCMe$_3$] was prepared by treatment of the otherwise very stable carbyne complex Tp*WCl$_2$(≡CCMe$_3$) with alumina. It was the precursor of an air- and moisture-stable ROMP catalyst.[742] Tp^{x}ML species (M = Mo, W) were part of a catalyst system for cyclic olefin polymerization.[743]

Tungsten metallacycles, exemplified by the structurally characterized **77**, and the complex **78**, were obtained through amine-induced coupling of carbonyl and nitrile ligands, starting with the cationic complex [Tp*W(CO)$_3$(MeCN)]$^+$.[744]

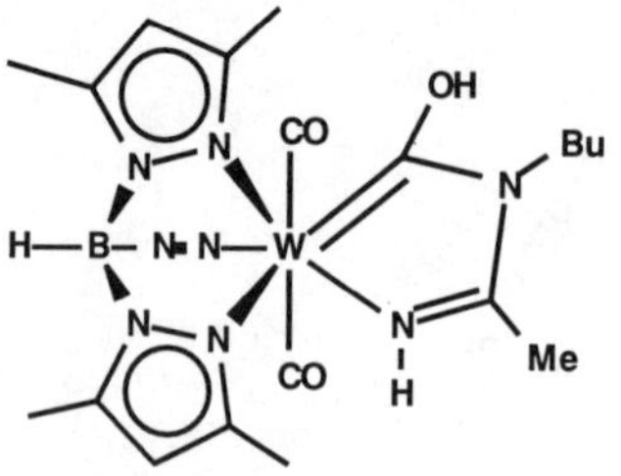

77 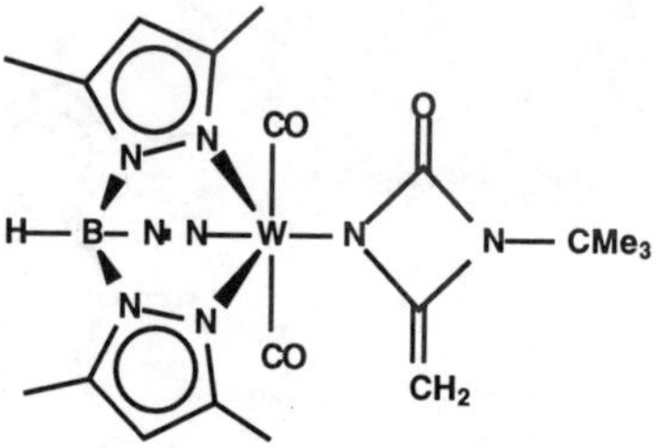   **78**

The complex TpW(=N-adamantyl)(=CHPh)(Br) underwent an unusual reaction with bromine. The structurally characterized product, showed a net insertion of an alkylidyne group into a tungsten-(pyrazolyl nitrogen) bond, as demonstrated by the structural feature W=C(Ph)(μ-pz)B.[745] Treatment of [TpW(CO)$_3$]$^-$ with 1,3-diiodopropane produced the structurally characterized η^2-acyl cyclopropyl complex TpW(CO)$_2$(η^2-COCpr), and the same type of product was obtained from the reaction with [TpMoCO)$_3$]$^-$. Decarbonylation of these complexes gave rise to the known TpM(CO)$_2$(η^3-allyl) compounds. When TpW(CO)$_2$(η^2-COCpr) was treated with CNBu^t, two products were obtained: the structurally characterized α-keto-η^2-iminoacyl complex TpW(CO)$_2$[η^2-C(COCpr)=NBu^t], and the seven-coordinate η^1-acyl complex, TpW(CO)$_2$(COCpr)(CNBu^t). The mechanism of these reactions was discussed in detail, and similar transformations where a methyl group was present in the place of cyclopropyl, were also investigated.[1554]

Unusual dinuclear complexes of structures **79** and **80** where the ligand L was (η^5-7,8-C$_2$B$_9$H$_{11}$) and E was S or Se, were synthesized, and were found to be

2.7 GROUP 6 COMPLEXES

readily cleaved by donor solvents, but were stable in dichloromethane or toluene. The structure of **79** was determined by X-ray crystallography.[673] Thioacyl complexes, TpMo(CO)$_2$(η^2-CSR), and related species were also reported.[1561]

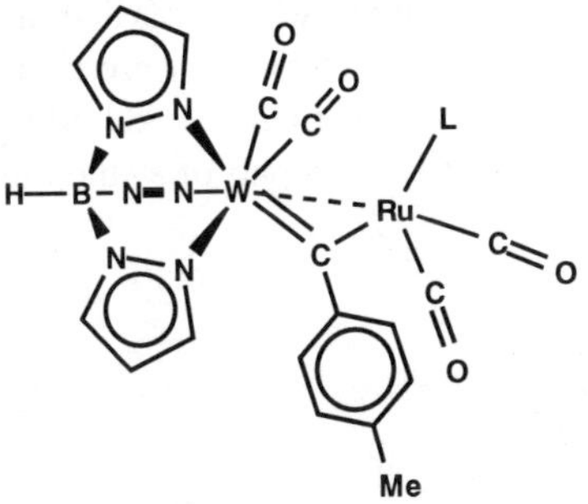

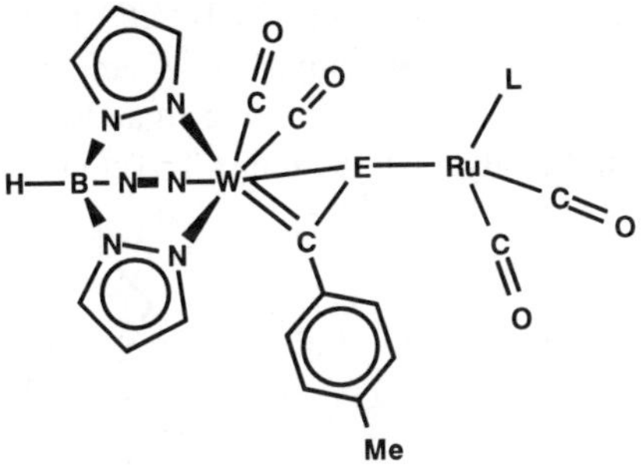

79   **80**

Numerous polymetallic clusters were reported, in which Tp^x ligands were capping a tungsten atom at a corner of the cluster. They were exemplified by the following species: TpWRu$_3$(μ_3-CTol)(CO)$_{11}$,[746] TpMoCo$_2$(μ_3-CR)(CO)$_8$,[747] TpWFe(μ–PPh$_2$)(CO)$_5$,[748] [TpWRh(μ–CMe)((μ–CO)(CO)Cp][PF$_6$]$_2$,[749] TpWFe(μ-CTol)(CO)$_5$,[750] TpWFeRh(μ_3-CTol)(CO)$_4$(μ-MeC$_2$Me)(η-C$_9$H$_7$),[751] and others.[752-755] Finally, the dinuclear tungsten complex [Tp*W(NO)(Cl)NH(CH$_2$)$_3$P(Ph)$_2$]$_2$CdI$_2$ was reported as a world recordholder for a compound with the most elements (ten).[756]

2.8 Group 7:   Mn, Tc, Re

2.8.1 Mn

The first reported homoscorpionate complexes of manganese included Tp$_2$Mn and pzTp$_2$Mn,[1,9] and their studies by spectroscopy,[13] Tp*$_2$Mn,[11] TpMn(CO)$_3$ and pzTpMn(CO)$_3$,[16,4] the structure of the latter complex,[403] and of Tp*$_2$Mn,[757] being determined more recently. Manganese(IV) complexes, [Tp$_2$Mn]$^{2+}$ and [Tp*$_2$Mn]$^{2+}$ were prepared by oxidation of the neutral precursors, and the structure of the [Tp*$_2$Mn][ClO$_4$]$_2$ salt was determined by X-ray crystallography.[757] Dinuclear complexes TpMn(μ-O)(μ-O$_2$CR)$_2$MnTp were synthesized as possible models for some manganese enzymes, and the structure of TpMn(μ-O)(μ-OAc)$_2$MnTp,[758] as well as its EPR spectra,[759] were determined. This compound was also found to cause chemiluminescence of luminol.[760]

2.8.2 Tc

The complexes TpTcOCl$_2$ and TpTcOBr$_2$ were the first technetium homoscorpionates reported, and the former was structurally characterized.[313,761] Subsequently both, high and low-oxidation states of Tc have been investigated, exemplified by the complexes TpTcO$_3$ and by TpTc(CO)$_2$(PPh$_3$), the structure of which was determined by X-ray crystallography.[762,763] Complexes TpTcO$_3$ and Tp*TcO$_3$ were prepared from Tc$_2$O$_7$ and the respective ligands,[764] and TpTcO$_3$ was found to add ethylene, forming the glycolate complex TpTcO(OCH$_2$CH$_2$O), **81**.[765] As part of a study of

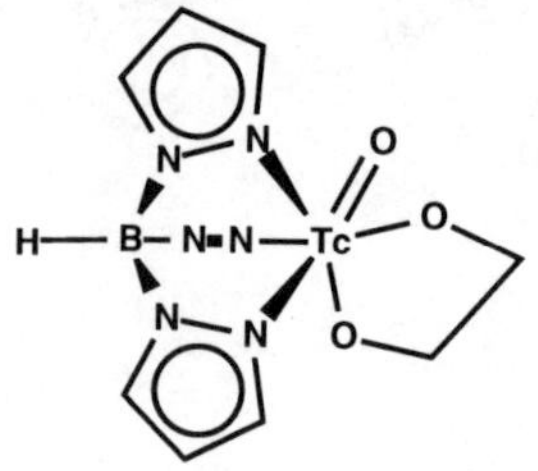

81

technetium chemistry, a series of TpM(CO)$_3$ and Tp*M(CO)$_3$ complexes (M = Mn, Tc and Re) was synthesized, and structurally characterized.[766] Irradiation of Tp*Tc(CO)$_3$ together with P(OMe)$_3$ produced Tp*Tc(CO)$_2$[P(OMe)$_3$].[767] The structurally characterized dinitrogen complex, [Tp*Tc(CO)$_2$]$_2$(μ-N$_2$) was obtained by irradiation of Tp*Tc(CO)$_3$ in a nitrogen atmosphere.[768] The first technetium complex containing a Tc=S bond, TpTc(=S)Cl$_2$, and the analogous Re complex, were synthesized by the reaction of TpM(O)Cl$_2$ with B$_2$S$_3$.[769,770]

2.8.3 Re

In addition to a few studies comparing analogous complexes of technetium and rhenium,[314,766,770] much research, especially that of Mayer, was devoted solely to rhenium. The structure of TpReO$_3$ was determined.[771] This compound was reduced by triphenylphosphine in the presence of Me$_3$SiX to yield TpReOX$_2$ species, which reacted with thiophenol to form either TpReOX(SPh) or TpReO(SPh)$_2$, both of which were structurally characterized, while chelating dithiols such as 1,2-ethanedithiol or 1,2-benzenedithiol yielded the appropriate chelate dimercaptides.[772] The reaction of TpReOCl$_2$ with toluidine gave the imido species, TpRe(=NTol)Cl$_2$, which upon treatment with ZnEt$_2$ was converted to the structurally characterized TpRe(=NTol)(Et)Cl.[773] The complex pzTpReO(OMe)$_2$ in which the ligand was

2.8 GROUP 7 COMPLEXES

82

tridentate, was converted to derivatives in which the ligand was κ^2: pzTpReO(OMe)(AcAc), **82**, the dinuclear [pzTpReO(AcAc)]$_2$O, [pzTpReO]$_2$(μ-O)(μ-pz^x)$_2$, and similar ones.[774] Analogous reduction of Tp*ReO$_3$ with PPh$_3$ yielded Tp*ReO(OH)Cl, from which several aryl thiolates, Tp*ReO(OH)(SAr), were prepared, as was the heterobimetallic complex Tp*ReO(OH)(SC$_6$H$_4$-4-O)Mo(NO)ClTp*.[775] The structurally characterized complex pzTpReO$_3$ was prepared similarly, and it was converted to pzTpReOCl$_2$.[776] The reduction of pzTpReO$_3$ with triphenylphosphine produced a compound formulated as [pzTpReO]$_2$O, which was a convenient starting material for the preparation of pzTpReO(LL) species, where LL could be glycolate, catecholate, (OR)$_2$ or (SR)$_2$. Structures of the diphenoxy, and dithiophenoxy derivatives were determined by X-ray crystallography.[777] Treatment of TpReOX(OTf), where X = Cl, Br, I, with pyridine N-oxide yielded the rare dioxo compounds TpReO$_2$X, which slowly disproportionated to TpReO$_3$ and TpReOX$_2$. The structure of TpReO$_2$Cl was determined by X-ray crystallography, and this complex was converted to several TpReO$_2$(OR) derivatives.[778] An improved procedure for the synthesis of Tp*Re(O)Cl$_2$ from Re$_2$O$_7$ has been reported.[1531]

Photochemical activation of TpReO(Cl)I in a variety of aromatic solvents resulted in the replacement of iodine by an aryl group. The yields were improved by the presence of pyridine.[779] This reaction was studied in detail and it was found that TpReO(Z)I (Z = Cl, I, Ph) species can be used in it, and that a variety of aromatics can replace the iodine to yield TpReO(Z)(Ar) derivatives. The structurally characterized complex TpReO(Ph)$_2$ was obtained from either TpReOI$_2$ or from TpReO(Ph)I.[780] When TpReOCl(Ph) was irradiated in the presence of pyridine or of other donor ligands, phenyl migration to the oxygen atom took place, yielding TpRe(OPh)(Cl)(L). It was shown that this is an intramolecular process, whereas a similar ethyl migration occurred via a radical pathway.[781] The structurally characterized oxalate complex, TpReO(C$_2$O$_4$), was prepared by treating TpReO(OCH$_2$CH$_2$O) with oxalic acid. It underwent photolysis to a very reactive four-coordinate rhenium(III) intermediate, TpReO, which was converted to TpReO$_3$ by oxygen, to TpReOCl$_2$ by chloroform, and to various 1,2-diolato derivatives by vicinal diones.[782] The structures of complexes **83** and **84** have been determined by X-ray crystallography.[783]

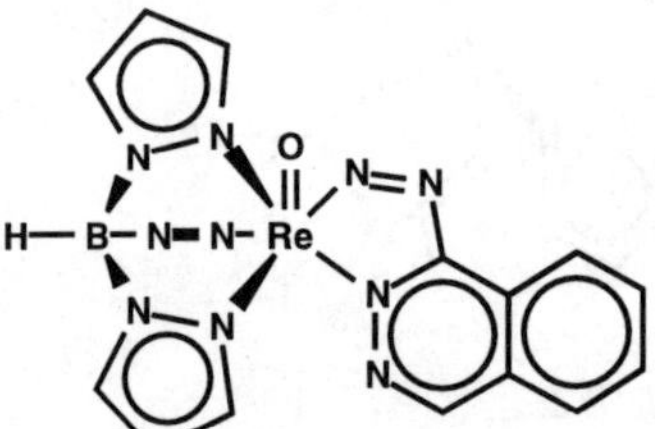

83 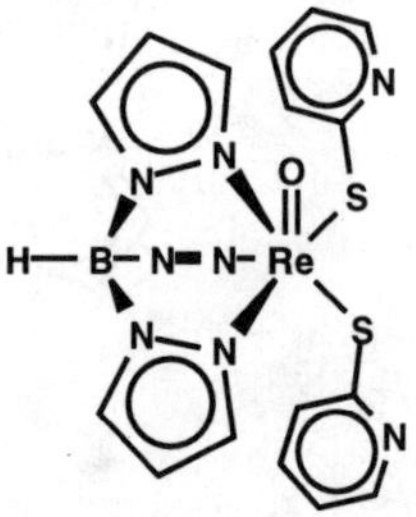   **84**

Irradiation of the oxalate complex TpRe(O)(C$_2$O$_4$) in acetonitrile along with hexafluoroacetone produced the unusual species, **85**, containing the elements of two hexafluoroacetone and one acetonitrile molecules.[784] The complex **86** was obtained on irradiation in MeCN of its precursor, which contained a κ^2-Tp ligand, and three carbonyl groups.[785] The Re^V complex TpReO(R)(OTf) was formed when TpReOCl$_2$

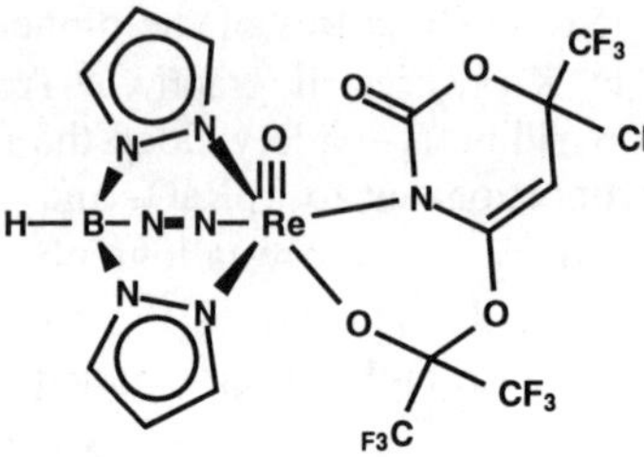

85 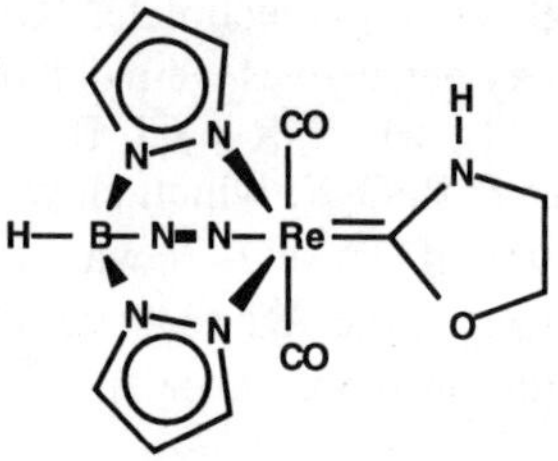   **86**

was treated with ZnR$_2$ and AgOTf, and the structure of this complex for R = Bu^n was determined by X-ray crystallography. Such complexes were readily oxidized by pyridine N-oxide or by DMSO to TpReO$_3$, and to the corresponding aldehyde, but in the presence of 2,6-lutidine, cis-2-butene was formed from the n-butyl derivative.[786] Treatment of TpReO(OR)$_2$ complexes, available from the reaction of TpReOCl$_2$ and alkoxides, with Me$_3$SiOTf, generated the reactive triflate species TpReO(OR)(OTf), which were readily oxidized to yield aldehydes and TpReO$_3$.[787] The structurally characterized pzTpReO(OMe)$_2$ was found to react with many vicinal diols forming the corresponding diolates.[788] Thermolysis of such diolates, derived from Tp*, was shown to produce Tp*ReO$_3$ and the appropriate olefin. The activation parameters for this reaction were determined, and compared with those for the Cp* analogs.[789]

2.8 GROUP 7 COMPLEXES

The reaction of TpReO(Ph)(OTf) with dimethylsulfoxide produced [TpReO(Ph)(DMSO)](OTf) which underwent phenyl-to-oxo migration, forming [TpReO(OPh)(DMSO)](OTf) and Me_2S. The complex [TpReO$_2$(Ph)](OTf) was an intermediate in this reaction. Kinetic and activation parameter data for this reaction have been determined.[790] The rhenium(V) oxo-hydride complexes TpReO(H)Cl, Tp*ReO(H)Cl, and Tp*ReO(H)$_2$ were prepared by treatment of the corresponding oxo-alkoxide pprecursors with THF·BH$_3$. The structurally characterized complex Tp*ReO(H)(O$_3$SCF$_3$) inserted olefins, and was oxidized by oxygen transfer agents to Tp*ReO$_3$.[791]

The carbyne/carbene complex TpRe(≡C-Bu^t)(=CH-Bu^t)(neopentyl), **87**, has been reported.[792] From a vibrational study of TpRe(CO)$_3$ and Tp*Re(CO)$_3$ complexes it was concluded that Tp^x ligands are better donors than Cp.[793] Irradiation of TpRe(CO)$_3$ in the presence of donor ligands resulted in replacement of one CO, with formation of TpRe(CO)$_2$(L) species, of which those with L = THF, and L = PPh$_3$, were structurally characterized.[794] The TpRe(CO)$_2$(THF) complex readily lost THF

87

and formed a number of adducts: with cyclopentene, an S-bonded one with thiophene, and binuclear ones with pyran, naphthalene, and N-methylpyrrole, in each case with a *trans* orientation of the TpRe(CO)$_2$ fragment. In the absence of other donors, a dinitrogen complex, [TpRe(CO)$_2$]$_2$(N$_2$) was obtained, in which the N—N bond length was close to that of free N$_2$.[795] Surprisingly, the bromination of TpRe(CO)$_3$ left the carbonyls untouched, instead introducing bromine at the 4-position, forming Tp*^{Br}Re(CO)$_3$.[796] The complexes TpReH$_6$ and TpReH$_4$(PPh$_3$) were the first polyhydrides stabilized solely by N-donor ligands.[797] Treatment of the structurally characterized pzTpReO(OMe)$_2$ with 1,2-, 1,3-, and 1,4-diols yielded the corresponding cyclic diolato complexes pzTpReO(diolato), and the synthesis of pzTpReCl$_3$ was also reported.[798] Reduction of κ^3-pzTpReO$_3$ with PPh$_3$ in the presence of donor ligands yielded rhenium(V) derivatives κ^2-pzTpReO$_2$L$_2$ (L = py, 4-Mepy, 4-(NMe$_2$)py and 1-Me-im), and κ^2-pzTpReO$_2$(P-P) (P-P = dmpe, dppe). Similar reactions also took place with κ^2-TpReO$_3$, and the structures of κ^2-pzTpReO$_2$(py-4-(NMe$_2$))$_2$, and κ^2-TpReO$_2$(dmpe) was determined by X-ray crystallography. By contrast, the reaction of [ReO$_2$(py)$_4$]$^+$ with KpzTp in dichloromethane resulted in partial ligand degradation, and formation of the structurally characterized complex **88**, which may be regarded as containing the novel ligand (HO)Tp. It contains two hydrogen bridges between pyrazolyl groups.[799]

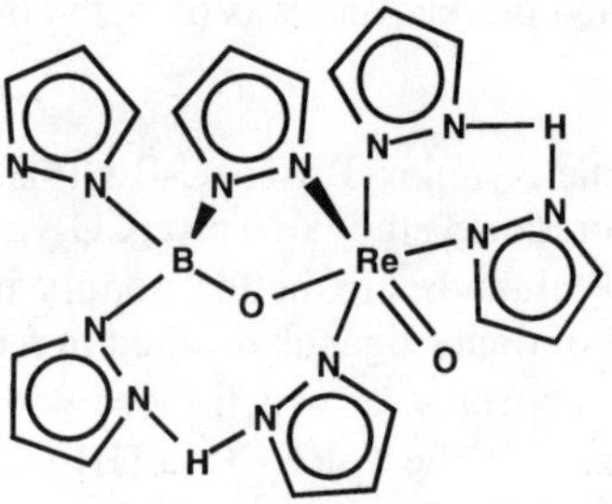

88

2.9 Group 8: Fe, Ru, Os

2.9.1 Fe

The early work with iron homoscorpionates included the synthesis of the octahedral Fe^{II} complexes, Tp$_2$Fe, pzTp$_2$Fe, and Tp*$_2$Fe,[9,11] the analysis of their spectral and magnetic properties,[13] and of the spin equilibria in octahedral Tp$^x{}_2$Fe complexes by means of Mössbauer spectroscopy.[800,801,14] In general, not much organometallic chemistry could be derived from Tp^{x}Fe species, as there was a strong tendency for the various organometallic intermediates to be converted into the very stable octahedral Tp$^x{}_2$Fe complexes. Nevertheless, the reaction of KTp with Fe(CO)$_3$I(η^3-allyl) produced, in addition to the ubiquitous Tp$_2$Fe, also two isolable novel organometallic iron species, namely, TpFe(CO)$_2$(COCH=CHCH$_3$) and TpFe(CO)$_2$(CH=CHCH$_3$).[802] In a similar reaction, TpFe(CO)$_2$(η^2-C(O)R) was obtained from Fe(CO)$_2$(η^2-C(O)R)I(PPh$_3$) and KTp,[803] but such reaction did not take place between KTp and Fe(η^2-OCNPr$^i{}_2$)I(CO)$_2$(PPh$_3$).[804] Thermolysis of TpFe(CO)$_2$(η^2-C(O)R) yielded Tp$_2$Fe.[803] From KTp, Fe$_2$(CO)$_9$, and MeI, the complex TpFe(CO)$_2$(COMe) was obtained in low yield.[805] Other reported organometallic iron compounds were TpFe(CO)(COMe)(PMe$_3$) and κ^2-TpFe(CO)(COMe)(PMe$_3$)$_2$.[806]

2.9 GROUP 8 COMPLEXES

Octahedral complexes Tp_2Fe and Tp^*_2Fe,[807] $pzTp_2Fe$,[808] $[Tp^*_2Fe](PF_6)$ and $[Tp^*_2Fe](TCNQ)$,[809] $[Tp_2Fe](NO_3)$,[810] $[Tp_2Fe][TpFeCl_3]$,[811] $Et_4N[TpFeCl_3]$ and $Et_4N[Tp^*Fe(N_3)_3]$,[812] and the unusual heteroleptic anionic complex of C_{3v} symmetry, $[TpFe(P_3O_9)]^-$,[813] were structurally characterized. The bond lengthening in going from low-spin to high-spin in Tp^*_2Fe was found to be one of the largest known.[807] The $[pz°Tp]_2Fe$ complex was also evaluated in a study of size-exclusion chromatography.[814] Tp_2Fe could be oxidized to Fe^{IV} in liquid SO_2, and was stable on the cyclic voltammetry scale.[815]

Spin equilibria in octahedral Fe^{II} homoscorpionates were studied in detail via partial molal volumes,[816] by means of the intersecting-state model,[817] through Mössbauer spectroscopy under different sets of conditions,[14,801,818-821] by means of gas-phase photoelectron spectroscopy,[822] via resonance and pulsed ultrasonic techniques,[823] and electrochemically.[824] The effect of pyrazole ring substitution on spin equilibria was investigated,[825] and a study of the temperature and pressure dependence of the electronic spin states of Tp_2Fe and Tp^*_2Fe (and of their cobalt(II) analogs) was carried out.[826] An unusual structurally characterized ferraoxetane, **89**, was obtained by the reaction of KTp with the carbamoyl complex $Fe(CO)_2(PPh_3)(CF_3)[CONPr^i_2]$.[827]

Considerable amount of research was devoted to binuclear Fe^{III} complexes of the general structure $[TpFe]_2(\mu\text{-}O)(\mu\text{-}Z)_2$, **90**, in which the μ-Z bridges were usually

89   **90**

carboxylato ligands, but could be phosphato, or other bridging anions. They were studied as structural and spectroscopic models for the spin-coupled diiron(III) centers in hemerythrin and ribonucleotide reductase. The basic model was **90** with Z = OAc, which was synthesized and structurally characterized.[828,829] It could be reversibly protonated, and the structure of the protonated cation was established.[830] The OAc bridges could be exchanged for their deuterated equivalents, and also for $(PhO)_2PO_2$ ligands.[831] Properties of $[TpFe]_2(\mu\text{-}O)(\mu\text{-}OAc)_2$ were studied by EXAFS,[832-834] by resonance Raman spectroscopy,[835] and were compared with those of the purple acid phosphatase from beef spleen,[836] with the 1,4,7-triazacyclononane analog,[837,838] and with other carboxylato Fe^{III} complexes.[839-842] A comparison of structures of **90**-type complexes bridged with Ph_2PO_2 or $(PhO)_2PO_2$ ligands, showed that the diiron Fe—O—Fe core is expanded in both compounds relative to the carboxylate-bridged analogs.[843] Hydroxo-bridged **90**-type complexes with carboxylato, Ph_2PO_2, and $(PhO)_2PO_2$ bridges, as well as the complex $[TpFe]_2[\mu\text{-}(PhO)_2PO_2]_3$ were structurally characterized, and compared to the oxo-analogs.[844] The complex $[TpFe]_2(\mu\text{-}O)(\mu\text{-}OAc)_2$ was studied in conjunction with excited-state electron transfer quenching involving the exchange coupled Fe^{III} ions.[845] The same complex was also found to catalyze the oxidation of alkanes with dioxygen,[846] while Tp*Fe(OOCCOPh), a model of α-ketoglutarate enzyme, was an olefin epoxidation catalyst.[847] The Tp ligand was used for capping $[Fe_4S_4]$[848,849] and $[VFe_3S_4]$ cores in various clusters.[850]

2.9.2 Ru

In contrast to iron, which had relatively little organometallic scorpionate chemistry, ruthenium exhibited a rich and varied one, mostly restricted to Ru^{II}. For instance, ruthenium provided the first examples of structurally characterized heteroleptic complexes such as $[pzTpRu(C_6H_6)][PF_6]$,[851,852] $TpRu(\eta^6\text{-p-cymene})$, and similar compounds with other aromatics.[853] In $TpRu(\eta^6\text{-}C_6Me_6)Cl$ the Tp ligand was κ^2 and Cl was covalent, but in the cationic species, Tp was κ^3.[854] The complex $[TpRu(\eta^6\text{-}C_6H_6)]^+$ reacted with anionic nucleophiles, X⁻, to yield $TpRu(\eta^5\text{-}C_6H_6X)$ species, of which the one with X = CN was structurally characterized, and found to contain a non-planar phenyl group.[855] The addition of nucleophiles to $[TpRu(\eta^6\text{-arene})]^+$, which gave rise to $TpRu(\eta^5\text{-cyclohexadienyl})$ species, has been studied by NMR, and the structure of the products has been probed by extended Hückel MO calculations, with emphasis on the rationalization of the experimentally observed rotational barriers of the carbocycles.[856] Tp^xRuCp species were prepared and studied in detail.[857-859] The structure of the unusual heteroleptic complex $TpRu(PPh_3)(B_3H_8)$ was established by X-ray crystallography.[860,861] Reaction of Ru^{IV} (aq) with TpK at a pH of 2 was studied in the presence of a variety of non-coordinating counterions, and it indicated replacement of coordinated water molecules in the tetranuclear Ru^{IV}(aq) core, with formation of species, exemplified by $H_4[(TpRuO_2)_4Hpz](NO_3)_4$.[862] Synthesis of the complex $TpRu(CO)(CH(CO_2Me)CH_2CO_2Me)(PPh_3)$ has also been reported.[863] A study of the protonation of the Ru—Ru bond in a series of $L_2Ru_2Me(CO)_2$ species, also included the complex $Tp_2Ru_2(CO)_4$.[865]

Dihydrogen complexes of ruthenium stabilized by homoscorpionate ligands, as exemplified by $TpRu(PR_3)(H_2)H_2$,[866] $TpRu(PR_3)(H_2)H$ and by $Tp*Ru(PR_3)(H_2)$ were synthesized.[867] Hydrogenation of Tp*RuH(COD) produced $Tp*Ru(H_2)_2H$ characterized by analytical and spectroscopic methods, and this compound was converted to deuteriated isotopomers.[868] Complexes $Tp^xRuH(H_2)$ and $Tp^xRuH(COD)$

2.9 GROUP 8 COMPLEXES

(Tp^x = Tp, Tp*) had good catalytic activity in reducing unactivated ketones, either by H_2 or by H_2 transfer from alcohols.[869] Cationic hydride and dihydrogen TpRu complexes, such as $TpRuH(PPh_3)_2$ and $TpRu(H_2)(PPh_3)_2$, have been reported. The hydrogen therein was readily replaced by neutral ligands such as CO, RNC, N_2, and by other related ligands.[870] The monohydrides $TpRu(H)(PR_3)$ were protonated to cations $[TpRu(H_2)(PR_3)]^+$. Neutral complexes $TpRu(H)(H_2)(PR_3)$, containing fluxional Ru-bonded hydrogens, were reported, as well as structures of the cations $[TpRu(N_2)(PEt_3)_2]^+$ and $[TpRu(H_2O)(PMePr^i_2)_2]^+$.[871] The complex TpRuCl(dippe), where dippe is 1,2-(diisopropylphosphino)ethane, reacted readily with 1-alkynes yielding vinylidene species, and structures of two such complexes $[TpRu=C(OMe)CH_2COOMe(dippe)][BPh_4]$ and $[TpRu(C\equiv CC(Ph)=CH_2)(dippe)]$, **91**, were determined by X-ray crystallography.[872] TpRuH(dippe), and also $TpRuH(PPh_3)_2$

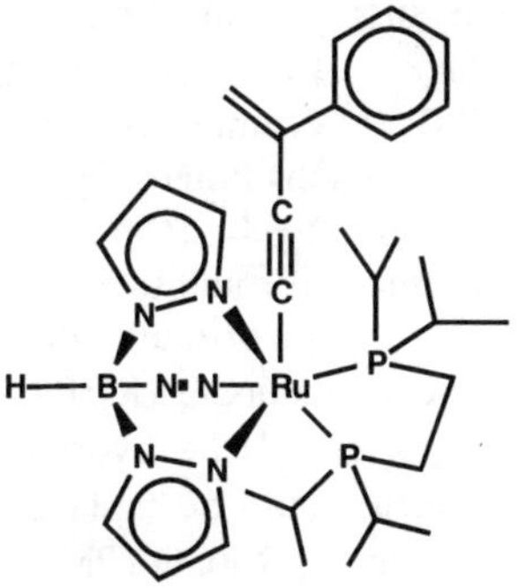

91

were converted to dihydrogen adducts on protonation. Of these two, the dippe derivative was more stable, and it was structurally characterized.[873] Protonation of $TpRu(CO)(H)(PR_3)$ converted the hydride ligand to (H_2), yielding the cation $[TpRu(CO)(H_2)(PR_3)]^+$.[874] Similarly, protonation of TpRuH(dppe) produced the monocationic dihydrogen complex $[TpRu(H_2)(dppe)]^+$. A comparison of the acidity of such monocationic and dicationic complexes, showed the dicationic complexes to be more acidic than their monocationic counterparts.[875] A Ru—Ru bonded dimer, $[TpRu(CO)_2]_2$, was structurally characterized.[876] Its electrochemical oxidation resulted in Ru-Ru bond cleavage, and the formation of cationic $[TpRu(CO)_2L]^+$ species, L being a solvent molecule.[877] Other complexes, such as $TpRu(CO)_2X$ (X = Cl, Br, I) have been reported,[878] as well as the acetyl complexes $TpRu(CO)(COMe)(PMe_3)$ and $TpRu(CO)(COMe)(PMe_3)_2$, the latter being κ^2.[806] The structurally characterized complex κ^2-$TpRuH(CO)(PPh_3)_2$ lost PPh_3 on heating, being converted to κ^3-$TpRuH(CO)(PPh_3)$, and the same happened with the CS analog. Treatment of κ^2-

TpRuH(CO)(PPh$_3$)$_2$ with a copper(I) salt produced cleanly the binuclear [TpRuH(CO)(PPh$_3$)][CuPPh$_3$], containing a Ru—Cu bond, and a Ru—H—Cu bridge.[879] The structurally characterized TpRu(S$_2$CNMe$_2$)(PPh$_3$) was converted by [TolS]$_2$ and Zn dust to the trinuclear [{TpRu(PPh$_3$)(μ-STol)$_2$}$_2$Zn.[880]

Complexes TpRu(X)(COD), where X was H or Cl have been readily synthesized.[881] Alkylation of TpRu(Cl)(COD) with Et$_3$Al, Et$_2$Mg, or EtLi produced mainly two products: the Ru-alkyl, TpRu(Et)(COD) and the hydride, TpRuH(COD). Using AlMe$_3$, the methyl derivative, TpRu(Me)(COD) was obtained.[882] In the complex Tp*Ru(Me)(COD), the Tp* ligand was coordinated in bidentate fashion, and the molecule contained an agostic B—H—Ru bond.[883] Complexes TpRuCl(PhCN)$_2$ and pzTpRuCl(PhCN)$_2$ were catalysts for olefin hydrogenation,[884] and also starting materials for numerous derivatives, including the cation [pzTpRu(CO)$_3$]$^+$.[885] Solid Tp$_2$Ru absorbed chlorine gas in redox-fashion, and released it on heating in vacuo.[886] Ruthenium(III) complexes of the Type Tp^{x}Ru(NO)Cl$_2$ were synthesized, and characterized by 2D NMR.[887] Numerous other complexes of structure TpRu(CO)(PPh$_3$)(R), where R was a substituted vinyl or aryl group, were prepared, and structures for R = p-tolyl, and for R = [-C(C≡CPh)=CHPh] were determined by X-ray crystallography.[888,889] An unusual route to Ru-alkyls entailed the reaction of TpRuCl(PPh$_3$)(MeCN) with NaBH$_4$ in primary alcohols, RCH$_2$OH, producing readily alkyl carbonyl complexes, TpRu(CO)(R)(PPh$_3$). The ruthenium hydride, TpRuH(H$_2$)(PPh$_3$), was also reported.[890] The hydride in TpRu(CO)H(PPh$_3$) was replaced by chloride which, in turn, could be replaced by CO, RNC, P(OMe)$_3$ or PMe$_3$, and the structure of the cation [TpRu(CO)(PMe$_3$)(PPh$_3$)]$^+$ was established by X-ray crystallography.[891] Ring opening metathesis polymerization of norbornene was effectively catalyzed by the complex TpRu(=C=CHPh)(Cl)(PPh$_3$).[892] Removal of chloride ion from ruthenium(II) alkylidenes TpRu(=CHPh)Cl(PCy$_3$) in the presence of donor ligands, yielded cationic species, which did not catalyze olefin metathesis reactions, but TpRu(=CHPh)Cl(PCy$_3$), **92**, catalyzed ring-closing metathesis in the presence of HCl, CuCl, or AlCl$_3$.[893] The unusual cyclopropene complex, **92a**, and some of its transformations have been reported.[1541]

92

92a

2.9 GROUP 8 COMPLEXES

Kirchner used TpRuX(COD) as an entry point to various TpRu derivatives, converting it first to cationic complexes [TpRu(COD)L]$^+$, from which COD could be replaced by chelating donor ligands. Some of the resulting complexes showed catalytic activity in the coupling of HC≡CPh with PhCOOH.[894] Carbon-carbon bond formation was achieved by the treatment of TpRu(COD)Cl with HC≡CR, which gave rise to η^3-butadienyl and η^2-butadiene complexes. Their chemistry, and the reaction mechanism of their formation were studied.[895] This coupling reaction was covered in more detail, employing complexes of the type TpRuCl(L)(L'), where L and L' were P, N, or O donors. Among them, the most efficient catalyst precursor for dimerization was TpRu(PPh$_3$)$_2$. In the absence of phosphine ligands polymerization occurred, with TpRu(COD)Cl being the most efficient catalyst precursor. The catalytically active species was assigned the structure TpRu(L)(-C≡C-R), and this species was trapped in one specific case as TpRu(PCy$_3$)(-C≡C-Bu)(CO), **93**.[896] A TpRu vinylidene complex, the structurally characterized [TpRu(tmen)=C=CHPh]$^+$, **94**, where tmen was Me$_2$NCH$_2$CH$_2$NMe$_2$, resulted from the treatment of the cation [TpRu(tmen)(solvent)]$^+$ with HC≡CPh. The type of bonding involved was analyzed

93   **94**

by extended HMOP calculations.[897] Similar cationic complexes were prepared using Ph$_2$PCH$_2$CH$_2$NMe$_2$ (= pn) instead of tmen.[898] Their reaction with HC≡CR (R = COOEt, Ph, CH$_2$Ph) led to novel coupling products, with loss of HNMe$_2$, and formation of complexes TpRu(Cl)[κ^3-(P,C,C)-Ph$_2$PCH=CHC(R)=CH$_2$)].[899] Regio- and stereoselective C—C coupling with terminal acetylenes was achieved through facile γ-C—H bond activation in phosphinoamine ligands of TpRuCl(L) complexes. The novel products thus obtained may be exemplified by TpRuCl[κ^3(P,C,C)-Ph$_2$PCH$_2$CH(NEt$_2$)CH=CHR)], and by related structures. These couplings took place, depending on the steric requirements of R, either at the internal or terminal carbon atom of the acetylene molecule.[900] A variety of vinylidene complexes of structure TpRuCl(PPh$_3$)(=C=CHR) was obtained from TpRuCl(PPh$_3$)(DMF) and diversely substituted 1-alkynes. The vinylidene moiety was labile, being easily replaced by CO, PR$_3$ and by other nucleophiles.[901] The complex reacted with several terminal alkynes forming vinylidene derivatives, such as TpRuCl(PCy$_3$)[=C=CHR], some of which catalyzed the dimerization of terminal alkynes.[902,903,1562] The formation of structurally characterized *trans*-diene complexes in TpRu moieties was achieved by first oxidizing TpRuCl(COD) to a Ru^{III} species, followed by addition of the diene in the presence of zinc.[904] 2-Acetamidopyridine replaced COD from TpRu(COD)Cl, and reaction of the resulting product with 1-alkynes yielded cyclic amido carbene complexes of general structure **95**.[905]

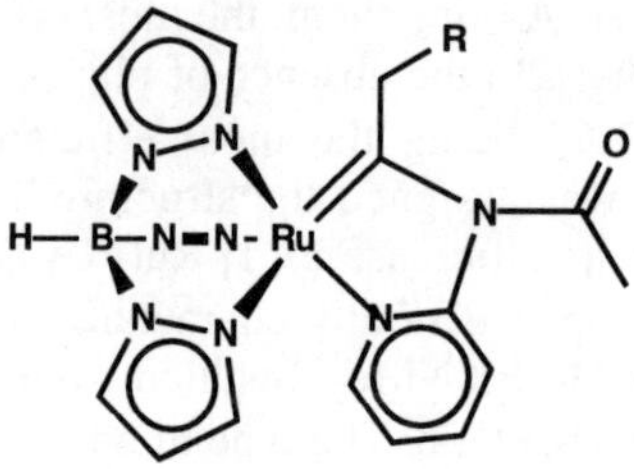

95

An example of ruthenium coordinating to a bis-bidentate pzTp ligand was the dinuclear complex (PPh$_3$)(CO)(MeCN)Ru[(pz)$_2$B(pz)$_2$]Ru(PPh$_3$)Cl(CO)(MeCN), in which one ruthenium was 5-coordinate and the other was six-coordinate.[906] Dinuclear complexes [TpRu]$_2$(μ-O)(μ-OAc)$_2$ and [TpRu]$_2$(μ-OH)(μ-OAc)$_2$, displaying a wide range of redox processes, were prepared and structurally characterized,[907] and the reactivity of the complex TpRuCl(PPh$_3$)$_2$ was explored in considerable detail.[1558]

2.9.3 Os

The scorpionate chemistry of osmium has been studied less than that of ruthenium. Earlier examples included κ^3-TpOsH(CO)(PR$_3$), κ^2-TpOsH(CO)(PR$_3$)$_2$, [κ^3-TpOs(CO)(PR$_3$)(κ^2-H$_2$)][PF$_6$] and its κ^2-HD analog,[908] as well as the structurally characterized complex TpOs($\equiv$CBu^t)(CH$_2$Bu^t)$_2$.[909,910] An additional example of a dihydrogen complex, [TpOs(H$_2$)(PPh$_3$)$_2$]$^+$, was prepared from TpOs(H)(PPh$_3$)$_2$ and HBF$_4$. It was more acidic than the Cp analog, but less so than its Ru equivalent.[911] Also reported were the dinuclear complexes [TpOs(CO)$_2$]$_2$ and [pzTpOs(CO)$_2$]$_2$, having Os-Os bonds, which were converted by bromine in good yield to mononuclear Tp^{x}Os(CO)$_2$Br species.[876] The complex κ^2-TpOs(Ph)(CO)(PPh$_3$)$_2$ lost PPh$_3$ on heating, yielding κ^3-TpOs(Ph)(CO)(PPh$_3$).[879] Similarly, κ^2-TpOsH(CO)(PPr$^i_3$)$_2$ lost one phosphine ligand, being converted to κ^3-TpOs(H)(CO)(PPr$^i_3$), which could be protonated to [κ^3-TpOs(H$_2$)(CO)(PPr$^i_3$)], from which H$_2$ could be displaced by

2.9 GROUP 8 COMPLEXES

acetone, forming κ^3-TpOs(κ^1-OCMe$_2$)(CO)(PPr$^i_3$).[874] The structure of the nitrido derivative Tp*Os($\equiv$N)Ph$_2$ was determined by X-ray crystallography.[912] Treatment of the nitrido complex TpOs($\equiv$N)Cl$_2$ with PhMgCl resulted in direct addition of the carbanion to the electrophilic nitride, producing [TpOs(=NPh)Cl$_2$]$^-$, which was protonated to the structurally characterized amido derivative TpOs(NHPh)Cl$_2$. When a large excess of PhMgCl was used, then compounds TpOs(NHPh)ClPh and TpOs(NHPh)(Ph)$_2$ were obtained.[913] Another unusual reaction of TpOs($\equiv$N)Cl$_2$ was the addition of triphenylborane to the nitrido nitrogen, producing TpOs([-N(Ph)BPh$_2$]Cl$_2$, in the structure of which one Cl was coordinated to boron, forming a four-membered ring. Hydrolysis of this compound yielded TpOs(NHPh)Cl$_2$. In an analogous reaction of TpOs($\equiv$N)Cl$_2$ with PhOBPh$_2$, the product was the structurally characterized complex, TpOs[-N(Ph)B-O-BPh$_2$]Cl$_2$.[914] Synthesis of the hydride complex, TpOsH(PPh$_3$)$_2$, from TpOsCl(PPh$_3$)$_2$ and KOH in 2-methoxyethanol has also been explored.[1558] Heating of κ^2 TpOs(CO)(CH=CHR)(PPh$_3$)$_2$ produced the tridentate complex κ^3 TpOs(CO)(CH=CHR)(PPh$_3$).[1361]

Amongst other reported osmium species was the mixed-valence compound TpOs^{III}Cl$_2$(N$\equiv$N)Cl$_2$Os^{II}Tp,[915] from which the rather unusual complex, [pz^{4Cl}Bp]OsCl$_2$(N$\equiv$N)Cl$_2$Os[pz^{4Cl}Bp]$^-$, was obtained and characterized by X-ray crystallography,[916] and the anionic species [TpOsCl$_3$]$^-$.[917] Electrochemical reduction of the Os^{VI} complex, TpOs(Cl$_2$)$\equiv$N, yielded the ammine species, TpOs^{III}(Cl$_2$)(NH$_3$), which could be reoxidized to the starting material.[918] The complex TpOsCl$_2$(NS) has also been prepared.[919] An interesting trimetallic species [TpOs(Cl$_2$)$\equiv$N]$_2$CoCp, **96**,

96

has been obtained from TpOs(Cl$_2$)$\equiv$N and the diene complex CpCo(C$_5$H$_5$-C$_6$F$_5$). Its structure showed both nitrido nitrogens bonded to cobalt, in a manner similar to other CpCoL$_2$ complexes, and it contained essentially linear Os-N-Co bonds. The complex TpOs(Cl$_2$)$\equiv$N also reacted slowly with PtCl$_2$(Me$_2$S)$_2$, displacing one Me$_2$S, and forming the adduct TpOs(Cl$_2$)$\equiv$N-PtCl$_2$(Me$_2$S), in which platinum was in a square planar environment, as was determined by X-ray crystallography.[920]

2.10 Group 9: Co, Rh, Ir

2.10.1 Co

Cobalt homoscorpionates were some of the earliest such complexes reported, and included Tp_2Co, $pzTp_2Co$,[1,9] and $Tp*_2Co$.[11] They were studied by spectroscopy,[13] their paramagnetic resonance spectra were recorded and analyzed,[921] as were their isotropic nuclear resonance shifts,[922] the theory of which was discussed.[923] Tp_2Co was the first homoscorpionate complex structurally characterized by X-ray crystallography.[15] The paramagnetism of this complex led to a number of studies concerned with its second-sphere coordination by pyridine or aniline,[924,925] by nitro compounds,[926] and even by saturated hydrocarbons.[927] Ligand field parameters for the Co^{III} complex, $[Tp_2Co][ClO_4]$ were determined.[415] The related homoleptic and heteroleptic complexes, $[Tp_2Co][PF_6]$, $[TpCo(tacn)][PF_6]_2$ and $[TpCoCl(dien)][PF_6]$ were prepared, studied by spectroscopy, and were structurally characterized.[928] The extremely unstable chlorine-bridged dimer, $[TpCoCl]_2$, was reported,[308] as was the complex $[Tp_2Co][Sn_2Co_5Cl_2(CO)_{19}]$,[929,930] and also the stable heteroleptic TpCo complexes with carbocyclic ligands, $[TpCoCp]^+$, **97**, and $TpCo[Ph_4C_4]$, **98**.[191] All

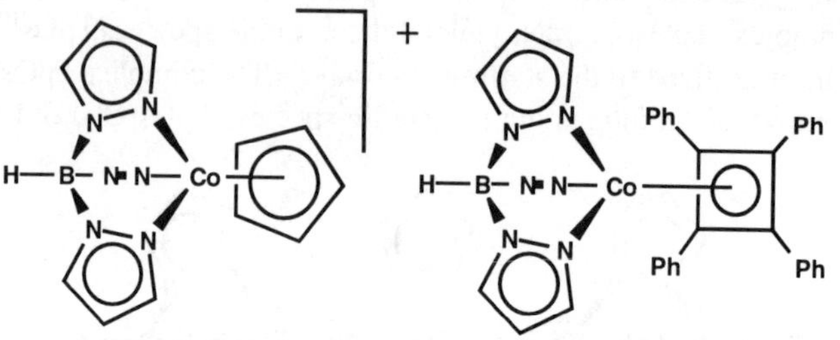

97   **98**

attempts to prepare a heteroleptic complex involving a carbollide ligand, produced the salt $[Tp_2Co][commo-3,1,2-Co(3,1,2-C_2B_9H_{11})]$.[931] In order to prepare synthetic approximations to a proposed active site in the cobalt(II)-substituted blue copper proteins, complexes $Tp*Co(L)$ (L = SC_6F_5, SC_6H_4-p-NO_2, O-ethylcysteinate) were synthesized, and studied by spectroscopy, while the structure of $Tp*Co(SC_6F_5)$ was determined by X-ray crystallography.[1546]

The only reported complexes containing alkyl groups on Tp^xCo moieties of the "first generation" homoscorpionate ligands were of the general structure $Tp^xCoCp(R_f)$ where R_f were various perfluoroalkyl groups. In all these complexes the Tp^x ligands were bidentate and Cp was η^5. In some cases separable, and in others

2.10 GROUP 9 COMPLEXES

inseparable, conformational isomers were present.[932] On the other hand, tetrahedral Tp^{x}CoR complexes were readily available from several homoscorpionate ligands of the "second generation", as shown in Chapter 3.

2.10.2 Rh

The first homoscorpionate complexes of rhodium were prepared by the reaction of the Tp^x ligand with RhCl(LL) dimers, where LL was a diolefin, (CO)$_2$ or a combination of other ligands,[4,933-936] and the structure of [pzTp]RhI$_2$CO was determined by X-ray crystallography.[937] Protonation of Tp*Rh(CO)$_2$ occured on the nitrogen, and the structure of the product [(κ^2-Tp*H)Rh(CO)$_2$][BF$_4$] was determined by X-ray crystallography,[938] as was that of TpRhI(PPh$_3$)Me.[939] The ^{103}Rh chemical shift anisotropy relaxation was studied for the compound [pzTp]Rh(COD),[940] and ^{103}Rh NMR studies were also done on a variety of [pzTp]Rh(diolefin) complexes, including those of COD, NBD and duroquinone.[941-943] A wide range of Rh^{III} alkyl complexes of general structure Tp*RhCl(CNCH$_2$Bu^t)R (R = Me, Pr, Pr^i, vinyl, cyclopropyl) was prepared by the reaction of RMgX with Tp*RhCl$_2$(CNCH$_2$Bu^t), and the complex Tp*RhCl$_2$(PMe$_3$) was structurally characterized.[944]

Graham studied the photochemistry of Tp*Rh(CO)$_2$ in aromatic or aliphatic solvents and found C—H bond activation, with oxidative addition of benzene or of

$$Tp*Rh(CO)_2 + R\text{-}H \text{ (or Ar-H)} \rightarrow Tp*Rh(CO)(H)(R \text{ or Ar}) + CO\uparrow \qquad 2.6$$

the alkane.[945] Tp*Rh(CO)(η^2-alkene) reacted thermally in benzene with displacement of alkene, and activation of the benzene C—H bond, forming Tp*Rh(CO)H(Ph).[946] This reaction could also lead to ethylene insertion, producing the structurally characterized Tp*Rh(CO)(Et)(Ph). Under carbon monoxide pressure, this complex was converted to Tp*Rh(CO)[C(O)Et](Ph).[947] The quantum efficiency for this reaction in n-pentane was calculated,[948] and the reaction was studied in detail, and found to be very efficient, and extremely wavelength dependent.[949-952] Studies of this reaction in argon and methane matrices at about 12 K indicated that it started with CO loss, followed by dechelation of one pz* arm of the Tp* ligand. In a nitrogen matrix, a N$_2$ complex was formed. Only at higher temperatures was C—H bond activation observed.[953] By using ultra-fast IR spectroscopy it was possible for Bergman to establish the femtosecond dynamics of the C—H activation process, which proceeded through loss of CO followed by a number of steps, and to assign the intermediate structures and the energy barriers for each specific reaction step.[954] The photochemical reaction of Tp*Rh(CO)$_2$ to activate C—H bonds was analyzed, and compared with that of the Cp and Cp* analogs.[955] When photolysis of Tp*Rh(PMe$_3$)(C$_2$H$_4$) was carried out in thiophene, the products were **99** and **100**, corresponding to C—H and S—C bond activation, respectively. At higher temperatures, complex **100** became the almost exclusive product.[956] A rather similar

99 **100**

photolytic reaction took place with Tp*Rh(CN-CH$_2$Bu^t)(PhN=C=N-CH$_2$Bu^t), which upon irradiation in alkanes or arenes yielded products arising from the elimination of the carbodiimide ligand, and formation of a C—H oxidative addition product.[957-959] The thermolysis of such a product, Tp*Rh(CN-CH$_2$Bu^t)(Ph)(H), in benzene in the presence of excess isocyanide produced Tp*Rh(CN-CH$_2$Bu^t)$_2$, and the mechanistic aspects of this reaction were studied in detail.[960] The structure of Tp*Rh(CN-CH$_2$Bu^t)$_2$ showed the Tp* ligand to be κ^2.[961] The 1,3-dipolar cycloaddition of phenyl azide to Tp*Rh(CNR)$_2$ gave rise to Tp*Rh(CNR)(η^2-PhN=C=NR), the photolysis of which in benzene produced Tp*Rh(CNR)(Ph)(H). These reactions, and related ones, were investigated,[962] with emphasis on the energetics of homogeneous intermolecular vinyl and allyl carbon–hydrogen bond activation by the 16-electron species Tp*Rh(CNCH$_2$CMe$_3$).[1528]

The thermal reaction of Tp*Rh(C$_2$H$_4$)$_2$ with CO, PMe$_3$, or Bu^{t}NC resulted in the replacement of one ethylene, and formation of Tp*Rh(C$_2$H$_4$)(L), while with acetonitrile or pyridine the ethyl/vinyl derivatives, Tp*Rh(Et)(L)(C$_2$H$_3$), were formed. They reacted thermally with benzene, producing Tp*Rh(Ph)(Et)(L). It was suggested that the mechanism of these transformations might involve Rh^{III} intermediates, although only Rh^I species seemed to be involved in the activation of benzene.[963] Theoretical studies of the relative stability of η^2-ethene and hydridovinyl Tp^{x}Rh(CO)(C$_2$H$_4$) complexes indicated that the favored species were κ^2-Tp^{x}Rh(η^2-C$_2$H$_4$).[964] Oxidation of Tp*Rh(CO)(PPh$_3$), which was prepared by heating κ^2-Tp*Rh(CO)$_2$ with PPh$_3$, yielded the stable Rh^{II} cation, [κ^3-Tp*Rh(CO)(PPh$_3$)]$^+$, the structure of which was determined by X-ray crystallography.[965] In the complex Tp*Rh(CO)(PMe$_3$) the Tp* ligand was found to be κ^2.[966] Several heteroleptic complexes, such as TpRh(C$_4$H$_4$BPh),[967] 3-TpRh-(3,1,2-C$_2$B$_9$H$_{11}$), closo-2-TpRh-(2,1,7-C$_2$B$_9$H$_{11}$),[931] and [TpRhCp*]$^-$,[191] were also reported.

2.10 GROUP 9 COMPLEXES

The reaction of TpRh(C_2H_4)$_2$ with PPh$_3$, afforded TpRh(C_2H_4)(PPh$_3$), in which there was an equilibrium between κ^3 and κ^2 coordination of the Tp ligand. Hydrogen displaced ethylene from this compound, producing TpRh(PPh$_3$)H$_2$.[968] The κ^2-κ^3 isomerism in Rh^I homoscorpionates was studied by IR spectroscopy for a series of Tp^{x}Rh(LL) complexes, including Tp*. Three types of structures were present in varying amounts: a κ^3-structure, a κ^2-structure with the third pz^x in axial position, exchanging with the two coordinated pz^x groups, and a κ^2-structure with the third pz^x group in the equatorial position, not exchanging with the coordinated pz^x groups. The last structure was favored when bulky (LL) and Tp^x ligands were present.[114] The bulkiness of the Tp^x and (LL) ligands played a crucial role in the catalytic activity of Tp^{x}Rh(LL) complexes toward stereoregular polymerization of phenylacetylenes, including those with a variety of substitutents in the *para*-position.[969] Experimental evidence was obtained for the presence of methane sigma-complexes prior to the dissociation of methane in the reductive elimination reaction from complexes of the type Tp*RhMe(H)(L).[1556]

A very useful starting material for TpRh chemistry was TpRh(PPh$_3$)$_2$, obtained readily from TpK and RhCl(PPh$_3$)$_2$. It reacted with dimethyl acetylenedicaboxylate to yield **101**, it formed a CS$_2$ complex TpRh(η^2-SCS)(PPh$_3$) which, upon the reaction with MeI in HCCl$_3$ produced the carbene **102**, and it was

101   **102**   **103**

converted by dimethylthiocarbamoyl chloride to **103**. The structures of these complexes were established by X-ray crystallography. Other transformations of TpRh(PPh$_3$)$_2$ included its reaction with oxygen, forming the complex TpRh(PPh$_3$)(O$_2$), the reaction with ethylene, which produced the compound TpRh(PPh$_3$)(CH$_2$=CH$_2$), and its conversion by COD to TpRh(COD).[970]

Cyclopropane added to the transient species Tp*Rh(CNCH$_2$Bu^t) yielding the hydrido cyclopropyl derivative **104**, which rearranged to the structurally characterized rhodiacyclobutane **105**, and this compound added stepwise isonitrile, CNR, being converted to **106**. The propylene complex, Tp*Rh(CNCH$_2$CMe$_3$)(H$_2$C=CHMe), was produced upon pyrolysis of **105**.[971]

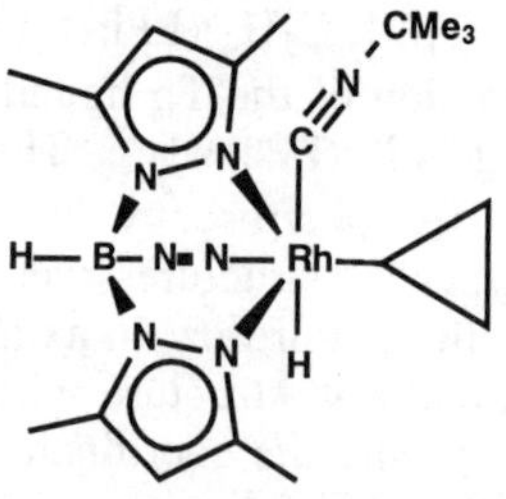

104 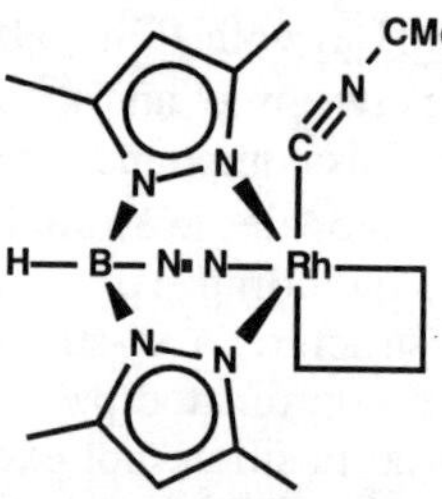   **105** 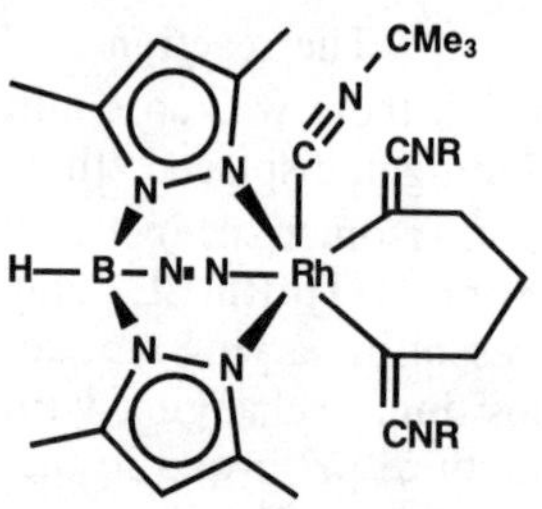   **106**

The nature of the various types of hydrogen bound to homoscorpionate rhodium complexes was studied using several techniques. The first "non-classical" polyhydrido complex stabilized by a nitrogen donor ligand, Tp*Rh(H$_2$)H$_2$, was reported by Venanzi,[972] and characterized via 2D-(^{1}H,^{103}Rh)-NMR. The nature of the Rh-H$_2$ bond in the above complex was investigated through inelastic neutron scattering spectroscopy. The H—H separation, and the barrier for rotation of the H$_2$ ligand were determined.[973] A density functional theory study established the dihydrogen-dihydride structure in Tp*Rh(H$_2$)H$_2$.[974] Protonation of the hydrogen complex Tp^{x}Rh(PPh$_3$)H$_2$ afforded the cation [TpRh(PPh$_3$)(H$_2$)H]$^+$, which was formulated as having a fluxional dihydrogen-hydride structure, as it exhibited a single hydrogen resonance in the ^{1}H NMR at all accessible temperatures.[975]

Oxidative addition of allyl bromide to Tp*Rh(L,L') complexes occurred with diminishing ease for complexes with L = MeCN, L' = cyclooctene; L = MeCN, L' = ethylene; and L,L' = COD. The initial product was the σ-allyl complex, which could be converted to the η^3-allyl derivative, Tp*RhBr(η^3-allyl). This complex reacted readily with nucleophiles, forming products exemplified by Tp*RhH(η^3-allyl) and Tp*RhMe(η^3-allyl), respectively.[976] Synthetic routes to complexes Tp^{x}RhCl$_2$(L) for Tp and Tp* (and for other Tp^x ligands), where L was MeOH or Hpz^x, were reported.[977]

2.10.3 Ir

The homoscorpionate chemistry of iridium resembles, up to a point, that of rhodium, and analogous Rh and Ir complexes were often prepared in pairs. Protonation, leading to TpIr(PR$_3$)H$_2$, proceeded as with the Rh analog, but the H—H bond lengths in the Ir complex were longer.[975] Triphenylphosphine displaced one ethylene from Tp*Ir(C$_2$H$_4$)$_2$, forming TpIr(C$_2$H$_4$)(PPh$_3$), which showed no exchange of the axial and equatorial pyrazolyl arms, while the complex TpRh(C$_2$H$_4$)(PPh$_3$) did show such an exchange on the NMR time scale.[968] A variety of five-coordinate, 18 electron

2.10 GROUP 9 COMPLEXES

complexes Tp*Ir(CH$_2$=CH$_2$)(L), **107**, for Tp and Tp* was prepared, and shown to be unstable with regard to their hydride-vinyl isomers, **108**. The hydrido vinyl complex

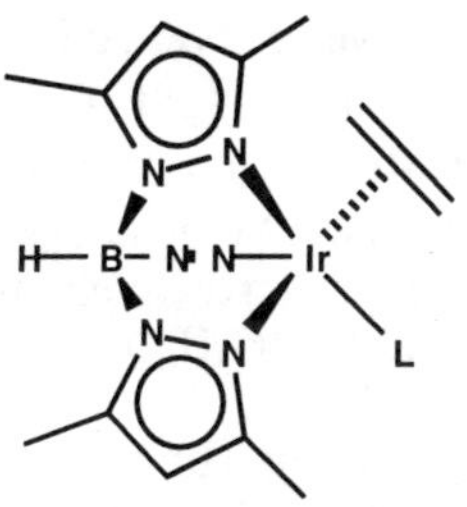

107 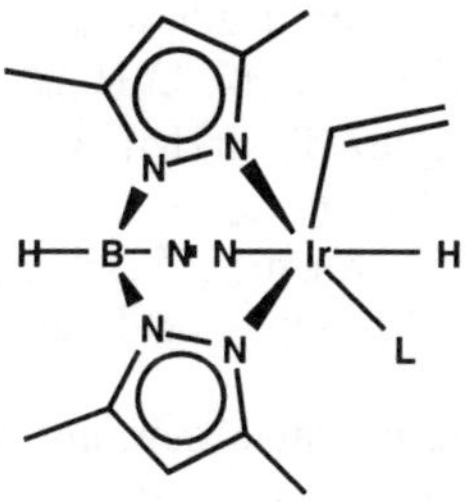   **108**

Tp*Ir(H)(CH=CH$_2$)(PMe$_2$Ph) was structurally characterized.[978] Protonation of Tp*Ir(CO)$_2$ took place at the metal, forming a cationic Ir^{III} hydride, in contrast to the Rh analog, where the uncoordinated nitrogen was protonated.[938] Ab initio quantum mechanical calculations, carried out for the reaction Tp^{x}M(CO)(C$_2$H$_4$) → Tp^{x}M(CO)(H)(C$_2$H$_3$), agreed with the experimental findings, that iridium favors the hydrido vinyl species, while rhodium favors the ethylene complex.[964] The equilibria between the κ^2 and κ^3 coordination modes of Tp, Tp*, and other Tp^x ligands in Tp^{x}Ir(COD) complexes were studied by NMR, and the structure of κ^3-TpIr(COD) was determined by X-ray crystallography.[979] Electrophilic cationic ethylidene complexes of Ir^{III} were prepared through the β-carbon protonation of Tp*Ir H(CH=CH$_2$)(PMe$_3$), itself obtainable through the thermal reaction of the Ir^I species, Tp*Ir(CH$_2$=CH$_2$)(PMe$_3$). The comparative α-migratory insertion processes for hydride and ethyl groups in the intermediate species, [Tp*Ir H(PMe$_3$)=CHMe]$^+$, were studied in detail.[980]

Compounds pzTpIr(COD) and pzTpIr(CO)$_2$ were prepared by the metathesis route from the corresponding chlorine bridged dimers.[981] Hydrogenation of Tp*Ir(COD) produced the stable, structurally characterized, Tp*IrH$_2$(COE) (COE = cis-cyclooctene), or Tp*IrH$_2$(COD), depending on the pressure used. The complexes Tp*Ir(C$_2$H$_4$)$_2$ and Tp*IrH$_2$(C$_2$H$_4$) were prepared similarly.[982] Protonation of TpIr(σ-C$_8$H$_{13}$)(η^2-C$_8$H$_{14}$) afforded the structurally characterized cation [TpIr(H)(σ-C$_8$H$_{13}$)(η^2-C$_8$H$_{14}$)]$^+$.[983] Photolysis in benzene of Tp*IrH$_2$(COE) in the presence of P(OMe)$_3$ produced Tp*IrH(Ph)[P(OMe)$_3$], while in ether, t-butyl acrylate gave rise to the complex Tp*IrH$_2$(CH$_2$=CHCOOBu^t), in each case the intermediate being the 16-electron Tp*IrH$_2$ species.[984] A more detailed study of the above reaction, and of the photolysis of Tp*Ir(COD) in methanol, produced a mixture of Tp*IrH$_4$ and Tp*Ir(CO)H$_2$ in each case, and it was shown that methanol is the source of the CO and hydride ligands, and that a common intermediate is involved for both reactions.[985] TpIr(C_2H_4)$_2$ reacted with carbon monoxide, forming either TpIr(C_2H_4)(CO),[986] or TpIr(CO)$_2$,[987] and with methyl acrylate (MA), to yield TpIr(C_2H_4)(MA). Upon irradiation this complex was converted to the hydrido vinyl species TpIrH(η^1-C_2H_3)(η^2-C_2H_4).[987] Thermolysis of Tp*Ir(C_2H_4)$_2$ produced the hydrido allyl complex Tp*IrH(η^3-$CH_2CHCHCMe$), formed through a hydrido vinyl intermediate.[988,989] That intermediate also activated the oxygen-adjacent hydrogens in, for instance, THF, forming Fischer-type carbene derivatives, also containing a hydrogen and butyl group on iridium, viz. Tp*IrH(Bu)(C_4H_6O).[990]

Carmona has also demonstrated that heating Tp*IrH(C_2H_3)(C_2H_4) under nitrogen in benzene produces the dinitrogen complex, Tp*Ir(Ph)$_2$(N_2), which under additional nitrogen pressure yields the structurally characterized dinuclear complex [Tp*Ir(Ph)$_2$]$_2$(μ-N_2). The reaction of Tp*Ir(Ph)$_2$(N_2) with CO or with PR$_3$ led to Tp*Ir(Ph)$_2$(L) complexes, while with THF the carbene Tp*IrH(Ph)(C_4H_6O) was obtained.[991] Thermolysis of Tp*Ir(C_2H_4)$_2$ in thiophene resulted in the replacement of both ethylene molecules by two 2-thienyl groups, and one coordinated intact thiophene molecule, producing Tp*Ir(2-thienyl)$_2$(SC_4H_4), which could be hydrogenated to Tp*IrH$_2$(SC_4H_4), and this complex on heating gave rise to the structurally characterized dinuclear **109**.[992] The above reaction was studied in more detail, being also extended to 2-methylthiophene and 3-methylthiophene, which were activated just like thiophene. The S-bonded thiophenes could be readily replaced with CO or with PMe$_3$, and the structure of Tp*Ir(2-SC_4H_3)CO was established by X-ray crystallography. Thermal activation of several substituted thiophenes by Tp*Ir(η^4-H_2C=(Me)C(Me)=CH_2) was also studied.[993] Tp*Ir(C_2H_4)$_2$ has also been converted to Tp*Ir(C_2H_3)(Et)(MeCN) which, in the presence of water, condensed to the iridapyrrole, **110**.[994]

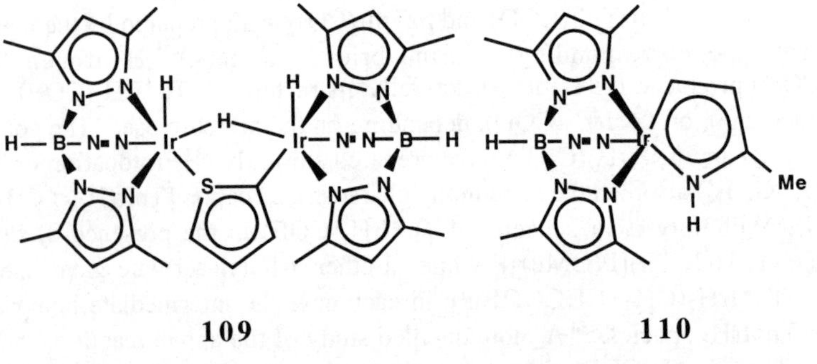

109   **110**

Proton donors such as ROH or even water, added to TpIr(CO)$_2$ producing complexes TpIr(H)(CO)(COOR) and TpIr(H)(CO)(COOH), respectively. The latter,

2.10 GROUP 9 COMPLEXES

on heating in MeCN, formed the dihydride TpIrH$_2$(CO).[995] Primary amines added in similar fashion, forming carbamoyl derivatives TpIrH(CO)(CONHR).[996] In an unusual reaction, Tp*Ir(C$_2$H$_4$)(PPh$_3$) reacted with excess PPh$_3$, yielding a product, with one pyrazolyl ring cyclometallated in the 5-position, in equilibrium with the starting material.[997]

A variety of diene complexes having the general structure Tp^{x}Ir(1,3-diene), with Tp^x being Tp or Tp*, and the dienes being butadiene, 2-Me-butadiene, 2,3-dimethylbutadiene, cyclopenta-1,3-diene and cyclohexa-1,3-diene, has been synthesized, and their chemistry was investigated in detail. Photolysis of Tp*Ir(H$_2$C=C(Me)C(Me)=CH$_2$) produced the η^3-allyl complex [Tp*Ir(H)(η^3-CH$_2$C(C(Me)=CH$_2$)CH$_2$]. On the other hand, thermal activation of benzene by Tp*Ir(H$_2$C=(Me)C(Me)=CH$_2$) yielded the N$_2$-bridged [Tp*IrH(Ph)]$_2$(μ-N$_2$).[998] The thermal reactions of TpIr(olefin)$_2$ and Tp*Ir(olefin)$_2$ leading to hydrido η^3-allyl species were very thoroughly examined, and a mechanism for the observed transformations has been proposed, involving sequential olefinic C—H bond activation and C—C coupling of alkenyl and olefin ligands, while maintaining the same iridium(III) oxidation state throughout all these processes.[999] The Tp*Ir(H$_2$C=(Me)C(Me)=CH$_2$) complex was also found to activate aldehydes producing, in the case of *p*-anisaldehyde, the complex **111**, which was obtained upon heating of the intermediate **112**.[1000]

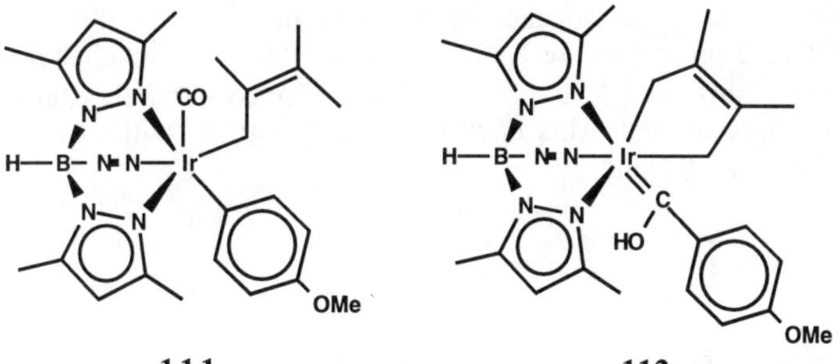

111   **112**

Various iridium polyhydrides, exemplified by TpIrH$_2$(PMe$_3$), by [TpIrH$_3$(PMe$_3$)]$^+$,[1001] and by Tp*IrH$_4$,[1002] were studied by NMR. It was established in this fashion that the structure of Tp*IrH$_4$, which was prepared in excellent yield upon hydrogenation of Tp*Ir(C$_2$H$_4$)$_2$, is best represented in solution as a distorted octahedron, which is capped with a hydride ligand, as shown in **113**. The related SiEt$_3$ derivative, **114**, was also synthesized, and it was structurally characterized by X-ray crystallography.[864]

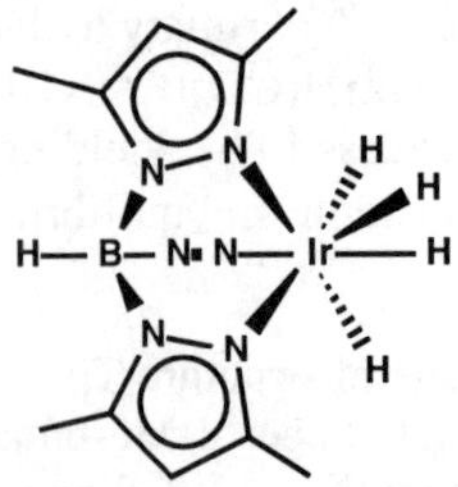

113 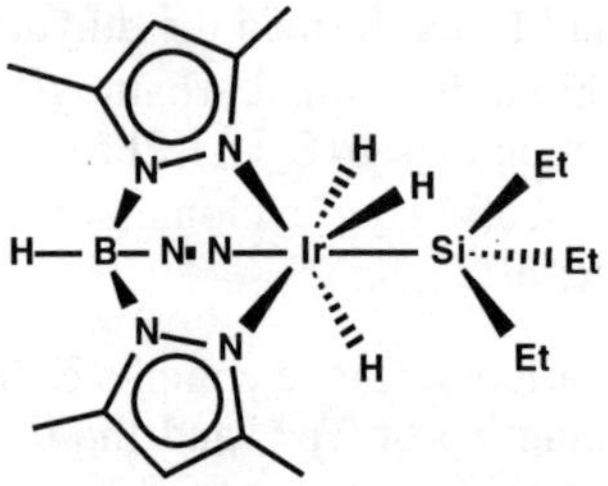 **114**

2.11 Group 10: Ni, Pd, Pt

2.11.1 Ni

Apart from the early reports of Tp$_2$Ni, pzTp$_2$Ni,[1,9] and Tp*$_2$Ni,[11] and spectral studies thereof,[13,922] the organometallic complex TpNi(η^3-allyl) was prepared and structurally characterized.[1003] A dinickel complex containing, in addition to other ligands, a Ni—Ni bond, with each Ni bonded also to a pz group of one Tp (the third pz group being uncoordinated), has also been reported.[1004] From NiX(R)(PMe$_3$)$_2$ and KTp or KTp*, the complexes Tp^{x}Ni(R)(PMe$_3$) were obtained, in which the Tp^x ligands were κ^2, although a distant Ni--N interaction with the third pz^x arm was observed.[1005] Also prepared were Tp*Ni(NO$_3$), Tp*Ni(O-ethylcysteinato) and the salt K[Tp*Ni(cysteine)], with the five-coordinate Tp*Ni(O-ethylcysteinato) being structurally characterized. This complex reacted rapidly with O$_2$, which led to cleavage of the Ni—S bond.[1534]

2.11.2 Pd

The first palladium homoscorpionate complexes were TpPd(η^3-allyl) and pzTpPd(η^3-allyl), in which all pz groups were equivalent in the NMR, spectrum, and thus rapidly exchanging.[4] The structure of TpPd(η^3-allyl) was established by X-ray crystallography.[1006] The complex pzTpPd(η^3-allyl) contained two coordinatively active pz groups, and formed the spiro cation [(η^3-allyl)Pd(μ-pz)$_2$B(μ-pz)$_2$Pd(η^3-allyl)]$^+$,[311] and related ones.[1007] A series of similar mono- and dinuclear Pd^{II} and Pt^{II} η^3-methallyl complexes was synthesized, and studied by NMR.[1008] The preparation of the complex [pzTp]Pd[η^3-CH$_2$C(CH$_2$Cl)CH$_2$], and its conversion to [pzTp]Pd[η^3-CH$_2$C(CH$_2$SO$_2$Ph)CH$_2$] has been reported.[1009] The structurally characterized Tp$_2$Pd complex contained square planar Pd, with bidentate Tp ligand.[1010] NMR studies

2.11 GROUP 10 COMPLEXES

indicated a similar structure for [pzTp]$_2$Pd.[1011] The complex TpPd(PPh$_3$)(COFc) and its pzTp analog were studied by NMR and electrochemically,[1012] while some Pd(II) complexes of pzTp were investigated by ^{1}H-^{13}C COSY NMR.[1013]

Many cyclopalladated and related organopalladium compounds, were derivatised as Tp or pzTp complexes in which the Tp^x ligand was bidentate, as was established for many of them by X-ray crystallography.[1014-1021] In the Tp^x derivatives of a 2-picolyl-bridged complex, both Tp^x ligands were bidentate, but pzTp retained the dimer structure in **115**. This dimer was cleaved by [Tp]$^-$, producing, in the presence of PPh$_3$, the complex **116**.[1022]

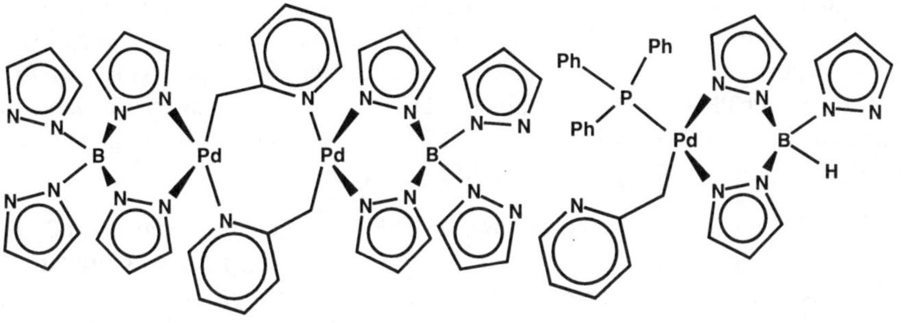

115   **116**

Different products were obtained from the reaction of dimers [LPdCl$_2$]$_2$ with pzTpK, including mono- or dinuclear, neutral or cationic species, depending on the nature of L.[1023] The structurally characterized dinuclear [TpPd]$_2$(μ-ArNC)$_2$, where Ar was mesityl, also contained a Pd-Pd bond.[1024] Complexes pzTpPd(Me)(PPh$_3$), pzTpPd(Ph)(PPh$_3$) and TpPd(Ph)(PPh$_3$) were also structurally characterized.[1006] Oxidation of the anion [TpPdMe$_2$]$^-$ with [PhCOO]$_2$ in the presence of PPh$_3$ yielded TpPdMe$_3$ and TpPdMe(PPh$_3$), while oxidation with [PhE]$_2$ gave rise to TpPdMe$_2$(EPh), where E was O or S.[1025]

Among the complexes of palladium(IV), the simplest ones, TpPdMe$_3$ and [pzTp]PdMe$_3$ were structurally characterized, as was the mixed complex TpPdMe$_2$Et. TpPdR$_3$ complexes containing various combinations of the three organyl groups on Pd (Me, Et, Ph, σ-allyl, benzyl, and 1,4-butylene) were also reported.[1026-1028] The palladacyclopentane complex, [TpPd(C$_4$H$_8$)]$^-$ was oxidized to palladium(IV) species by water, by H$_2$O$_2$, or by halogens, and the structures of TpPdMe$_2$(OH) and TpPd(C$_4$H$_8$)(OH) were established by X-ray crystallography.[1029,1030] The hydrogen bonding of the aquapalladium(IV) group in TpPd(C$_4$H$_8$)(OH) to *m*-cresol and to pentafluorophenol was studied by X-ray crystallography.[1031]

2.11.3 Pt

Homoscorpionate ligands stabilized five-coordinate Pt^{II} complexes of general structure Tp^xPtL, where L was an olefin, a fluoroolefin, an allene, CO, CNR, or an acetylene.[1032-1034] These olefin complexes were stereochemically rigid at room temperature, but the CO complex was fluxional, and five-coordinate in solution (^{195}Pt coupling was observed to all H-3 and H-4 protons), yet four-coordinate in the crystal.[1035,1036] The ^{13}C upfield shift of the olefinic carbon resonances was correlated to the π back-bonding.[1037] Stereochemical nonrigidity in $Tp^xPtMe(L)$ complexes was studied by NMR.[1038] The Tp ligand was found to be $κ^3$ in the acetylenic complex $TpPtMe(F_3CC≡CCF_3)$,[1039] but it was $κ^2$ in $TpPtMe(CNBu^t)$.[1040] Mono- and dinuclear $Pt(η^3$-methallyl) complexes, based on pzTp, were also reported.[1028] The reaction of $[Pt(PEt_3)Br_2]_2$ with $[pzTp]^-$ afforded the five-coordinate $pzTpPt(PEt_3)Br$, but $[Tp]^-$ formed the structurally characterized dinuclear complex $(PEt_3)Br_2Pt(μ-pz)BH(μ-pz)_2Pt(PEt_3)Br$.[1041] Oxidation of $[TpPtMe_2]^-$ by $(RE)_2$ produced stable Pt^{IV} complexes $TpPtMe_2(ER)$, where E was PhCOO, O or S, and R was Ph or Me, while the reaction with $SnBrMe_3$ yielded $TpPtMe_2SnMe_3$.[1025]

The simplest Pt^{IV} homoscorpionate, $TpPtMe_3$, could be prepared either from $[PtIMe_3]_4$, or from $CpPtMe_3$.[1042] The related $Tp*PtMe_3$, **117**, was prepared in similar fashion. Its reaction with bromine left the methyl groups unaffected, and afforded $Tp*^{Br}PtMe_3$, **118**, as the only product.[1043] The reaction of $[TpPtMe_2]^-$ with a proton afforded $TpPt(H)Me_2$.[1044] Related complexes, $Tp*PtR_2(X)$ (R = Ph, Me; X

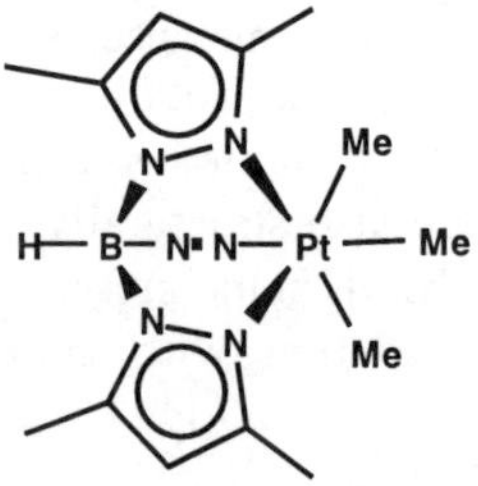

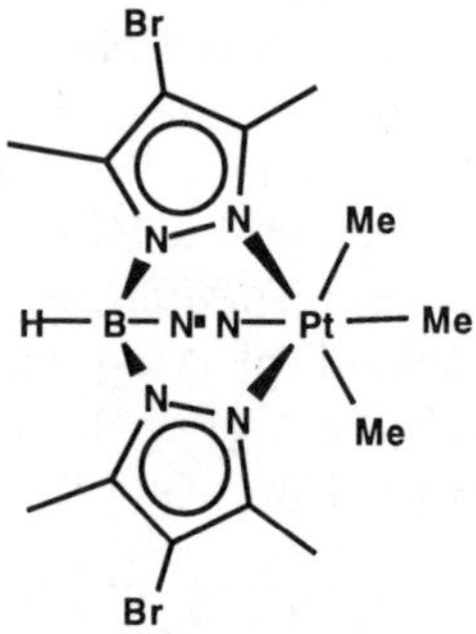

117   **118**

= I, Cl, Me), were also synthesized, and the structure of $Tp*PtMe_2H$ was established by X-ray crystallography.[49] Complexes $TpPtR_2(OH)$ and $pzTpPtR_2(OH)$ were obtained upon the oxidation of Pt^{II} precursors with water or with protic acids, and structures of $TpPtMe_2(OH)$, $pzTpPtMe_2(OH)$ and $TpPt(Tol)_2(OH)$ were established by X-ray crystallography.[1045] Another approach to such compounds involved demethylation of [Tp*PtMe$_2$]$^-$ with B(C$_6$F$_5$)$_3$, and oxidative addition of R—H to the resulting [TpPtMe] species.[1046]

2.12 Group 11: Cu, Ag, Au

2.12.1 Cu

Some of the first homoscorpionate complexes of copper were Tp$_2$Cu, [pzTp]$_2$Cu,[9] and Tp*$_2$Cu,[11] which were studied by spectroscopy.[13] The compounds Tp$_2$Cu,[1047] and Tp*$_2$Cu,[1048,1049] were structurally characterized, and excited-state distortions in Tp*$_2$Cu were determined from Resonance Raman intensities and a normal coordinate analysis.[1050] The first stable Cu^I carbonyl complex, TpCu(CO), **119**, was prepared by Bruce, along with [pzTp]Cu(CO) and Tp*Cu(CO),[1051-1053] and it was structurally characterized by Churchill.[1054] In [pzTp]Cu(CO) all pz groups were identical by NMR.[1055] The carbon monoxide was readily replaced by phosphines, arsines, or isonitriles. Heating converted Tp^{x}Cu(CO) complexes to dimers [Tp^{x}Cu]$_2$, which on further heating disproportionated to Tp^x$_2$Cu and metallic copper.[1052] Reaction with PEt$_3$ produced, in addition to TpCu(PEt$_3$), a dinuclear species Tp$_2$Cu$_2$(PEt$_3$)$_2$ in which all pz and Et groups were NMR-equivalent.[1056] In the course of a broader study of Tp^{x}Cu(PR$_3$) and Tp^{x}Cu(PAr$_3$) complexes, using four different Tp^x ligands, the structures of TpCu[PCy$_3$] and Tp*Cu[PCy$_3$],[1057] and of TpCu(PPh$_3$),[1058] were determined by X-ray crystallography. In the dimers [TpCu]$_2$,[1055] and [Tp*Cu]$_2$,[1059] the scorpionate ligands exhibited different modes of ligation. The complex pzTpCu[1,2-(Me$_2$As)$_2$C$_6$H$_4$] has also been reported.[1060] The carbonyl complex TpCuCO was converted to TpCu—MoH$_2$Cp$_2$, **120**,[1061] and was also used as sensitizer for the valence isomerization of norbornadiene to quadricyclene.[1062]

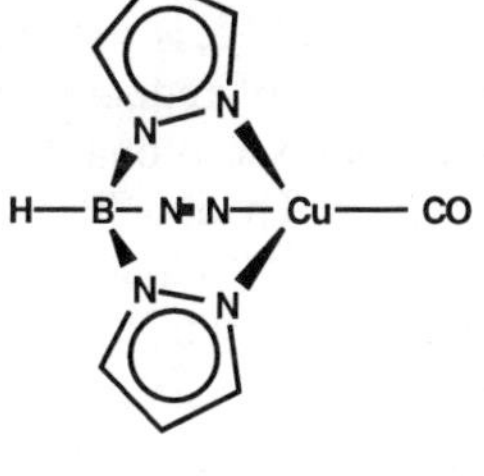

119   **120**

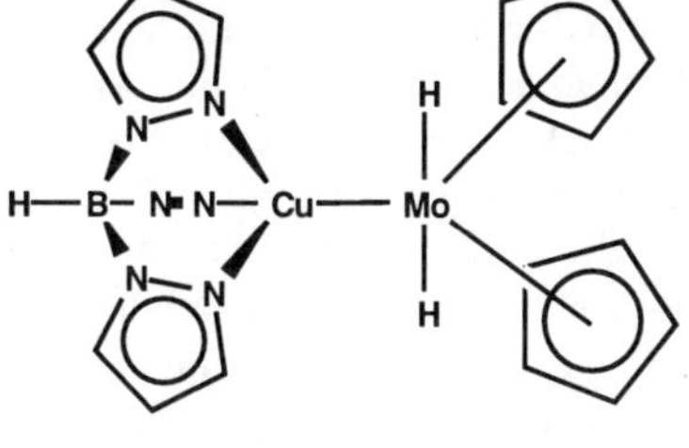

The compounds K[Tp*CuSAr] and Tp*Cu(SAr) were synthesized, aiming for models for the active sites in the blue copper proteins, as were complexes

Tp*CuL (L = SR, OAr or Me$_2$NCS$_2$).[1063,1064] Various olefin complexes of structure TpCu(olefin)·CuCl, were synthesized, and structures of Tp*Cu(CH$_2$=CH$_2$) and Tp*Cu(CH$_2$=CH$_2$)·CuCl were determined by X-ray crystallography. While Tp* was κ^3 in Tp*Cu(CH$_2$=CH$_2$), it was κ^2 in the second complex, with the third pz* arm coordinating to CuCl.[1065,1066] Various aspects of copper-dioxygen chemistry, including the synthesis, properties and reactions of copper(II)-superoxide, and -peroxide compounds were discussed.[1547] The copper(I) complex, Tp*Cu(CH$_2$=CH$_2$), was a catalyst for carbene and nitrene transfer to form cyclopropanes, cyclopropenes, and aziridines.[1067] Dioxygen from the Tp*Cu(O$_2$) complex could be displaced by CO, MeCN, or by ethylene.[1068] Although TpMX complexes were quite unstable, being rapidly converted to Tp$_2$M and MX$_2$, the dimeric complex, [TpCuCl]$_2$, containing two chloride bridges, could be isolated and structurally characterized.[308]

Dinuclear Cu(II) complexes were studied as possible models for oxyhaemocyanin. For instance, [Tp*Cu]$_2$O was prepared, and found to react with PPh$_3$ forming Tp*Cu(PPh$_3$) and PPh$_3$O,[1069] and with H$_2$O$_2$ producing a peroxo-bridged dinuclear complex,[1070] which was studied as an oxidizer for various phenols under aerobic and anaerobic conditions.[1071] Other dinuclear complexes of this general type were the structurally characterized [Tp*Cu](μ-OH)$_2$, and [Tp*Cu](μ-CO$_3$) obtained from the former compound upon exposure to CO$_2$.[1072] The energetics of oxygen binding, and the core isomerization between the Cu$_2$(μ-O)$_2$ and Cu$_2$(μ-η^2:η^2-O$_2$) core structures, were studied for TpCu complexes by gradient-corrected density functional methods.[1073] An uncommon trinuclear Tp*$_2$Cu$_3$ complex, for which the structure HB(pz*-Cu-pz*)$_3$BH was proposed, was converted to the structurally characterized Tp*Cu(μ–N$_3$)$_2$Cu(μ–N$_3$)$_2$CuTp*.[1075]

2.11.2 Ag

In general, homoscorpionates containing a B—H bond formed less stable complexes with silver(I) than with copper(I) due to the relative ease of reducing silver(I) to metallic silver. Accordingly, relatively few silver homoscorpionates have been reported. They were of structure Tp^{x}AgL for Tp, pzTp and Tp*, with L = phosphines, phosphites, arsines, or isonitriles.[1076-1078] This reaction was investigated in considerable detail, and the structures of TpAg(PPh$_3$), TpAgP(o-Tol)$_3$, TpAgP(Bz)$_3$ and Tp*AgP(p-Tol)$_3$ were determined by X-ray crystallography.[1079,1080] Other structurally characterized silver complexes were of the type pzTpAgPR$_3$ (R = Ph, m-Tol, o-Tol; PR$_3$ = PMePh$_2$, PEtPh$_2$, P(C$_6$H$_{11}$)$_3$,). With the exception of [pzTp]Ag[P(o-Tol)$_3$], in which the pzTp ligand was κ^3, in all the other complexes the [pzTp] ligand was κ^2, and the silver ion was in a three-coordinate environment. In such cases it was possible to coordinate an additional small ligand to the Ag ion.[1081] The structure of [Tp*Ag]$_2$ showed it to be an asymmetric dimer.[1568]

2.11.3 Au

Most of the reported homoscorpionate gold complexes were those of gold(III). These included Tp^{x}AuCl$_2$,[1082,1083] and TpAuMe$_2$.[317] In the structurally characterized pzTpAuMe$_2$ the ligand was κ^2, and the geometry was square planar.[316] The only example of an Au^I complex was Tp*Au(PPh$_3$)$_2$, prepared from KTp* and AuCl(PPh$_3$) in the presence of excess PPh$_3$.[1080]

2.13   Group 12: Zn, Cd, Hg

2.13.1   Zn

Simple homoscorpionates of zinc have been long known,[9,11] and they were studied by spectroscopy.[13] The structures of Tp$_2$Zn,[1084] and of Tp*$_2$Zn,[1085] were determined by X-ray crystallography. Both complexes were octahedral, although zinc was usually tetrahedral in solution, and zinc homoscorpionates could be used as Tp^x ligand transfer agents. The complex TpZn(NO$_3$) contained the nitrate ligand bonded to zinc in anisobidentate mode, the two Zn-O bonds being 1.981(2) Å and 2.399(3) Å, respectively.[1086] An *ab initio* comparison of the bonding mode of nitrate and bicarbonate ligands in TpZnL and Tp^{tBu}ZnL was carried out.[1087] The structurally characterized Tp*ZnMe was prepared from Tp*Tl and ZnMe$_2$, and was found to be converted to Tp*$_2$Zn in most of its reactions.[1088] Thiolate complexes, Tp*ZnSR (R = Et, CH$_2$Ph) were synthesized from Tp*ZnCl and the corresponding sodium thiolates, and the structure of Tp*ZnSEt was determined by X-ray crystallography. These complexes produced the corresponding thioethers upon reacting with methyl iodide or dimethyl sulfate.[1089]

2.13.2   Cd

Cadmium homoscoprionates Tp*CdI and Tp*$_2$Cd were synthesized, and the latter was structurally characterized.[1090] The complex Tp*Cd(BH$_4$), **121**, contained κ^3-Tp* and a κ^2-borohydride ligand. The presence of a Cd—Cd bond in the unusual dimer [Tp*Cd]$_2$, **122**, was inferred from ^{113}Cd NMR.[46] Complexes Tp$_2$Cd, [pzTp]$_2$Cd, Tp*$_2$Cd, and heteroleptic ones, such as Tp*Cd[pzTp] were studied by ^{113}Cd NMR, and structures of the octahedral Tp*$_2$Cd and [pzTp]$_2$Cd were determined by X-ray crystallography.[1091] CP/MAS studies were carried out on Tp*$_2$Cd and [pzTp]$_2$Cd complexes, and ^{113}Cd shielding tensors were obtained.[1221] Heteroleptic complexes with five-coordinate Cd, such as Tp*CdL (L = Bp*, AcAc, S$_2$CNR$_2$, and other ligands) were also prepared, studied by ^{113}Cd NMR, and the structure of Tp*Cd(S$_2$CNEt$_2$) was determined by X-ray crystallography.[1092] A number of octahedral heteroleptic compounds Tp^{x}CdL was prepared, studied by ^{113}Cd NMR, and the compound Tp$_2$Cd was structurally characterized.[1091,1093] Also studied by ^{113}Cd NMR were the organocadmium compounds Tp*CdR, with R = Me, Et, Pr, *i*-Pr, *i*-Bu, t-Bu and Ph.[1094]

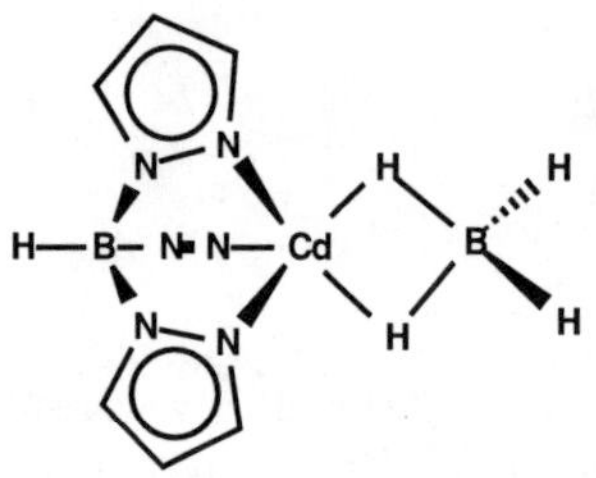

121

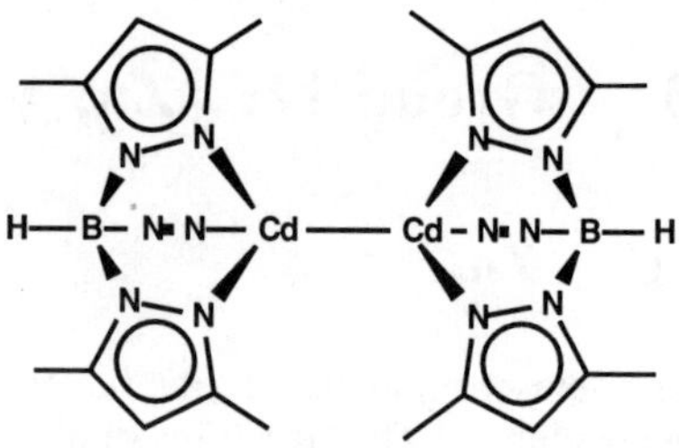

122

2.13.3 Hg

The crystal structure of [pzTp]HgMe showed three-coordinate mercury, and bidentate pzTp, with unequal N—Hg bonds.[1095] The synthesis of mercury compounds Tp*$_2$Hg, [pzTp]$_2$Hg, Tp^{x}HgX and [Tp*Hg]$_2$ was reported, and the pzTp derivatives were found to be more stable than those of Tp*.[1096] Organomercury compounds TpHgMe,[253] and Tp*HgR (R = alkyl, aryl, ferrocenyl) have been prepared, as were those of the type pzTpHgR. Many of these complexes were fluxional, and were thought to contain two-coordinate Hg.[1097,1223] Solid state and solution NMR spectra,

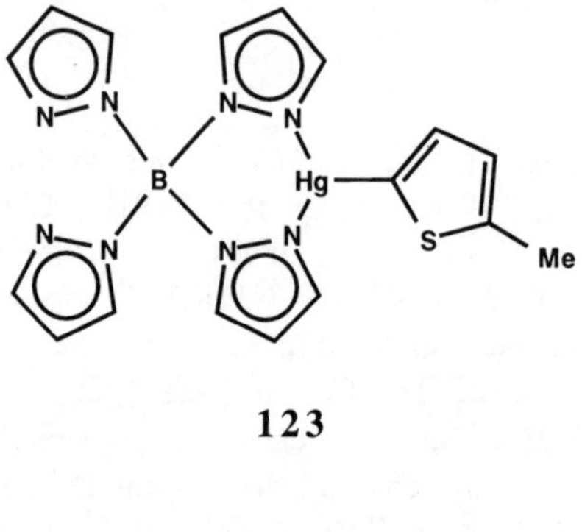

123

including ^{199}Hg, have been determined for Tp*HgMe, [pzTp]HgMe, Tp*HgSR and [pzTp]HgSR complexes.[1098] Numerous complexes of general structure Tp^{x}Hg(2-thienyl) have been synthesized (Tp^x = Tp*, pzTp), studied by spectroscopy and by NMR, and the structure of [pzTp]Hg(5-methylthien-2-yl), **123**, was determined by X-ray crystallography. The mercury ion was in a roughly planar, T-shaped configuration, and the ligand was bidentate.[1099] Similar studies were conducted on a variety of Tp^{x}HgCN complexes, among which the tetrahedral Tp*HgCN was structurally characterized.[1100]

2.14 Group 13: B, Al, Ga, In, Tl

2.14.1 B

The only homoscorpionate complexes of boron were the cation [HB(μ-pz*)$_3$BH]$^+$, which can be regarded as the [Tp*]$^-$ complex of [HB]$^{2+}$, and also the related [Tp^{Me}BEt]$^+$ cation.[105,1101] The pyrazabole heterocycles, which contain a pyrazolyl substituent in the 4- or 8-positions, (pz)HB(μ-pz)$_2$BH(pz) may be also regarded as boron derivatives of Tp ligands.

2.14.2 Al

Reaction of KTp with AlCl$_3$ afforded the salt [Tp$_2$Al][AlCl$_4$],[1102] while Tp*AlMe$_2$ was obtained from KTp* and AlMe$_3$,[1103] and it could be hydrolyzed in controlled fashion to Tp*Al(OH)$_2$.[1104]

2.14.3 Ga

An octahedral cation of gallium, [Tp*$_2$Ga]$^+$, was obtained upon the reaction of GaCl$_3$ with an equimolar amount of Tp*K. This led to the isolation of the salt [Tp*$_2$Ga][GaCl$_4$], the structure of which was determined by X-ray crystallography.[1102] Similar stable gallium cations based on Tp and on pzTp were also prepared, while from GaMeCl$_2$ and KTp* the complex Tp*GaMeCl was obtained. The solid state structure of this compound showed it to be four-coordinate, and to contain a bidentate Tp* ligand. It was, however, fluxional in solution, and its NMR displayed equivalence of all the pz* rings down to –90 °C. The reaction of GaMe$_2$Cl with the Tp^x ligands mentioned above produced in each case Tp^{x}GaMe$_2$. All of these compounds were studied by NMR, and the structures of Tp*GaMe$_2$ and of [pzTp]$_2$GaMe were established by X-ray crystallography. The Tp* ligand was bidentate in Tp*GaMe$_2$, and the structure tetrahedral, while in [pzTp]$_2$GaMe one pzTp ligand was tridentate, and the other bidentate, and the molecular geometry was octahedral.[1105] The organogallium complex [pzTp]GaEt$_2$ has also been reported.[1106] When Na$_2$Fe(CO)$_4$ was treated with Tp*GaMeCl, the product was Tp*Ga—Fe(CO)$_4$, **124**, which contained a Ga—Fe bond, as was established by X-ray crystallography.[1107]

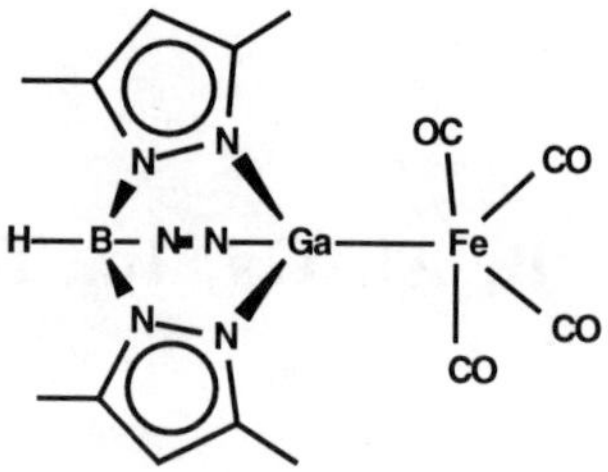

124

2.14.4 In

Numerous homoscorpionate complexes of indium have been prepared, in all of which indium tended toward six-coordination, as for instance in Tp*InCl$_2$(THF). In the complex [pzTp]$_3$In, all pzTp ligands were coordinated in κ^2 fashion, while in [pzTp]$_2$InCl, and in the structurally characterized [pzTp]$_2$InMe one pzTp ligand was κ^2, and the other was κ^3. The related complex Tp$_2$InCl was ionized in polar solvents, producing the cation [Tp$_2$In]$^+$. This cation could be obtained as the structurally characterized [TpInCl$_3$]$^-$ salt by the reaction of 3 KTp with 2 InCl$_3$.[1108] Many heteroleptic compounds of structure Tp*In(L)X were obtained from Tp*InCl$_2$(THF). L was Bp, Bp*, Tp, pzTp, Et$_2$NCS$_2$, OAc, various β-diketonates, catecholates, and

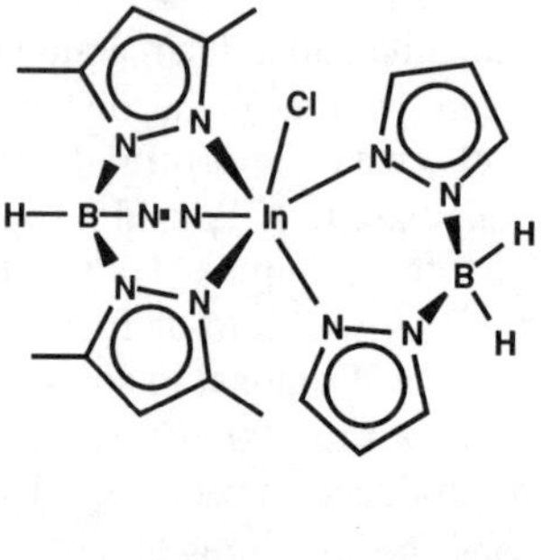

125 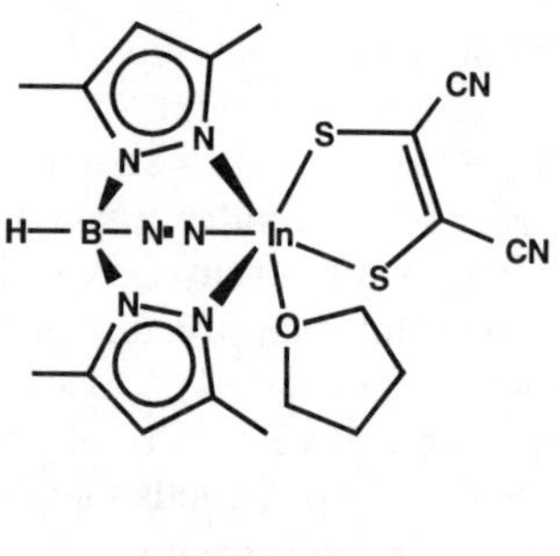   **126**

2.15 GROUP 14 COMPLEXES

[$S_2C_2(CN)_2$], while X could be Cl, THF or Hpz*. Structures of Tp*In(Bp)Cl, **125**, and Tp*In[$S_2C_2(CN)_2$](THF), **126**, were determined by X-ray crystallography.[1109] The reaction of Tp*InCl$_2$(THF) with Na$_2$Fe(CO)$_4$ and Na$_2$W(CO)$_5$ led to the complexes Tp*In—Fe(CO)$_4$ and Tp*In—W(CO)$_5$, respectively, the structures of which were established by X-ray crystallography,[1110] while the reaction of Tp*InCl$_2$(THF) with K$_2$S$_5$ gave rise to Tp*In(S$_4$)(Hpz*).[1111] The structure of Tp*$_2$InI has also been determined.[1112]

2.14.5   Tl

Tl^I derivatives of most of the homoscorpionates have been prepared, mainly to facilitate the reaction of these ligands in organic solvents, but no studies specifically devoted to the Tp^{x}Tl complexes were done. The structure of TpTl has been determined by X-ray crystallography.[1113] The reported organometallic complexes, pzTpTlR$_2$ (R = Et, Bu) are the only examples of thallium(III) scorpionates.[1114]

2.15   Group 14:   C, Si, Ge, Sn, Pb

2.15.1   C

Although "first-generation" homoscorpionate derivatives of carbon have not been reported, the dications [RB(μ-pz)$_3$CH]$^{2+}$, which may be formally regarded as derived from [RTp]$^-$ replacing three chloride ions from HCCl$_3$, should be obtainable by the reaction of HC(pz)$_3$ with RBX$_2$, where X is a good leaving group (halide, OTs, etc.).

2.15.2   Si

The complex pzTpSiCl$_3$ was obtained from the reaction of KpzTp with SiCl$_4$.[1106]

2.15.3   Ge

Treatment of GeCl$_4$ with KpzTp produced the complex pzTpGeCl$_3$.[1106] The germanium(II) complex, [Tp*Ge]Cl, the first example of a Ge^{II} homoscorpionate, was obtained from Tp*K and GeCl$_2$(dioxane), and it was converted to the iodide, and to [PF$_6$]$^-$ salts. The structure of the salt [Tp*Ge]I was determined by X-ray crystallography. There was no interaction between Ge and I, the shortest Ge····I distance being over 4 Å.[1115] Treatment of Tp*GeCl with sodium azide produced the azido analog, Tp*GeN$_3$, and structures of both complexes were determined by X-ray crystallography. In the complex Tp*GeCl the halogen occupied one axial site of a distorted pseudo trigonal bipyramid, the other site being occupied by the lone pair. The azido complex, Tp*GeN$_3$, had a very similar structure.[1115,1116]

2.15.4 Sn

Among the group 14 elements, most homoscorpionate derivatives were those of tin. The first reported, and structurally characterized, tin(II) complexes were Tp*$_2$Sn and Tp*SnCl. In the first one, one Tp* ligand was κ^3 and the other κ^2, while in the second complex, the structure was roughly trigonal bipyramidal.[1117] Tp$_2$Sn had a structure similar to that of its Tp* analog.[1118] In the complex [pzTp]$_2$Sn both ligands were κ^2, but exchanging rapidly on the NMR time scale.[1119] Other reported complexes were Tp$_2$Sn, pzTp$_2$Sn and Tp*SnCl, which were studied by ^{119}Sn NMR.[1120] Solid state and solution ^{119}Sn NMR studies were also done on Tp*$_2$Sn.[1121] The reaction of SnCl$_2$ with KTp and NaCpCo[P(O)(OEt)$_2$]$_3$ yielded the unusual heteroleptic complex TpSn{CpCo[P(O)(OEt)$_2$]$_3$}.[1122]

Sn^{IV} chemistry was investigated to a greater extent than that of Sn^{II}. Complexes pzTpSnCl$_3$,[1106] the not very stable pzTp$_2$SnR$_2$ (R = Me, Et, Bu),[1123] TpSnMe$_3$, TpSnMe$_2$Cl and TpSnMeCl$_2$,[1124] and the series Tp^{x}SnR$_n$X$_{3-n}$, where R was Me, Et, Bu or Ph, X was Cl or Br, and Tp^x was Tp,[1125] pzTp,[1126] and Tp*,[1127] were reported, as was the complex TpSnCl$_2$CH$_2$(COOMe)CHCH$_2$COOMe,[1128] and the structurally characterized TpSn(NCS)$_2$CH$_2$CH$_2$COOMe.[1129] The reaction of TpSnCl$_3$ with [Co(CO)$_4$]$^-$ yielded, among other species, the [Tp$_2$Co]$^+$ cation.[930] Complexes TpSnMe$_3$,[1130] and Tp*SnCl$_3$ were structurally characterized, and the latter one was used as a diluent in EPR studies of Tp*Mo(E)X$_2$ complexes.[612] The structurally characterized compounds TpSnPh$_2$Cl, Tp*SnPh$_2$Cl and TpSnCl$_3$, and related species, were also studied by Mössbauer spectroscopy,[1131] and Tp^{x}SnR$_n$X$_{3-n}$ complexes (Tp^x = Tp or Tp*) by X-ray absorption spectroscopy.[1132] Some Sn^{IV} compounds of general structure TpSnR$_3$ were investigated for antimutagenic activity,[1133] while Tp*SnBu$_2$Cl was found to be a useful transfer agent for the Tp* ligand to Zr, Nb, and Ta compounds.[356]

2.15.5 Pb

Although the simple complexes Tp$_2$Pb and [pzTp$_2$]Pb were reported long ago,[9] they were structurally characterized, along with Tp*$_2$Pb, much later.[1134] While in Tp$_2$Pb both ligands were κ^3, and the molecular structure was a capped octahedron with the lone pair at the capping position, in Tp*$_2$Pb both ligands were also κ^3, but the

2.16 GROUP 15 COMPLEXES

structure was octahedral. The compound [pzTp]$_2$Pb had pseudo-trigonal-bipyramidal geometry, with both ligand being κ^2. The complex TpPbCl has also been reported,[1057,1134] as were the heteroleptic complexes TpPb{CpCo[P(O)(OEt)$_2$]$_3$}, Tp*Pb(κ^1-NO$_3$), and also the five-coordinate Tp*PbBp. Structures of Tp*Pb(NO$_3$), Tp*PbCl(Hpz*) and of the cationic complex [Tp*Pb(Hpz*)$_3$]Cl were determined by X-ray crystallography.[1122] A mixed-ligand cationic complex {TpPb[HC(3,5-Me$_2$pz)$_3$]}$^+$, and its Tp* analog, were also synthesized, and the latter was structurally characterized.[1135] Early work on Sn^{II} and Pb^{II} scorpionate chemistry has been briefly reviewed.[1136]

2.16 Group 15: N, P, As, Sb, Bi

2.16.1 P

While there are no examples of nitrogen homoscorpionate complexes, a phosphorus complex, Tp*P-Fe(CO)$_4$, has been reported, and structurally characterized. It contains a κ^2 Tp* ligand.[1137]

2.16.2 As

The only arsenic derivative reported, and characterized by NMR, was pzTpAsMe$_2$.[1106]

2.16.3 Sb, Bi

Although no Tp^x derivatives of either Sb of Bi have been reported, there no reason why they should not be capable of existence.

2.17 Lanthanides

Some of the first trivalent lanthanide homoscorpionates reported were those of the type Tp$_3$Ln and pzTp$_3$Ln, along with Tp$_n$ErCl$_{3-n}$.[1138] The structures of Tp$_3$Pr and Tp$_3$Nd showed nine-coordination,[329] while in Tp$_3$Yb eight-coordination prevailed,[1139] and this complex was stereochemically rigid by NMR.[1140] Scorpionate complexes of Tb and Eu were studied by spectroscopy,[1141] as were those of Nd and Eu.[1142] It was found in a series of Tp*$_2$Ln(OTf) complexes that those of La, Ce, Pr and Nd were seven-coordinate in the solid state, as was determined by X-ray crystallography for the Nd complex, but those of Sm, Eu, Gd, Dy, Ho and Yb were six-coordinate and ionic in the solid state, as indicated by the structure of the Yb compound.[330] Several series of heteroleptic lanthanide complexes of general structure Tp$_2$M(L), where L was typically some bidentate ligand, most often an aliphatic β-diketonate, were synthesized. Those structurally characterized were Tp$_2$Yb(Bu^{t}C(O)CHC(O)Bu^t), **127** (R = t-Bu),[1143] and among the acetylacetonates Tp$_2$M(AcAc) (M = La, Ce, Pr, Nd, Sm, Eu, Tb, Dy, Ho, Er, Yb, Tm and Lu), the cerium and ytterbium complexes,

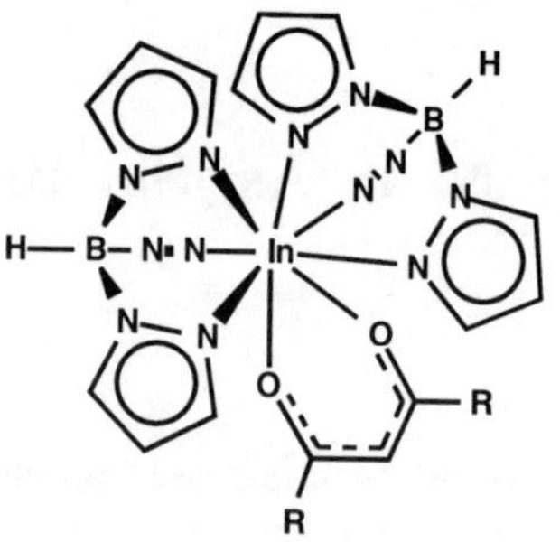

127

127, (R = Me).[331,335] Exchange of the diketonato ligands in the above complexes was studied.[338] Heteroleptic lanthanide complexes, similar to β-diketonates, were those of general structure Tp$_2$M(L), where L was picolinate-N-oxide (M = Y, Eu, Gd, Tb, Er, Yb, Lu), and the complex with M = Tb was structurally characterized.[1530] Other examples were Tp$_2$M(OOCPh) for M = Sm, Eu, Yb, Lu, Tp$_2$M(OAc) for M = Yb and Lu,[331] and the oxalates [Tp$_2$M]$_2$(C$_2$O$_4$).[335] Also reported were the related lanthanide compounds Tp$_2$M(salicylaldehydates) and their 5-methoxysalicylaldehydate analogs,[339] the tropolonates Tp$_2$M(O$_2$C$_7$H$_5$) for M = La, Ce, Pr, Nd, Sm, Eu, Tb and Lu, including the structure determination of the Yb complex,[333] and the complex Tp$_2$LaCl(H$_2$O).[340] The synthesis of Tp$_2$MCl and [Tp$_2$M(OAc)]$_2$ complexes of Sm, Tb and Er was reported, and the structure of the dimeric complex, [Tp$_2$Sm(OOCPh)]$_2$, was determined by X-ray crystallography.[1144] The separation of lanthanides, by means of their extraction from aqueous into organic phase, using a combination of Tp or pzTp ligands and β-diketonates was investigated.[1145] Trivalent lanthanides, PrCl$_3$ and NdCl$_3$, reacted with KTp* to yield dimeric [Tp*MCl(μ-Cl)(Hpz*)]$_2$ complexes, of which the seven-coordinate Pr derivative was structurally characterized. On the other hand, YbCl$_3$ produced the octahedral Tp*YbCl$_2$(THF), which underwent intermolecular rearrangement reactions to produce [Tp*YbCl$_2$(Hpz*)](THF) and [Tp*YbCl$_3$][H$_2$pz*], both of which were structurally characterized.[1146] Lanthanide(II) complexes Tp*$_2$M (M = Sm and Yb) have also been prepared in good yield.[1493]

2.18 ACTINIDE COMPLEXES

Complexes of the type $Tp_2MCl(L)$ were synthesized for M = Lu, Nd, Y and Yb,[1147] with the structure of $Tp_2NdCl(Hpz)$ determined by X-ray crystallography,[1148] as was that of the related $Tp_2SmCl(Hpz)$.[1149] Such samarium complexes were converted to various $Tp_2Sm(\beta$-diketonate) complexes, and to Tp_2SmBp, which was structurally characterized, and was found to contain an agostic B—H—Sm bond.[1150,1151] Air-sensitive Tp* complexes of divalent lanthanides, $Tp*_2M$ (M = Sm, Eu and Yb) were synthesized.[1152] The structures of a redox-related pair of lanthanide complexes, $Tp*_2Yb$ and $[Tp*_2Yb][O_3SCF_3]$, were compared, and a shortening of the M—N distance by 0.16 Å was observed in the reduced species.[1153] The structurally characterized $Tp*_2Sm(\kappa^2-O_2)$, obtained by the reaction of Tp_2Sm with oxygen, was the first example of a lanthanide superoxo complex.[1154] A similar structure was found in the analogous azobenzene adduct, $Tp*_2Sm(PhN=NPh)$, which contained both azobenzene nitrogens symmetrically bonded to Sm.[1155] Unusual cyclooctatetraene complexes $Tp^xM(COT)$, where M was Ce, Pr, Nd and Sm, and Tp^x was Tp or Tp*, have been reported.[1156] In the pair of compounds with the same composition, $[Tp*_2La(O_3SCF_3)(MeCN)]MeCN$ and $[Tp*_2Nd(MeCN)_2](O_3SCF_3)$, the former was a neutral molecule, while the latter adopted an ionic structure.[1157] The synthesis and fluorescence studies of $Tp^x_2Eu(THF)_2$, where Tp^x = Tp and pzTp, as well as the structures of $[pzTp]_2Eu(THF)_2$ and of $pzTp_3Yb$, were reported. In the Yb complex two ligands were tridentate and one was bidentate.[1158] Neodymium complexes $Tp_2NdCl(H_2O)$ and $Tp*Cl_2(L)$, where L was 4,4'-$(Bu^t)_2$-2,2'-bipyridine were prepared, and structurally characterized. In each instance the Tp^x ligands were tridentate.[1159] Trinuclear complexes **128** and **129** were structurally characterized,[1160] and some of the Tp^x-lanthanide chemistry was presented in a mini-review.[1161]

128 Ln = Yb, Z = $[-CH_2-]_3$
129 Ln = Lu, Z = $[-CH_2-]_2$

2.18 Actinides

Bagnall synthesized the first homoscorpionate actinide complex, $TpUCpCl_2$,[1162] and numerous other Tp, pzTp and Tp* complexes of uranium and thorium, also containing Cp, substituted Cp, and halide ligands.[1163-1168] The compounds $Tp*U(NCS)_2(THF)$ and $Tp*Th(NCS)_2(THF)$ have also been reported,[1169] and various

TpNp complexes were studied by ^{237}Np Mössbauer spectroscopy.[1170] Alkoxide and aryloxide uranium derivatives of general structure $Tp_2UCl_n(OR)_{2-n}$ were prepared from Tp_2UCl_2,[1171] as was $Tp_2U(SBu^t)_2$.[1172] The structures of Tp_2UCl_2 and Tp_2ThCl_2,[1173] $Tp_2UCl(OBu^t)$ and $Tp_2UCl(OPh)$,[1174] $Tp_2U(OPh)_2$,[1175] and of $Tp_2U(SPr^i)_2$,[1176] were determined by X-ray crystallography. The reaction of UI_4 with 2 KTp in methylene chloride yielded Tp_2UI_2, but in THF the product was $Tp_2UI[O(CH_2)_4I]$.[1177] Scorpionate complexes of U, Th and Np were studied by NMR.[1178,1179]

The reaction of KTp* with UCl_4 or $ThCl_4$ provided the corresponding Tp*MCl$_3$ complexes,[1180] which formed adducts with THF, pyridine, nitriles, etc.,[1181] or with Hpz*,[1182] and which served as starting materials for numerous derivatives, obtained by nucleophilic displacement of chloride ions. Thus, their sequential treatment with alkoxides, followed by [Cp]$^-$, afforded the asymmetric Tp*MCl(Cp)(OR) species,[1183] while the reaction with alkoxides or aryloxides produced Tp*MCl$_n$(OR)$_{3-n}$ compounds,[1184,1185] and carboxylate anions produced Tp*M(OOCR)$_3$ complexes, of which Tp*U(OAc)$_3$ was structurally characterized.[1186] Reactions with simple lithium alkyls or aryls yielded ill-defined mixtures, but the more bulky reagents produced isolable complexes such as Tp*UCl$_2$(2-CH$_2$C$_6$H$_4$NMe$_2$) and Tp*UCl$_2$(2-C$_6$H$_4$CH$_2$NMe$_2$).[1187] The heteroleptic compounds Tp*MCl$_2$Cp, Tp*MCl$_2$N(SiMe$_3$)$_2$ and Tp*M(NPh$_2$)$_3$ (M = U, Th) were also prepared.[1188] From UCl$_3$ the complex Tp*UCl$_2$ was obtained,[1189] and the structures of Tp*UCl$_3$ and Tp*UCl$_2$Cp were determined by X-ray crystallography.[1190] Organometallic derivatives of actinide scorpionates were briefly reviewed,[1191] and haloactinide complexes with Tp and Tp* ligands were studied by laser desorption Fourier transform mass spectrometry.[1192] An extraordinary uranium(III) structure was found in Tp*$_2$UI, which contained one pz* ring of the Tp* ligand bonded to U through both of its nitrogen atoms, as shown in **130**.[1198]

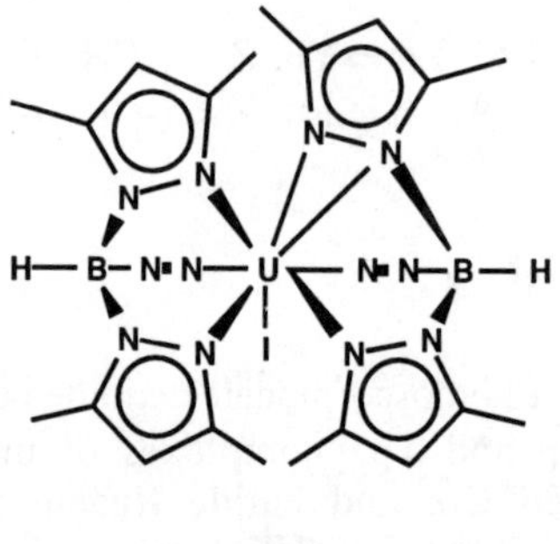

130

2.18 ACTINIDE COMPLEXES

The complex Tp*UCl$_2$CH$_2$SiMe$_3$ underwent an insertion reaction with acetone, forming Tp*UCl$_2$(OCMe$_2$CH$_2$SiMe$_3$), which upon reaction with additional acetone yielded the aldolate Tp*UCl$_2$(OCMe$_2$CH$_2$C(O)Me). However, the related compound Tp*UCl$_2$CH(SiMe$_3$)$_2$ only formed the aldolate with excess acetone. Such reactions were investigated in detail.[1193] Treatment of uranium metal with iodine in THF, followed by KTp*, produced Tp*UI$_2$[O(CH$_2$)$_4$I].[1194] Enthalpies of uranium-ligand bond dissociation in complexes of the general structure Tp*UCl$_2$L, were determined by means of solution calorimetry measurements.[1195,1196] Sodium naphthalenide reduced Tp*UCl$_3$ to Tp*UCl$_2$.[1197]

The only fully characterized plutonium(III) compound was the dimeric [Tp*PuCl(μ-Cl)(Hpz*)]$_2$, which was obtained by the reaction of KTp* with PuCl$_3$. It was isomorphous with the analogous Pr and Nd complexes. The synthesis of another plutonium(III) complex, Tp$_3$Pu, was also mentioned, but no details were given.[1146]

Chapter 3

Homoscorpionates — Second Generation

3.1 General Considerations

This chapter deals mainly with the "second generation" Tp^x and pz^oTp^x ligands, that is, those characterized by relatively large 3-R substituents, as well as with those containing various combinations of substituents in the 3-, 4-, and 5-positions, and on boron. It includes di- and trisubstituted pyrazole rings, plus those containing fused benzo- and naphtho-moieties. Most of these ligands were reported after 1986,[3] although a few have been synthesized much earlier. A large number of "firsts" has been achieved with such Tp^R ligands containing large R substituents. Three issues are of relevance to many ligands within this group:

1. The regiochemistry in ligand synthesis.

2. The quantification of the steric effects in coordination to a metal ion.

3. Ligand rearrangement.

3.1.1 Regiochemistry in Ligand Synthesis

In Chapter 2, there was no issue of regiochemistry, since both, pyrazole (leading to the parent Tp and pzTp ligands) and 3,5-dimethylpyrazole (leading to Tp*), had identical 3 and 5 substituents, their anions were of C_{2v} symmetry, and therefore the attachment of boron to either of the two pyrazole nitrogen atoms would generate an identical product. The same holds true for 4-substituted pyrazoles, since only Tp^{4R} can be formed. A different situation prevails when Tp^x ligands are prepared from 3(5)-monosubstituted pyrazoles or, more generally, from pyrazoles whose anions are not of C_{2v} symmetry. Restricting ourselves to the tris(pyrazolyl)borate ligands, and starting with a 3(5)R-pyrazole, the products could be either [HB(3Rpz)$_3$]$^-$ (**131**) or [HB(5Rpz)$_3$]$^-$ (**132**), as shown below, or even a mixture of ligands having the general structure [HB(3Rpz)$_n$(5Rpz)$_{3-n}$]$^-$ which may contain, in addition to the two isomers **131** and **132**, also two other isomers: [HB(3Rpz)$_2$(5Rpz)]$^-$ and [HB(3Rpz)(5Rpz)$_2$]$^-$,

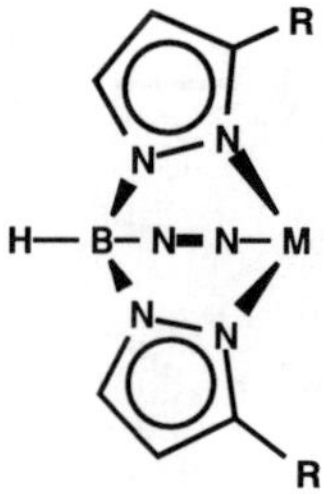

131 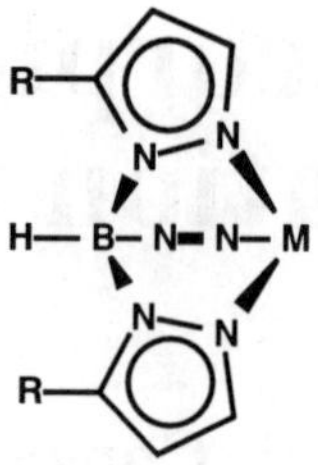 **132**

arising from random bonding of boron to N1 and N2 of the asymmetric pyrazole. The same considerations apply when Tp^x ligands are prepared from 3,4-disubstituted pyrazoles, or from 3,5-disubstituted pyrazoles containing different 3R and 5R' substituents. However, it has been found in practice that mixtures are rarely obtained, and that 3(5)R-monosubstituted pyrazoles led to essentially one product, which was [HB(3Rpz)$_3$]$^-$ (= Tp^R), as well as to the related ligands [H$_2$B(3Rpz)$_2$]$^-$ and [B(3Rpz)$_4$]$^-$ (= pz^{o}Tp^R). This regioselectivity was demonstrated first in the reaction of 3(5)-methylpyrazole, leading to [Bp^{Me}]K,[132] and to [Tp^{Me}]K.[79] Later on, ligands Tp^{Ph}, Tp^{tBu},[2,3] and other homoscorpionates of structure Tp^R were prepared. The probable reason for this regioselectivity is that boron-nitrogen bond formation, involving a concerted loss of hydrogen, proceeds through a less sterically encumbered transition state, when bonding occurs to the less hindered N1 rather than to N2. The same holds true for 3,5-disubstituted pyrazoles, when there is substantial size disparity between the 3 and 5 substituents (for instance, Me and *t*-Bu, or Me and Ph), so that the larger substituent R ends up in the 3-position of the Tp$^{R,R'}$ ligand, next to the coordinated metal, and the smaller R' in the 5-position, remote from the coordinated metal, but affording some steric protection to the B—H bond. Pyrazoles with 3,4-substituents (for instance, 3-isopropyl-4-bromopyrazole, 3-phenyl-4-methylpyrazole, and other similar pyrazoles) also tend to yield ligands with boron being bonded to the least hindered nitrogen atom. The only exception is benzopyrazole (indazole) and its derivatives containing alkyl or aryl substituents in the 3-, 4-, 5-, or 6-positions, or any combinations thereof, which produce scorpionate ligands where boron is bonded to the more hindered nitrogen, as in **133**. This happens, presumably, because electronic effects outweigh steric ones. On the other hand, indazoles with a 7-alkyl or

3.1 REGIOCHEMISTRY IN LIGAND SYNTHESIS

aryl substituent, as in **134**, or a 6,7-fused benzo ring, as in **135**, give rise to scorpionate ligands with boron bonded to the less hindered nitrogen.

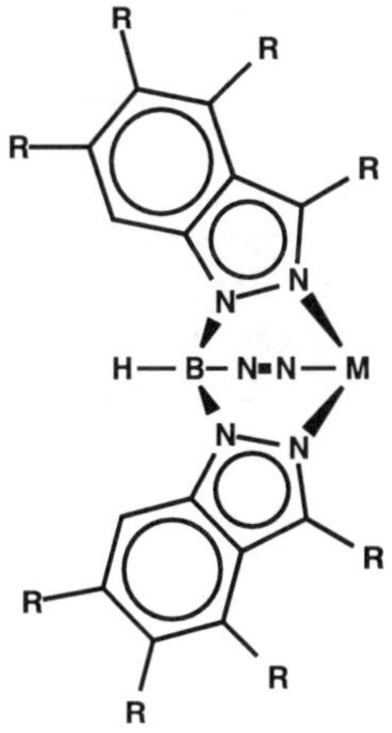

133 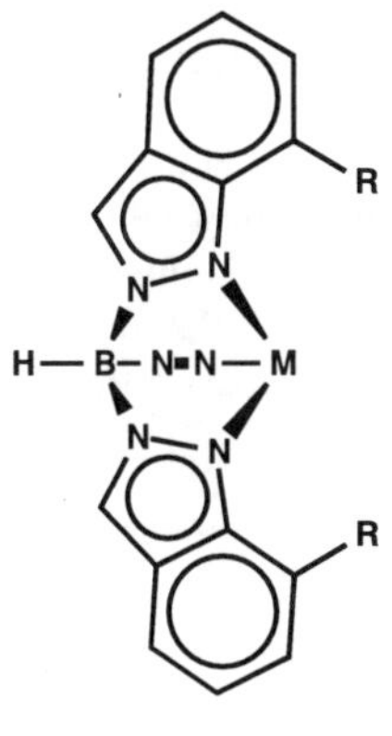   **134** 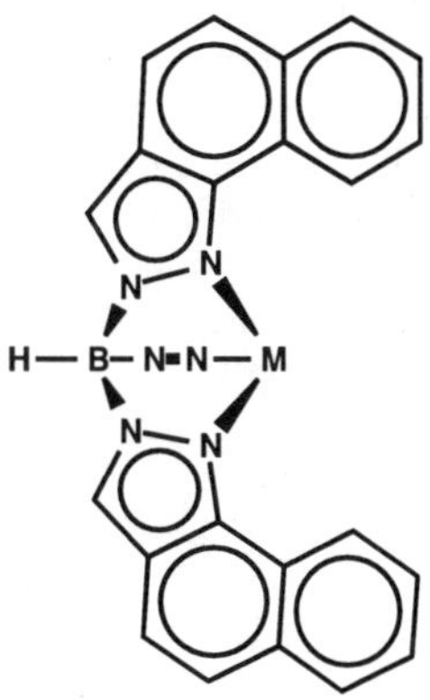   **135**

While the formation of Tp^R ligands is the expected norm, in a few instances there is concurrent formation of an isomeric ligand, [HB(3Rpz)$_2$(5Rpz)]$^-$ (= Tp^{R*}) which, being of lesser symmetry, is usually more soluble than Tp^R, and can be separated from it by crystallization. This occurs, for instance, in the reaction of 3-neopentylpyrazole with KBH$_4$, when both, Tp^{Np} and Tp^{Np*} are formed, although Tp^{Np} is by far the dominant product.[80] A similar situation prevails in the case of 3-(9-anthryl)pyrazole.[89] Only with 3-mesitylpyrazole is Tp^{Ms*} the major reaction product, although it can be thermally isomerized to Tp^{Ms}.[93] Even in the case of Tp^{x*} ligands, the majority (two out of three) of pyrazolyl groups is still bonded to boron through N1. From 3,5-disubstituted pyrazoles containing one t-butyl group and a smaller substituent, such as methyl,[86] 2-thienyl,[83] and even isopropyl,[92] the Tp^{tBu,R} ligands are obtained as the only product. On the other hand, when the substituents are similar in size, such as methyl and ethyl, or ethyl and propyl, inseparable mixtures are inevitably produced.[1199] Even with 3-isopropyl-5-methylpyrazole, where the two substituents are fairly different in size, one obtains in addition to the desired Tp^{iPr,Me} ligand, about 20% of the 3,3,5-isomer.[106] The only instant in Tp^x chemistry of ligand rearrangement going beyond the Tp^{R*} stage was found in the case of Tp^{Me}Ir(COD), which rearranged at first to Tp^{Me*}Ir(COD), and then to [HB(5-Mepz)$_2$(3-Mepz)]Ir(COD).[979]

3.1.2 The Quantification of Steric Effects in Tp^x Ligands

In contrast to the cyclopentadienyl ligand, where the bonds to the Cp substituents (H, Me, other alkyl or aryl groups) are pointing outward in the plane of the cyclopentadienyl ring, and away from the coordinated metal, the bonds in a Tp^x ligand emanate at an angle which makes the substituents protrude in space past the metal, enveloping it, and forming a protective pocket of varying size and shape. As the substituents become larger, the pocket becomes tighter, limiting both, frontal, and side-on access to the coordinated metal by other ligands. This can be seen from figures **136** and **137** below, which show frontal and side-on accessiblity to the metal in the presence of spherical 3-R substituents in the Tp^{x}M fragment. The smaller the cone angle, and the larger the wedge angle, the easier it is for other ligands to coordinate to the metal. Because of this feature, the proper choice of 3-R

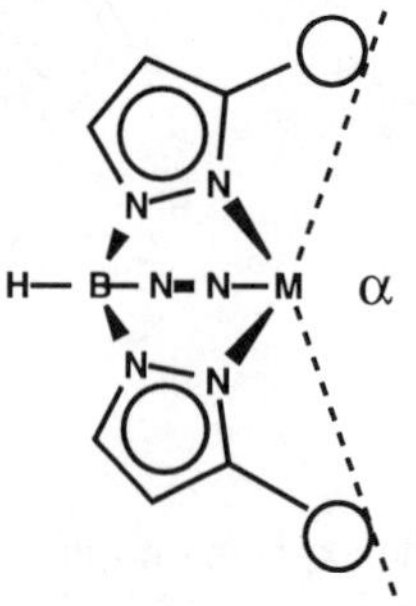

136
cone angle = (360-α)
side view

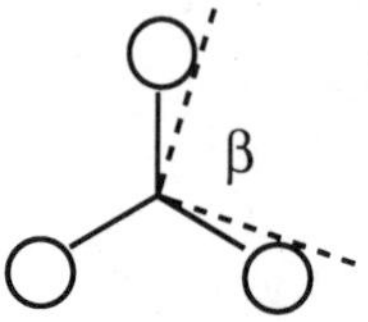

137
wedge angle = (β)
view along the B-M axis

substituents does adjust the steric accessibility of the coordinated metal, in this fashion controlling the coordination chemistry of the Tp^{x}M species. Some of the approaches to quantify the steric characteristics of the Tp^x ligands entailed the calculation of the cone and wedge angles of such ligands containing different 3-R substituents. Indeed, the trends in the values of these angles were consistent with the trends in the coordination chemistry of the Tp^x ligands. Three main categories of Tp^x ligands can be distinguished:

1. Those of low steric hindrance (small cone angle), characterized by a strong tendency to form Tp$^x{}_2$M complexers with divalent first row transition metals, and the inability to form stable Tp^{x}MX species. Apart from the parent Tp, such ligands are exemplified by those where 3R is methyl, cyclopropyl, or 2-thienyl.

3.1 STERIC EFFECTS IN HOMOSCORPIONATES 103

2. Those with large steric hindrance (large cone angle), characterized by predominant formation of tetrahedral Tp^{x}MX complexes, and by the inability to form Tp$^x{}_2$M species. The classic example here is Tp^{tBu} plus its 5-substituted analogs (Tp^{tBu,Me}, Tp^{tBu,Tn}, Tp^{tBu,iPr}), as well as Tp^{Ms}, Tp^{Ant} and Tp^{Trip}.

3. Those of intermediate steric hindrance, which are capable of forming both, Tp$^x{}_2$M and Tp^{x}MX species, although not necessarily with equal ease. Examples of such ligands are Tp^{Np} and Tp^{Ar}, where Ar is phenyl, or a number of variously substituted phenyl groups.

Without going into any calculations, it is quite clear that the metal gets progressively more and more encased in the ligand cavity, as one proceed from the least hindered parent ligand, Tp (**138**), through Tp^{Me} (**139**), Tp^{tBu} (**140**) to the extremely hindered Tp^{4Bo,7tBu} (**141**), as shown schematically below. Structure **141** approximates the steric effect of a neopentyl 3-substituent, frozen in its most sterically blocking conformation.

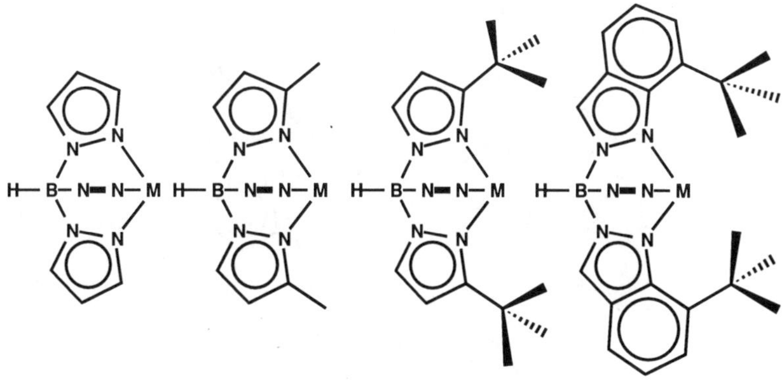

138   **139**   **140**   **141**

The cone and wedge angles obviously depend not only on the ligand itself, but also on the length of the N—M bond. For instance, the cone angle calculated for Tp^{3Bo,7tBu}Tl (**141**, M = Tl) was 261°, while for the Co complex of the same ligand this value increased to 315°.[102] The most recent compilation of some cone and wedge angles was based on first row transition metals (Co, Fe, Ni). In each case the lines defining the cone angle were drawn from existing X-ray structures of known complexes, connecting the center of the metal atom to the outermost point of the R-group, taking into account its van der Waals radii, and based on either the hydrogen or the carbon atoms.[81]

Several problems exist in calculating the cone and wedge angles. For simple, symmetrical 3-R substituents, such as H, Me or t-Bu, this is relatively easy, but in the case of asymmetric ligands, such as i-Pr, Ph and other planar heterocycles, the cone angle depends on the orientation of such substituents. On the one hand, a freely rotating isopropyl group can sweep out a space equivalent to that of a *t*-butyl group. However, a static isopropyl group can adopt a position with both of its methyl groups straddling the pyrazolyl plane, and pointing away from the metal, in which case its steric effect effect in terms of the cone angle would approximate that of a methyl group, although its wedge angle would be smaller. Conversely, if its methyl groups were pointed at the metal, the effect would resemble that of a *t*-butyl group. There are examples of both of these isopropyl group orientations being adopted by different Tp^{iPr} complexes in the solid state. It also matters whether we are dealing with octahedral, $Tp^x{}_2M$ complexes, or with tetrahedral Tp^xMX ones. Furthermore, some of the Tp^xMX complexes are five-coordinate dimers as, for instance, $[Tp^{iPr,4Br}NiNCS]_2$, **142**.[86]

142

The choice of a "standard" metal for calculating the cone and wedge angles has to satisfy the requirement that all Tp^xM complexes be isostructural. This would exclude transition metals, because of the variety of structures they can adopt with Tp^x ligands. Moreover, no stable structure, whether octahedral, tetrahedral, or five-coordinate can be achieved with all the known Tp^x ligands (see above). For these reasons Tp^xTl complexes seemed to be the best choice. Except for the tetrameric $[Tp^{Cpr}Tl]_4$,[81] all other known Tp^xTl complexes are monomeric and isostructural. Furthermore, many of them have already been structurally characterized. While the absolute cone and wedge angles derived from Tl(I) complexes cannot be directly transferred to other metals, they can be used to establish a relative steric hierarchy for the various Tp^x ligands. The cone and wedge angles of Tp^xTl complexes, the structures of which have been established by X-ray crystallography, are presented below in Tables 1 and 2, respectively. These were averaged values, based on the outermost hydrogen atoms of the 3-R substituent (except when the 3-R sustituents

3.1 STERIC EFFECTS IN HOMOSCORPIONATES

outermost hydrogen atoms of the 3-R substituent (except when the 3-R sustituents were perfluoromethyl or bromo), and taking into account the appropriate van der Waals radii.[1542]

Table 2. Tp^{X}Tl Cone Angles.

Complex	Cone angle (in degrees)	
TpTl	183	
Tp^{Cpr}Tl	223	
Tp^{Cbu}Tl	234	
Tp^{Br3}	234	(Br-based)
Tp^{4Bo,3Me}	235	
Tp$^{(CF_3)_2}$	237	(F-based)
Tp*Tl	239	
Tp^{iPr,4Br}Tl	243	
Tp^{tBu,Me}Tl	243	
Tp^{tBu}Tl	251	
Tp^{Cpe}Tl	253	
Tp^{Cy,4Br}Tl	273	
Tp^{3Bo,7tBu}	277	
Tp^{Cy}Tl	281	

As can be seen from Table 1, the range of cone angles covers about 100°. The complexes tabulated excluded planar aromatic 3-substituents, but included alicyclic ones. Even with accurately calculated cone angles, the feasibility of forming certain complexes depends on the other ligands involved. For instance, the ligand **177** (see p. 135), with a very large cone angle, but a small wedge angle (the three fused rings being planar), readily forms homoleptic Tp$^x{}_2$M complexes, but it is incapable of forming a κ^3 Tp^{x}Mo(CO)$_2$(η^3-methallyl) complex, because one of the coordinating arms would crash into the η^3-methallyl ligand, and therefore a κ^2 structure with an agostic B—H—Mo bond was adopted.[103] The ligand **187** (see p. 153), with an even larger cone angle (the tighter bite being the result of the 5-Me substituents), would still yield octahedral Tp$^x{}_2$M species, but could not form even a κ^2 Tp^{x}Mo(CO)$_2$(η^3-methallyl) complex, like **177** did.[101] At the same time the ligand Tp^{Ms} with a modest cone angle, but a very small wedge angle, would not form homoleptic Tp$^x{}_2$M species, but readily produced the structurally characterized κ^3 Tp^{Ms}Mo(CO)$_2$(η^3-methallyl) complex.[93]

Table 3.   Tp^{X}Tl Wedge Angles.

Complex	Wedge angle	
TpTl	70	
Tp^{Cpr}Tl	68	
Tp^{4Bo,3Me}	68	
Tp*Tl	67	
Tp^{Br3}	60	(Br-based)
Tp^{Cy}Tl	53	
Tp^{Cbu}Tl	51	
Tp$^{(CF_3)_2}$	49	(F-based)
Tp^{Cpe}Tl	46	
Tp^{Cy,4Br}Tl	46	
Tp^{3Bo,7tBu}	33	
Tp^{tBu,Me}Tl	31	
Tp^{tBu}Tl	29	
Tp^{iPr,4Br}Tl	28	

3.1.3 LIGAND REARRANGEMENTS

The total range of wedge angles, about 40 degrees, is considerably smaller than that of the cone angles, and wedge angles play a somewhat lesser role in restricting the formation of certain types of complexes. Still, as shown above in the example of Tp^{Ms}, their influence is not to be discounted.

Other ways to quantify the steric effects of Tp^x ligands were by means of the "bite size" (i.e. the distance between the coordinated nitrogens)[808] and also by means of the "ligand profile" measurement, which was applied to several second generation Tp^x ligands.[1200] Another attempt to compare the pocket size in Tp^{Ar} ligands involved comparing the areas of triangles formed by connecting the midpoints of the 3-phenyl rings in copper(I) complexes.[1538] None of these approaches have been widely used.

3.1.3 Ligand Rearrangments

The reaction of borohydride ion with 3-monosubstituted pyrazoles leads, in general, to regiochemicaly pure Tp^R complexes, except for occasional mixtures of Tp^R and the isomeric Tp^{R*}. There are situations, however, when a regiochemically pure ligand Tp^R undergoes rearrangement to Tp^{R*} during the course of complex formation. Such rearrangements were also observed with Bp^x ligands, but never with pz^{o}Tp^R ligands. The first instance of Tp^x rearrangement was established in the case of the ligands Tp^{iPr} and Tp^{iPr,4Br} which formed octahedral cobalt(II) complexes with rearrangement of the original ligands to Tp^{iPr*} and to Tp$^{(iPr,4Br)*}$, respectively.[5] This rearrangement was facilitated by the presence of polar solvents, and was explained in terms of the combined effects of

1. the driving force to form an octahedral complex of cobalt(II), and

2. the inability of six isopropyl groups to be accomodated in the equatorial belt of an octahedral complex.

Similar rearrangments were also found in the case of the regiochemically pure Tp$^{Np}{}_2$M complexes, which upon heating were converted to [Tp^{Np*}]$_2$M,[80] and in Tp^{tBu}AlR$_2$, **143**, which was converted to Tp^{tBu*}AlR$_2$, **144**, (R = Me, Et).[1104,1201] Other examples of of such rearrangements were exemplified by Tp^{Ph} → Tp^{Ph*} in the cation [Tp$^{Ph*}{}_2$Al]$^+$,[85] and in the 3,5-disubstituted system Tp^{iPr,Me}.[106] The only reverse isomerization, from Tp^{x*} to Tp^x occurred in the case of Tp^{Ms*}, the major product from the reaction of KBH$_4$ with 3-mesitylpyrazole, which was cleanly converted at the melting point of its Tl salt to TlTp^{Ms}.[93]

The mechanism of such rearrangements has been thought to be "borotropic", proceeding by detachment of one of the pyrazolyl groups from the metal, the migration of boron from N1 to N2 of the detached pz^x group, followed by the re-coordination of the detached pz^x group to the metal, but now through N1. The consensus was that this is a sterically-driven process, leading to a more comfortable arrangement of the pz^x and pz^{x*} groups in the complex. Nevertheless, one could also envisage a "metallotropic" mechanism leading to the same result. Here, a bridging

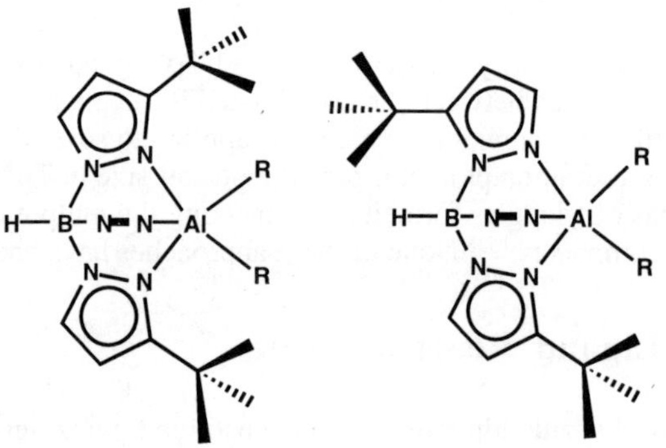

143   **144**

pz^x group in B(pz^x)M would first become detached from boron (probably with assistance by a polar solvent), and undergo a N2 to N1 metallotropic rearrangement on the metal, which would proceed through a η^2-pyrazolato intermediate. Such η^2-bonding of pyrazolato groups has now been found to be not that uncommon, and applicable not only to the *f*-group elements,[1202-1209] but also to the first row transition metals.[1210] Of particular significance is the structurally characterized complex Tp*$_2$UI, which contains one pz* from a Tp* ligand bonding η^2, that is side-on, to the uranium atom while still connected to boron,[1198] and thus lending support to the possibility of a metallotropic mechanism in Tp^x ligand rearrangements. Finally, one could envisage a rearrangement mechanism in which the pz^x group goes through a transition state in which it is endobidentate bonded to both, B and M, leading in the end to the Tp^{x*} isomer. An example of related bis-endobidentate bridging between two metal ions has been recently reported.[1211]

The only ligand rearrangement going beyond the Tp^{x*} stage, was found in the case of Tp^{Me}Ir(COD). This initially pure complex, rearranged in solution first to Tp^{Me*}Ir(COD) and then, on heating, to [HB(5-Mepz)$_2$(3-Mepz)]Ir(COD).[979]

3.2   Individual Ligands

What follows is a coverage of individual ligands, some of which led to the most exciting developments in the homoscorpionate area. They are arranged starting with RTp^x ligands, followed by monosubstituted, disubstituted, and trisubstituted (including fused benzo and naphtho rings) Tp^x species. In most cases the boron-substituted [RTp^x]$^-$ ligands are based on plain pyrazole but, obviously, 3-substituted

3.2 LIGANDS RTp^X

pyrazoles should also be accessible through the synthetic schemes employed, and a few such ligands are included.

3.2.1 B-Substituted Ligands, RTp^X (R ≠ pz^X)

3.2.1.1 MeTp

This ligand has not been reported, but the related ligand MeTp^{Me} ligand was synthesized, and converted to MeTp^{Me}Rh(NBD).[77]

3.2.1.2 EtTp

The EtTp ligand was not isolated as such, but only as part of the boronium cation [EtB(μ-pz)$_3$BEt]$^+$, which may be regarded as [EtTp(BEt)]$^+$,[124,1101] and this cation was structurally characterized as the [PF$_6$]$^-$ salt.[1212]

3.2.1.3 iPrTp

Derivatives of iPrTp included iPrTpZrCl$_3$,[355] the chlorine atoms of which could be sequentially replaced with tert-butoxy substituents.[360]

3.2.1.4 BuTp

BuTp was first prepared from butylboronic acid, and was converted to octahedral BuTp$_2$M complexes for M = Mg, Mn, Fe, Co, Ni, Cu and Zn.[11] In the zirconium complex BuTpZrCl$_3$, the chloride ligands could be replaced with OBu^t groups.[360]

3.2.1.5 MeS(CH$_2$)$_3$Tp and MeS(CH$_2$)$_3$Tp^R

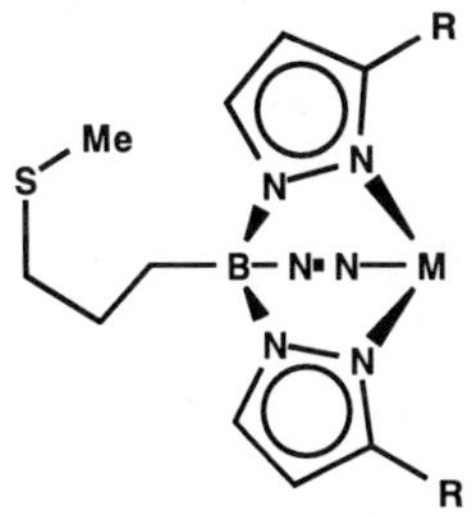

145

3.2.1.5 MeS(CH$_2$)$_3$Tp and MeS(CH$_2$)$_3$Tp^R

Synthesis of the cobalt complex [MeS(CH$_2$)$_3$Tp]$_2$Co, and its ^{13}C NMR was reported.[78] This ligand, and its MeS(CH$_2$)$_3$Tp^R analogs, **145**, were synthesized from the commercially available MeS(CH$_2$)$_3$BH$_2$ and [pz^R]$^-$ + 2 Hpz^R. The coordination chemistry of the MeS(CH$_2$)$_3$Tp^R ligands was very similar to that of their Tp^R counterparts, and there was no evidence of the thioether functionality coordinating to the transition metal ion.[83]

3.2.1.6 PhTp and PhTp^{tBu}

The first PhTp$_2$M complexes for M = Mn, Fe, Co, Ni, Cu and Zn were prepared from the free acid, [PhTp]H, obtained in situ from the direct reaction of PhBCl$_2$ and excess pyrazole.[11] Complexes PhTpMo(CO)$_2$(η^3-allyl) and PhTpMo(CO)$_2$(η^3-C$_7$H$_7$) were structurally characterized. They contained κ^3 PhTp, and η^3-C$_7$H$_7$ (rather than κ^2 PhTp, and η^5-C$_7$H$_7$).[1213] Mössbauer studies were carried out on the complex PhTp$_2$Fe, and its structure was determined by X-ray crystallography.[808]

The related, but more hindered ligand, PhTp^{tBu}, was prepared from PhBH$_2$ and Hpz^{tBu} in standard fashion. It was converted to PhTp^{tBu}FeMe (via the PhTp^{tBu}FeCl precursor). This complex provided an interesting array of derivatives, including the structurally characterized 15-electron complex PhTp^{tBu}FeCO, **146**, the 17-electron PhTp^{tBu}Fe(NO)$_2$, **147**, in which the PhTp^{tBu} ligand was found to be κ^2, and the dinuclear [PhTp^{tBu}Fe(OH)]$_2$, containing two hydroxy bridges, in which the ligand was also κ^2.[1215]

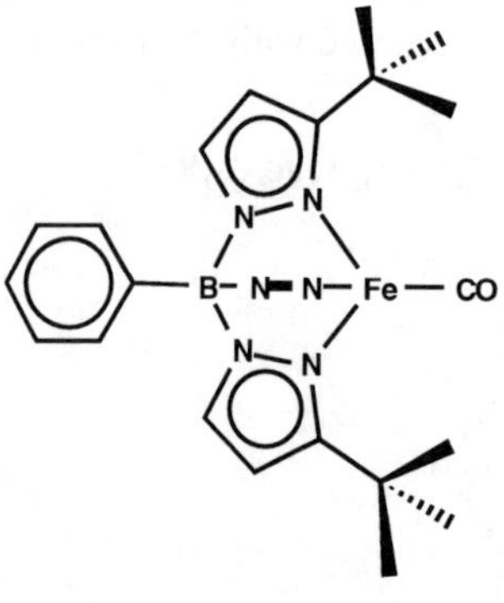

146 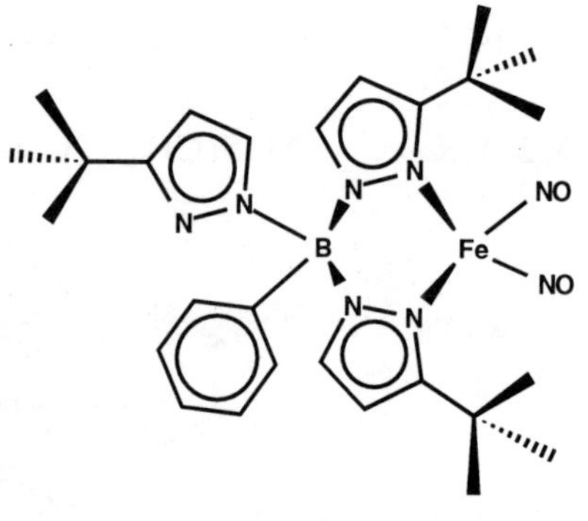   **147**

3.2 LIGANDS RTp^x

3.2.1.7 TolTp (mixture of m- and p-)

The above ligand was prepared from the appropriate mixed dichlorotolylboranes, and the Tp$^x{}_2$Co complex of the above isomeric ligand mixture was used in a study of proton-proton dipolar coupling in ultra-high field NMR.[126]

3.2.1.8 (p-BrPh)Tp

The similarly prepared p-BrPhTp ligand, permitted the use of the bromine substitutent in its octahedral Co^{II} complex, to obtain by means of standard organic reactions, the Li, D, Bu, COOH and COOMe derivatives.[127]

3.2.1.9 (C$_6$D$_5$)Tp

The phenyl-perdeuterated analog of PhTp was synthesized and converted to the complex [(C$_6$D$_5$)Tp]$_2$Co, which was used for direct measurment of the electron susceptibility anisotropy, using high-field deuterium NMR,[125,1557] and for the determination of multiple quantum ^{2}H spectra.[1216]

3.2.1.10 FcTp, Fc(Tp)$_2$, FcTp^{Me} and FcTp^{Ph}

Treatment of ferrocenyldibromoborane with dimethylamine yielded FcB(NMe$_2$)$_2$ which produced, upon reaction with [pz]$^-$ + 2 Hpz, the ligand FcTp. This ligand was converted to the trinuclear [FcTp]$_2$Fe complex, and also to the analogous complex based on FcTp^{4SiMe3}, the electrochemistry of which was studied. Derivatives [FcTpMo(CO)$_3$]$^-$, FcTpZrCl$_3$, and FcTpMo(CO)$_2$(η^3-methallyl) were prepared from FcTp, and from the related difunctional Fc(Tp)$_2$, which was synthesized from 1,1'-bis(dibromoboryl)ferrocene. The complex FcTpMo(CO)$_2$(η^3-methallyl) was structurally characterized.[75,128] The related ligands, FcTp^{Ph} and FcTp^{Me}, as well as the difunctional Fc(Tp^{Me})$_2$ and Fc(Tp^{Ph})$_2$ were synthesized by an analogous route, starting with the appropriate 3-substituted pyrazoles, and characterized as their Tl salts. The structure of [FcTp^{Ph}]Tl was determined by X-ray crystallography.[1217]

3.2.1.11 Me$_2$NTp

The free acid of this ligand, [Me$_2$NTp]H, was readily obtained from (Me$_2$N)$_3$B and pyrazole.[129,1218] It could be converted directly to the molybdenum derivative, [Me$_2$NTp]Mo(CO)$_2$(η^3-allyl), **148**, in which the ligand was κ^3-bonded through two pyrazolyl rings, and through the NMe$_2$ group, with the third pz group remaining uncoordinated, as was indicated by NMR.[130]

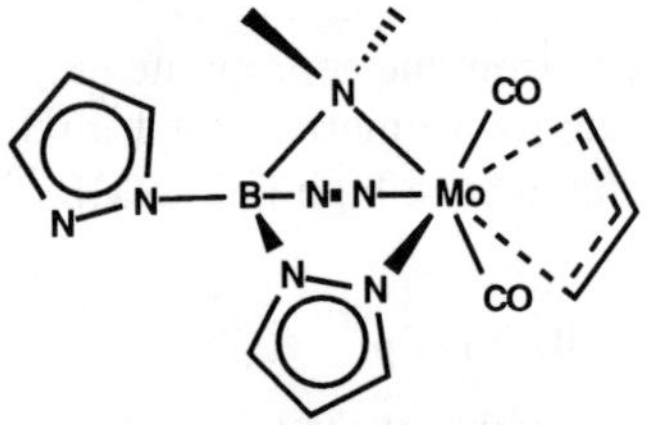

148

3.2.1.12 Tp-Tp

This simplest bis-Tp ligand was readily available from $(Me_2N)_2B-B(NMe_2)_2$. It formed intractable linear $[(Tp-Tp)M]_n$ polymers with divalent first row transition metals.[83] This ligand was structurally characterized as the dinuclear palladium complex $(\eta^3$-allyl$)Pd(pz)_3B-B(pz)_3Pd(\eta^3$-allyl$)$.[131]

3.2.1.13 Tp^{Me}-Tp^{Me}

This bifunctional homoscorpionate ligand was prepared just like Tp-Tp, but starting with Hpz^{Me} instead of Hpz. It was converted to $[Tp^{Me}$-$Tp^{Me}][Mo(CO)_2NO]_2$, **149**, and also to $[Tp^{Me}$-$Tp^{Me}][MoCl_2NO]_2$, from which the complex $[Tp^{Me}$-$Tp^{Me}][MoCl(py-tBu)NO]_2$ was obtained. It contained two 17-electron paramagnetic centers, which were electrochemically active.[1219]

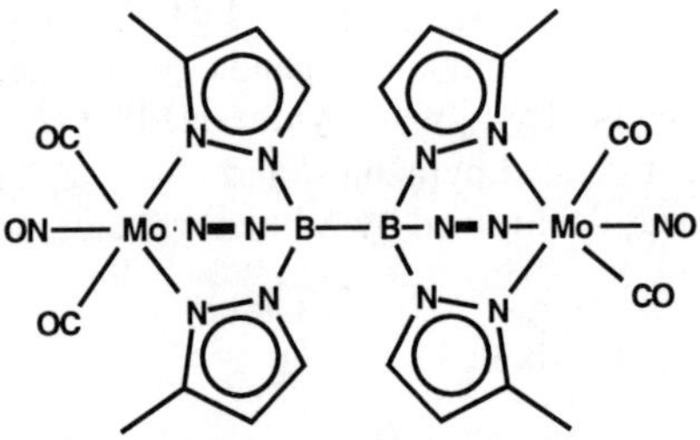

149

3.2.1.14 Tp^{Py}-Tp^{Py}

This potentially dodecadentate (bis-hexadentate) ligand, was prepared in the same manner as Tp-Tp, but using Hpz^{Py} instead of Hpz. Its dodecadenticity was

3.2 LIGANDS Tp^R

convincingly demonstrated by the synthesis of the gadolinium complex [Tp^{Py}-Tp^{Py}][Gd(NO$_3$)$_2$]$_2$, the structure of which was established by X-ray crystallography, showing each Gd to be in a ten-coordinate environment.[1219]

3.2.2 3-Monosubstituted Ligands, Tp^R

The placement of a substituent at the 3-position of a Tp ligand has the most telling effect on the coordination chemistry of the resulting Tp^x ligand. This is because the 3-substituent is closest to the coordinated metal ion, and it defines the size of the cavity harboring the metal, as expressed by cone and wedge angles, or by some other means. The effect of a 3-substituent is dominant, and any additional substitution at the 4- or 5-positions, while at times of some significance, is of secondary importance.

3.2.2.1 Tp^{Me} and pz^{o}Tp^{Me}

The simplest of 3-substituted Tp ligands, Tp^{Me}, resembled Tp* in its coordination chemistry but, unlike Tp* it also had its tetrakis-ligand pz^{o}Tp^{Me} readily available, which coordinated just like Tp^{Me}. Tp^{Me} was synthesized in 1985,[79] and it was converted to Tp^{Me}$_2$M complexes for M = Mn, Fe, Ni, Cu, Zn, Cd and Pb, which were structurally similar to their Tp* analogs,[1220] as well as to Tp^{Me}HgSR,[1223] and to Tp^{Me}Hg(2-thienyl) derivatives.[1099] The structurally characterized Tp^{Me}$_2$Fe was studied by Mössbauer spectroscopy,[821] and together with the Fe^{III} derivative [Tp^{Me}$_2$Fe][PF$_6$] it was studied electrochemically, spectroelectrochemically and by X-ray absorption spectroscopy,[824] while pz^{o}Tp^{Me}$_2$Cd was studied by means of ^{113}Cd NMR.[1221] The salt [Et$_4$N][Tp^{Me}Mo(CO)$_3$] was structurally characterized,[1222] as was Tp^{Me}Rh(NBD), in which the ligand was found to be κ^3-bonded.[114]

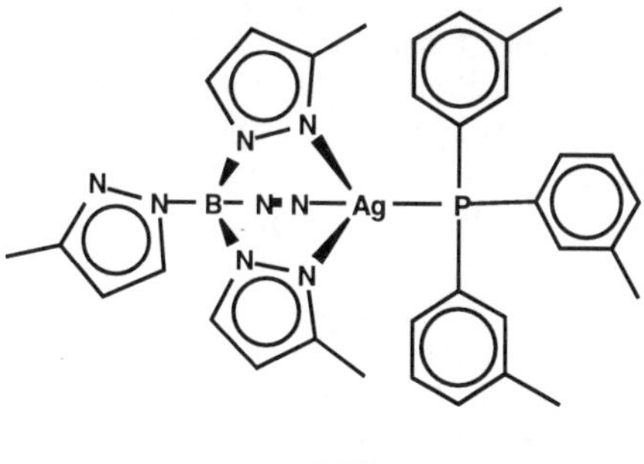

150 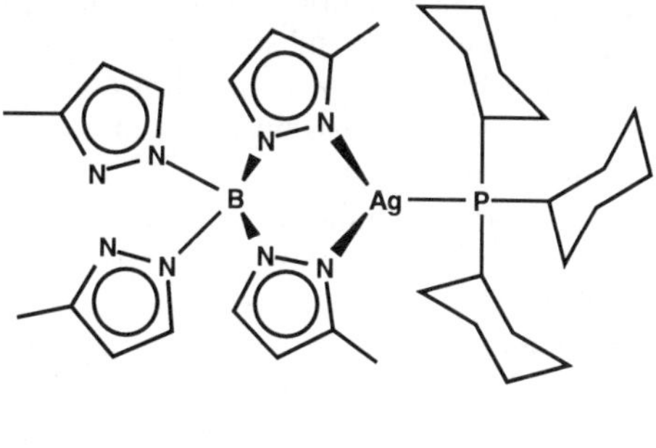   **151**

The complex Tp^{Me}Ir(COD) underwent rearrangement in solution first to Tp^{Me*}Ir(COD) and then, on heating, to [HB(5-Mepz)$_2$(3-Mepz)]Ir(COD).[979] The structures of several complexes of the type pz^{o}Tp^{Me}AgPR$_3$ (R = m-Tol, PMePh$_2$, P(C$_6$H$_{11}$)$_3$, CH$_2$Ph) were determined by X-ray crystallography. Except for pz^{o}Tp^{Me}Ag[P(m-Tol)$_3$], in which the pz^{o}Tp^{Me} ligand was κ^3, **150**, it was κ^2 in all the other complexes, as exemplified by **151**. In those cases where the pz^{o}Tp^{Me} ligand was κ^2, and the silver ion was three-coordinate, it was possible to coordinate a fourth ligand, such as imidazole.[1081]

Mössbauer and NMR studies were carried out on Tp^{Me} derivatives of tin(IV), and the structure of Tp^{Me}SnPhCl$_2$ was determined by X-ray crystallography.[1224] Other complexes of Tp^{Me} ligand reported were Tp^{Me}Cu(PPh$_3$), Tp^{Me}Cu(PTol$_3$),[1058] Tp^{Me}Cu(PCy$_3$),[1057] as well as [pz^{o}Tp^{Me}]$_2$Sn and pz^{o}Tp^{Me}SnCl.[1120] The reaction of [Tp^{Me}]$^-$ with EtBX$_2$ produced the cation [Tp^{Me}BEt]$^+$.[1101]

3.2.2.2   Tp^{Perfluoroalkyl} (Tp^{CF_3}, Tp^{C_2F_5}, Tp^{C_3F_7})

The ligand Tp^{CF_3} was prepared as the sodium salt, and it was converted to Tp^{CF_3}AgPPh$_3$,[94] as well as to Tp^{CF_3}Mn(CO)$_3$, and to the structurally characterized Tp^{CF_3}CuCO, which were studied by cyclic voltammetry.[95] The other ligands with longer perfluoroalkyl chains in the 3-position, Tp^{C_2F_5}, Tp^{C_3F_7}, were also prepared and converted to the corresponding Tp^{x}CuCO complexes. The sodium salt of Tp^{C_2F_5} crystallized with one molecule of 3-pentafluoroethylpyrazole, and its structure was determined by X-ray crystallography, as were the structures of Tp^{C_2F_5}CuCO and of Tp^{C_3F_7}CuCO.[96]

3.2.2.3   Tp^{iPr}, pz^{o}Tp^{iPr} and Tp^{iPr*}

The ligands Tp^{iPr} and pz^{o}Tp^{iPr} were the first representatives of the Tp^x sub-group, containing a 3-isopropyl substituent. Later on, variants of these ligands with additional 4- and 5-substituents were reported. Tp^{iPr} formed tetrahedral complexes Tp^{iPr}MX readily, but octahedral complexes were obtained only with, presumably sterically driven, rearrangement of the Tp^{iPr} ligand to Tp^{iPr*}, as was proven by NMR studies of the paramagnetic [Tp^{iPr*}]$_2$Co complex, and by X-ray crystallography. This was the first example of homoscorpionate ligand rearrangement from Tp^R to Tp^{R*}.[5] The same rearrangement to Tp^{iPr*} was also observed in the iridium complex Tp^{iPr}Ir(COD), which was converted to Tp^{iPr*}Ir(COD).[979] The ligand pz^{o}Tp^{iPr} provided the first example of a scorpionate ligand yielding an exclusive series of isomorphous tetrahedral complexes, **152**, with Mn, Fe, Co, Ni, Cu, and Zn, all exhibiting dynamic D$_{2d}$ symmetry, i.e. rapid ring inversion, but without exchange of the coordinated and uncoordinated pz^{iPr} arms. The structure and bonding in these complexes was investigated by X-ray absorption spectroscopy,[1225] and the complex [pz^{o}Tp^{iPr}]$_2$Fe by Mössbauer spectroscopy.[1226] Also reported, and studied by EPR were compounds Tp^{iPr}MoO$_2$X (X = Cl, OMe, OEt, SPh), of which Tp^{iPr}MoO$_2$(OMe)

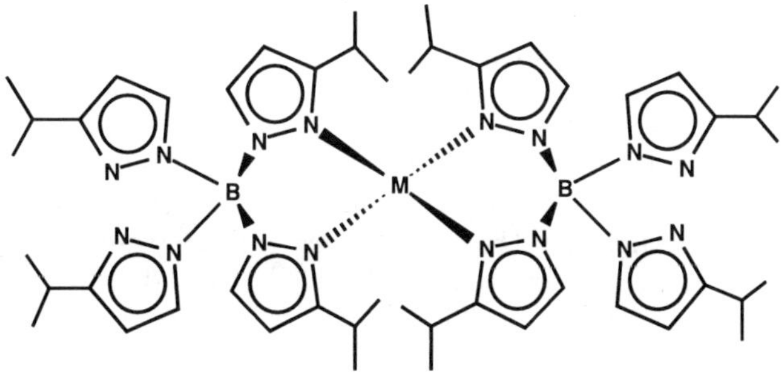

3.2 LIGANDS Tp^R

152

was structurally characterized.[624] The reaction of KTp^{iPr} with MoO(S$_2$PR$_2$)$_2$ resulted in ligand rearrangement, affording Tp^{iPr*}MoO(S$_2$PR$_2$), which was converted by boron sulfide to Tp^{iPr*}MoS(S$_2$PR$_2$). Ferrocenium oxidation of Tp^{iPr*}Mo^{IV}E(S$_2$PR$_2$) species (E = O or S) yielded the cations [Tp^{iPr*}Mo^{IV}E(S$_2$PR$_2$)]$^+$. The complex Tp^{iPr}MoO$_2$Cl did not react with sulfiding agents to yield the analogous thio derivative.[1227] The reaction of [Tp^{iPr}W(CO)$_3$]H with propylene sulfide, produced the structurally characterized dinuclear[Tp*W(CO)$_2$]$_2$S.[689] Treatment of [Tp^{iPr}W(CO)$_3$]$^-$ with iodine yielded Tp^{iPr}W(CO)$_3$I,[683] and the related tricarbonyl species Et$_4$N[Tp^{iPr}W(CO)$_3$] and [Tp^{iPr}W(CO)$_3$]H were studied by IR and NMR.[680] A side-on bonded κ^2-N,C nitrile ligand was present in Tp^{iPr}W(I)(CO)(MeCN).[731] Titanium imido-complexes, exemplified by Tp^{iPr}Ti(=NBu^t)(Cl)(4-tBu-py), were found to be fluxional, and the appropriate activation parameters have been determined.[350]

3.2.2.4 Tp^{Cpr}, pz^{o}Tp^{Cpr} (Cpr = cyclopropyl)

In contrast to the moderately sterically hindered Tp^{iPr} and pz^{o}Tp^{iPr}, the cyclopropyl analogs, Tp^{Cpr} and pz^{o}Tp^{Cpr} were remarkably unhindered. Thus, they formed readily

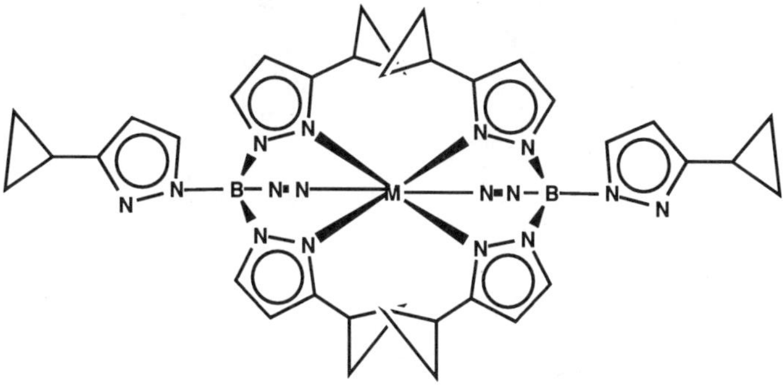

153

octahedral complexes $Tp^{Cpr}{}_2M$ without rearrangement to Tp^{Cpr*}, and displayed a chemistry akin to that of the parent Tp, or to Tp^{Me}.[82] For instance, unlike pz^oTp^{iPr}, which only yielded tetrahedral complexes, such as **152**, with first row divalent metals, pz^oTp^{Cpr} formed only octahedral complexes, exemplified by **153**. One remarkable feature of Tp^{Cpr} was the structure of its Tl salt, which was a monomer in solution, but crystallized as a tetramer containing a perfect Tl_4 tetrahedron, capped at each apex with a Tp^{Cpr} ligand.[81] The only instance of ligand rearrangement to Tp^{Cpr*} was encountered during the reaction of $Tp^{Cpr}Tl$ with $Nb(MeC≡CPh)Cl_3$. The resulting product, $Tp^{Cpr*}Nb(MeC≡CPh)Cl_2$, was structurally characterized by X-ray crystallography.[1542]

3.2.2.5 Tp^{tBu}, pz^oTp^{tBu}

Some of the most exciting results obtained with the second generation homoscorpionates, were those involving the Tp^{tBu} ligand (and its variants, containing additional substituents at the 5- or 4-positions), often employed as its structurally characterized Tl salt.[1228] First of all, it provided a series of stable tetrahedral $Tp^{tBu}MX$ complexes for the first row transition metals, thus earning the monicker "tetrahedral enforcer",[2,3] although it was later shown that five-coordination is possible with certain compact bidentate ligands. The tetrakis-ligand, pz^oTp^{tBu}, produced similar compounds.[3] The reaction of $Tp^{tBu}MCl$ (M = Fe, Co) with $AgBF_4$ resulted in fluoride abstraction, giving rise to $Tp^{tBu}MF$ complexes.[1229] Octahedral coordination of first-row transition metals was never encountered, although it was possible with second row metals, such as Mo and W.

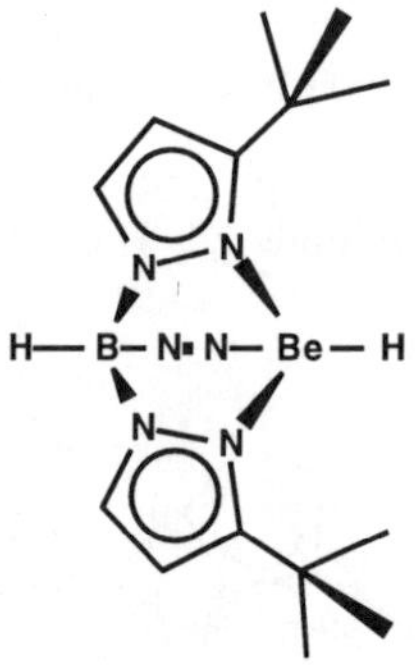

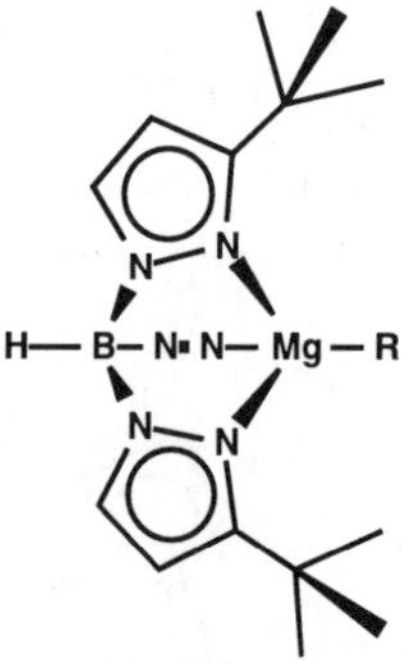

154   **155**

Monomeric haloberyllium complexes, $Tp^{tBu}BeX$ (X = Cl, Br) were readily prepared from $Tp^{tBu}Tl$ and BeX_2, and $Tp^{tBu}BeBr$ was structurally characterized. Its reaction with $LiAlH_4$ produced $Tp^{tBu}BeH$, **154**, which could be converted to

3.2 LIGANDS Tp^R

Tp^{tBu}BeSH by treatment with either sulfur or H$_2$S, and to Tp^{tBu}BeI by the reaction with I$_2$ or with MeI. These compounds were studied by ^{9}Be NMR.[1230] Although Tp^{tBu}BeMe could not be prepared by treating Tp^{tBu}BeX with MeLi, it was obtained by the reaction of Tp^{tBu}Tl with BeMe$_2$. Tp^{tBu}BeMe reacted with iodine and with H$_2$S, forming Tp^{tBu}BeI and Tp^{tBu}BeSH, respectively.[32]

The first examples of stable, monomeric magnesium alkyls, Tp^{tBu}MgR, **155**, were obtained from the reaction of MgR$_2$ with Tp^{tBu}Tl; the reaction of Tp^{tBu}Tl with RMgX reagents was less clean, leading in some cases to Tp^{tBu}MgR, and in others to Tp^{tBu}MgX.[33,1103,1231] The R substituents included primary, secondary and tertiary alkyl groups, CH=CH$_2$, CH$_2$SiMe$_3$ and phenyl. Several of these complexes were structurally characterized.[33] Reactivity studies on these compounds showed facile replacement of the alkyl groups with protic reagents Z-H, leading to Tp^{tBu}MgZ complexes (Z = SH, SR, OR, C≡CR, NHPh, Cl, OOBu^t), while from methyl ketones, RCOMe, the enolate derivatives Tp^{tBu}Mg-O-(CR=CH$_2$) were obtained. Insertion of carbon dioxide into the Mg—Me bond produced the acetato complex, Tp^{tBu}Mg(OAc), while the reaction with O$_2$ led to the formation of peroxo species Tp^{tBu}MgOOR.[325,1232] The oxygen reaction took a different course in the case of Tp^{tBu}MgCH$_2$SiMe$_3$, producing Tp^{tBu}MgOSiMe$_3$.[1233] The complex Tp^{tBu}MgOEt was found to be a catalyst for lactide ring-opening polymerization.[1234] Ligand degradation occurred during the reaction of Tp^{tBu}K with VCl$_3$,[1235] and also during the attempted preparation of Tp^{tBu}Rh(NBD).[1236] On the other hand, complexes Et$_4$N[Tp^{tBu}W(CO)$_3$] and [Tp^{tBu}W(CO)$_3$]H were synthesized, and studied by IR and NMR.[680] A trinuclear complex anion, **156**, was synthesized by the reaction of Tp^{tBu}CoCl with [Fe(CN)$_6$]$^{3-}$, and the spectral and magnetochemical properties of this multispin (S = 7/2, at 292 K) species were determined.[1237] The organocobalt complex, Tp^{tBu}CoMe was structurally characterized.[1529]

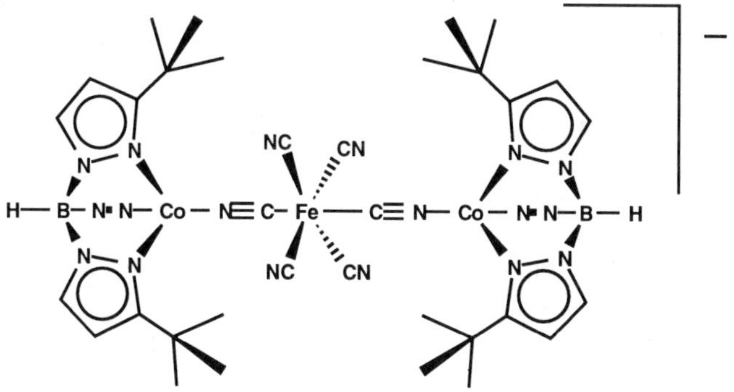

156

Aiming at a model for the substrate adduct of copper nitrite reductase, the complex $Tp^{tBu}CuNO_2$, containing five-coordinate copper with κ^2 nitrite, was prepared and studied by epr spectroscopy, as was $Tp^{tBu}CuOSO_2CF_3$, both of which were structurally characterized.[38] The first mononuclear copper nitrosyl complex, $Tp^{tBu}CuNO$, **157**, was also prepared, and structurally characterized, as was the dimer $[Tp^{tBu}Cu]_2$, **158**, from which the nitroso complex was made by treatment with NO.[37] This dimer, and related $[Tp^xCu]_2$ dimers were studied by electrochemistry.[1238] The chemistry of $Tp^{tBu}CuNO$, and of related complexes, was elaborated in more detail, and they were studied by EPR spectroscopy. $Tp^{tBu}CuNO$ reacted with oxygen to form a bidentate nitrate complex, and was converted by CO to $Tp^{tBu}CuCO$.[1239]

The organozinc chemistry based on the Tp^{tBu} ligand was also developed to a considerable extent. Stable organozinc compounds, $Tp^{tBu}ZnR$, were readily synthesized from ZnR_2 and $Tp^{tBu}Tl$,[34,1088] and could also be prepared from $Tp^{tBu}ZnCl$ and RLi.[1240] The hydride, $Tp^{tBu}ZnH$ was obtained similarly from ZnH_2. It reacted with active hydrogen compounds, REH, (E = O, S) eliminating H_2 and forming $Tp^{tBu}ZnER$ species, and its hydride was also replaced by halide in reactions with, for instance, CCl_4, I_2, or $CHBr_3$, while CO_2 inserted into the Zn-H bond to produce the bicarbonate derivative. The structure of $Tp^{tBu}ZnH$ was determined by X-ray crystallography.[1241,1088] $Tp^{tBu}ZnR$ complexes underwent similar reactions. X-ray crystallographic investigations on solid solutions of pairs of complexes $Tp^{tBu}ZnMe$,

157   **158**

$Tp^{tBu}ZnCl$ and $Tp^{tBu}ZnI$ showed that the presence of a co-crystallized impurity results in an apparent intermediate bond length for Zn-X.[1242] Similar phenomena were also encountered in mixtures of halide complexes $Tp^{tBu}ZnX$ with $Tp^{tBu}ZnCN$,[1243] and these findings were discussed in a more general review.[1214] Other $[Tp^{tBu}Zn]$ derivatives have also been reported,[1244] and ligand degradation was observed in working with some $Tp^{tBu}ZnOH$ complexes.[1245] The structure of $Tp^{tBu}Zn(OAc)$ contained a monodentate acetate ligand.[1246] $Tp^{tBu}CdI$ was prepared,

3.2 LIGANDS Tp^R

along with other Tp^{x}CdI complexes, and its structure was determined by X-ray crystallography, as were those of Tp^{tBu}CdMe and Tp^{tBu}Cd(κ^2-O$_2$NO),[1247] Tp^{tBu}Cd(OAc),[1272] and of Tp^{tBu}SnCl.[1248] The reaction of CdCl$_2$ with KBH$_4$ or LiBHEt$_3$, followed by Tp^{tBu}Tl, yielded the stable hydride Tp^{tBu}CdH, characterized by NMR.[46]

The complex Tp^{tBu}AlMe$_2$ was prepared from Tp^{tBu}Tl and AlMe$_3$, and was assigned a tetrahedral structure with κ^2 Tp^{tBu} ligand.[1104] The ethyl analog was prepared in similar fashion, or through the reaction of Tp^{tBu}Tl with AlClEt$_2$, and it was structurally characterized. It contained four-coordinate, tetrahedral Al, with a κ^2 Tp^{tBu} ligand. Over time, this complex isomerized to Tp^{tBu*}AlEt$_2$, which was also structurally characterized, and was found to contain the rearranged 5-Bu^{t}pz arm uncoordinated.[44,1201] The In(I) complex, Tp^{tBu}In was readily prepared from Tp^{tBu}Tl and InCl, and its structure was determined by X-ray crystallography.[1249] However, the reaction of Tp^{tBu}Tl with InI$_3$ produced, unexpectedly, Tp^{tBu}In-InI$_3$(Hpz^{tBu}), **159**, a dinuclear complex containing an In-In bond, a rearranged Tp^{tBu*} ligand, plus an additional coordinated Hpz^{tBu}. Its formation involved reduction of In^{III} to In^I, along with some degradation of the Tp^{tBu} ligand.[1250] Tp^{tBu} was decomposed upon reacting with GaI$_3$, yielding the polynuclear species Ga$_4$(OH)$_6$(Hpz^{tBu})$_{10}$I$_6$.[1251]

The Tp^{tBu} ligand provided the rare exception to the empirical rule that Tp^x ligands are always at least bidentate. In the sterically very congested complex Tp^{tBu}Ni(Tol)(PMe$_3$), structurally characterized by X-ray crystallography, the Tp^{tBu} ligand was found to be κ^1, as was also the case with a Pd^{II} analog.[1252]

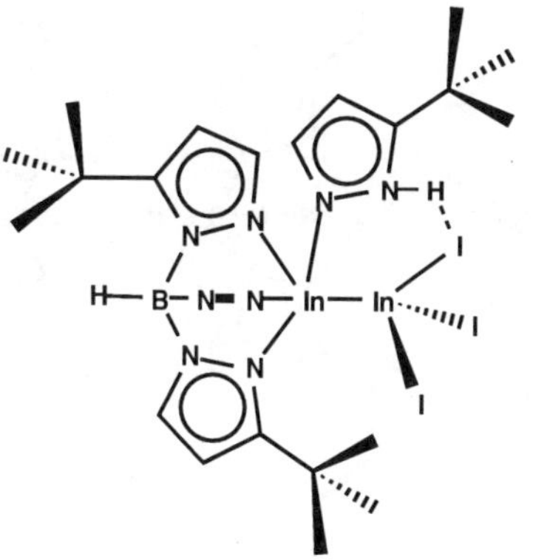

159

3.2.2.6 Tp^{Cbu}, pz^oTp^{Cbu} (Cbu = Cyclobutyl)

In terms of its coordination chemistry, the ligand Tp^{Cbu} was intermediate in steric requirements between cyclopropyl and isopropyl. It was isolated as the structurally characterized $Tp^{Cbu}Tl$, and, like Tp^{Cpr} was converted to an octahedral $[Tp^{Cbu}]_2Co$ complex without rearrangement, meaning that six cyclobutyl groups could be accomodated in the equatorial belt of the molecule. Such structure persisted also in solution, as evidenced by the sharp NMR of this paramagnetic complex, which showed only one type of pz^{Cbu} present. Like Tp^{iPr}, the Tp^{Cbu} ligand formed a tetrahedral $Tp^{Cbu}CoCl$ complex, which could be converted to the heteroleptic, octahedral $Tp^{Cbu}CoTp$.[83]

3.2.2.7 Tp^{Np} and Tp^{Np*} (Np = Neopentyl)

The presence of a methylene spacer between the C3 and the *tert*-butyl group, as in the case in the Tp^{Np} ligand, permitted high flexibility of this bulky substituent via rotation of the 3C-CH$_2$ bond, so that the *tert*-butyl group could either crowd in on the coordinated metal, or be far removed from it. During the synthesis of this ligand a small amount of Tp^{Np*} was also obtained, the major product being Tp^{Np}. Structurally characterized octahedral complexes, Tp^{Np}_2M were formed, but on heating they rearranged to Tp^{Np*}_2M. Tetrahedral species $Tp^{Np}MX$ were readily synthesized

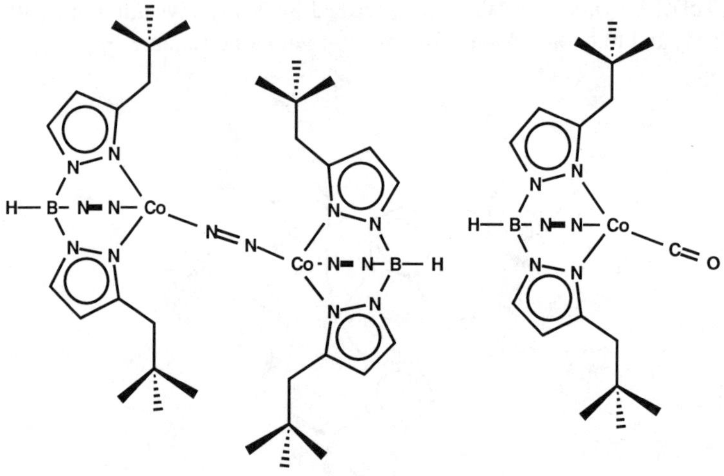

160   **161**

without rearrangement of the ligand.[80] They proved to be very useful for the preparation of heteroleptic Co^{II} complexes for NMR studies. A dinuclear complex $[Tp^{Np}Co]_2(\mu\text{-}N_2)$, **160**, was prepared by reducing $Tp^{Np}CoI$ with magnesium in a

3.2 LIGANDS Tp^R

nitrogen atmosphere, and its structure was determined by X-ray crystallography. The end-on bridging N_2 molecule was bent away from the B-Co axis by about 36°. The nitrogen complex was readily converted to Tp^{Np}CoCO, **161**, in which the CO ligand was also bent away from the three-fold axis by about 27°. On the basis of EHT and density functional theory (DHT) analysis, such bent structures should be favored in d^8 Tp^{x}CoL complexes.[1253]

3.2.2.8   Tp^{Cpe} and pz°Tp^{Cpe} (Cpe = Cyclopentyl)

When 3-R was cyclopentyl, the Tp^{Cpe} ligand was too hindered to form [Tp^{Cpe}]$_2$M complexes, but readily yielded Tp^{Cpe}MX species, which could be converted to octahedral, heteroleptic compounds, Tp^{Cpe}CoTp^x, with unhindered Tp^x ligands. The structures of Tp^{Cpe}Tl and of Tp^{Cpe}CoTp were established by NMR and by X-ray crystallography.[83] The tetrakis-ligand, pz°Tp^{Cpe}, resembled pz°Tp^{iPr}, in forming a purple tetrahedral [pz°Tp^{Cpe}]$_2$Co complex, and similar tetrahedral complexes with other first row divalent transition metals, as well as with zinc.[83]

3.2.2.9   Tp^{Cy} and pz°Tp^{Cy} (Cy = Cyclohexyl)

The coordination chemistry of the Tp^{Cy} ligand resembled that of Tp^{iPr}, but this ligand had a much more sizeable hydrophobic pocket. It formed tetrahedral Tp^{Cy}MX species, which could be converted to the heteroleptic octahedral Tp^{Cy}MTp^x complexes with relatively unhindered Tp^x ligands.[83] A rare example of a complex of a Tp^x free acid, [Tp^{Cy}H][CuCl$_2$], has beeen structurally characterized. It contained a hydrogen bond between the protonated pz^{Cy} arm and one of the chlorine atoms, and it was converted by sodium bicarbonate to the dinuclear complex [Tp^{Cy}Cu]$_2$(μ-CO$_3$), **162**, in which the carbonate ion was coordinated in bidentate fashion to each copper

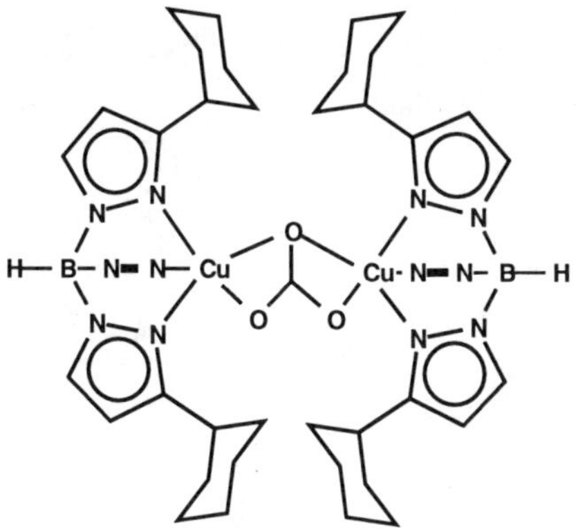

162

atom.[84] The pz°Tp^{Cy} ligand formed a series of tetrahedral [pz°Tp^{Cy}]$_2$M complexes, containing κ^2 ligands, with no exchange between the coordinated and free pz^{Cy} arms, as was indicated by NMR.[83] The pz°Tp^{Cy} ligand was also converted to the κ^3 complex pz°Tp^{Cy}Mo(CO)$_2$NO, the structure of which showed all cyclohexyl substituents to be in the 3-positions of the pyrazolyl ring.[1254]

3.2.2.10   Tp^{Trip} (Trip = Tripticyl)

This ligand, containing a 3-triptycyl substituent, has been reported as the Tl salt, but nothing has been published about its coordination chemistry.[97]

3.2.2.11   Tp^{Menth} and Tp^{Mementh} (162 and 163)

In an effort to prepare Tp^x ligands containing a chiral cavity, it was possible to synthesize the optically active ligands hydrotris(7(R)-isopropyl-4-(R)-methyl-4,5,6,7-tetrahydroindazolyl)borate, **163**, abbreviated as Tp^{Menth},[1255] and also the related hydrotris(7(S)-*tert*-butyl-4(R)methyl-4,5,6,7-tetrahydroindazolyl)borate, **164**, (= Tp^{Mementh}),[1256] both ligands being isolated as the thallium(I) salts. Their NMR spectra were assigned using COSY and HETCOR methods, and their absolute configurations and steric properties were established by X-ray crystallography. In terms of coordination chemistry, these ligands were intermediate between Tp^{iPr} and Tp^{tBu}. A large number of complexes was prepared from **163**. They included Tp^{Menth}MCl (M = Mn, Fe, Co, Ni, Cu, Zn), Tp^{Menth}M(NO$_3$) (M = Cu, Ni), and the corresponding acetates. The nitrate and acetate ions were coordinated in bidentate fashion. An equilibrium between four- and five-coordination was observed in Tp^{Menth}Rh(CO)$_2$. Interestingly, the Tp^{Menth} ligand was isomerized to Tp^{Menth*} during the reaction with TiCl$_4$ in THF, to produce Tp^{Menth*}TiCl$_3$. The same reaction done

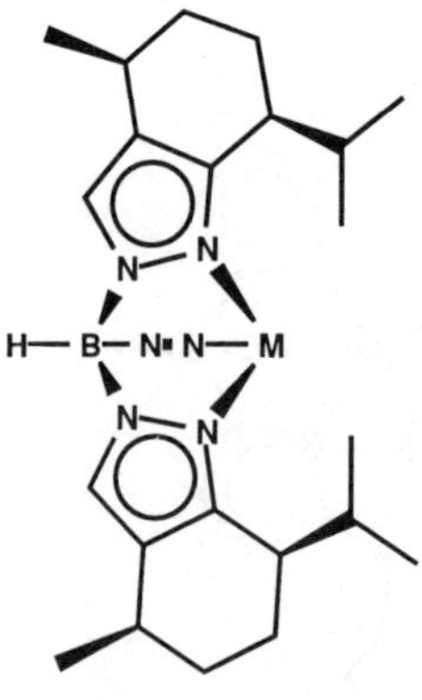

163 (Tp^{Menth}) 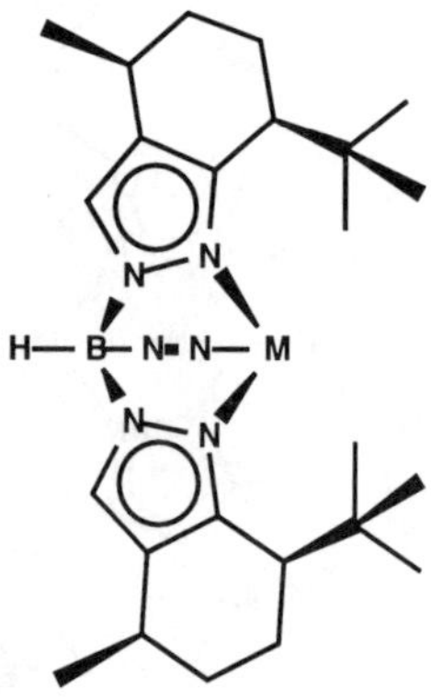   **164** (Tp^{Mementh})

3.2 LIGANDS Tp^R

in methylene chloride yielded about 90% of the unrearranged Tp^{Menth}TiCl$_3$. Structures of the complexes Tp^{Menth}ZnCl, Tp^{Mementh}ZnCl, Tp^{Menth}Ni(OAc) and Tp^{Menth*}TiCl$_3$ were determined by X-ray crystallography.[98]

3.2.2.13   pz^{o}Tp^{Camph} (166)

Also related to the optically active **163**, **164** and **165** ligands, was the ligand tetra123, methyl-2H-4,7-methanoindazolyl)borate, structure **166**, which was abbreviated as pz^{o}Tp^{Camph}, and was derived from camphorpyrazole. The complex pz^{o}Tp^{Camph}CuCO was prepared and characterized. This complex was used as a catalyst for the cyclopropanation of styrene with ethyl diazoacatate, yielding a mixture of *cis/trans* isomers, with an enantiomeric excess of the *trans* isomer.[1258]

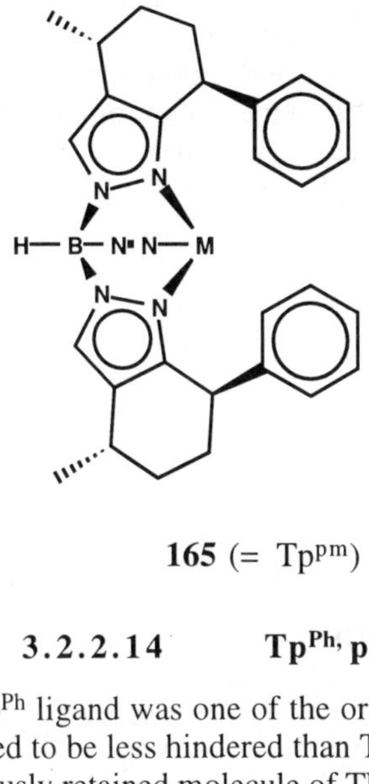

165 (= Tp^{pm})   **166** (= pz^{o}Tp^{Camph})

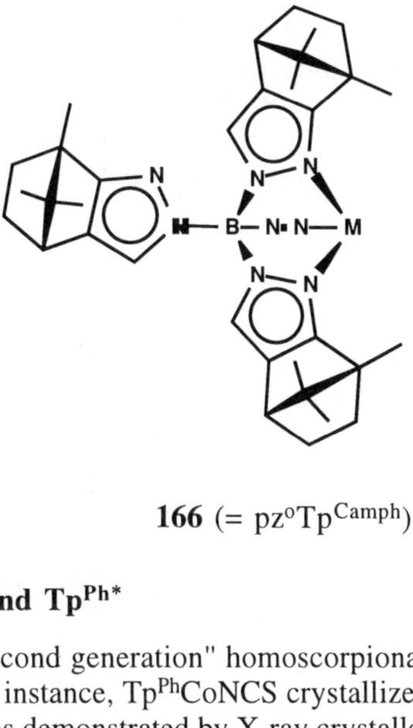

3.2.2.14   Tp^{Ph}, pz^{o}Tp^{Ph} and Tp^{Ph*}

The Tp^{Ph} ligand was one of the original "second generation" homoscorpionates, and it proved to be less hindered than Tp^{tBu}. For instance, Tp^{Ph}CoNCS crystallized with a tenaciously retained molecule of THF, as was demonstrated by X-ray crystallography, whereas Tp^{tBu}CoNCS was strictly unsolvated. Nonetheless, unsolvated Tp^{Ph}MX species could also be prepared (M = Co, Ni, Zn; X = NCO, NCS, N$_3$).[2,3] Moreover, unrearranged Tp^{Ph}$_2$M complexes (M = Fe, Mn) were synthesized, structurally characterized, and studied by electrochemistry.[1259] By contrast, the Tp^{Ph}$_2$Co complex contained one tridentate, and one bidentate Tp^{Ph} ligand, with the sixth coordination site occupied by an agostic B—H—Co bond.[1200] A still different structure was adopted by Tp^{Ph}$_2$Zn, which was tetrahedral, with κ^2 Tp^{Ph} ligands.[1260] Tetrahedral Tp^{Ph}CoX complexes with X = Cl, Br and NCS, and heteroleptic ones of structure Tp^{Ph}M(carboxylato) for M = Co and Zn, were also reported.[1200,1260] The rearranged

Tp^{Ph*} ligand was documented in the structure of the cation [Tp$^{Ph*}_2$Al]$^+$, which was obtained as the [AlCl$_4$]$^-$ salt, by the reaction of regiochemically pure Tp^{Ph}K with AlCl$_3$.[85]

Numerous complexes of the [Tp^{Ph}]$^-$ ligand, such as Tp^{Ph}Mo(CO)$_2$NO, Tp^{Ph}Mo(CO)$_2$(η^3-allyl), Tp^{Ph}Mo(CO)$_2$(η^3-methallyl), Tp^{Ph}Pd(η^3-methallyl) and Tp^{Ph}Pd(η^3-phenallyl) have also been reported,[3] as were the compounds [Et$_4$N][Tp^{Ph}W(CO)$_3$], [Tp^{Ph}W(CO)$_3$]H,[680] as well as the carbyne complex Tp^{Ph}W($\equiv$CTol)(CO)$_2$.[693] Various rhodium species, exemplified by Tp^{Ph}Rh(COD), Tp^{Ph}Rh(NBD) and Tp^{Ph}Rh(COD)$_2$ were studied by NMR, and were structurally characterized.[1261] Among copper complexes, Tp^{Ph}Cu(AcAc) and Tp^{Ph}CuX(Hpz^{Ph}) were reported,[1262] as was Tp^{Ph}CuCO,[1053] and the structurally characterized Tp^{Ph}Cu(pterin), synthesized in an approach towards a model for the metal site in phenylalanine hydroxylase.[1263] The structure of the cationic copper complex **167** (R = Ph) was established by X-ray crystallography, and it was studied by UV/VIS, by EPR, and by cyclic voltammetry.[90] Stereoregular copolymerization of ethylene with carbon monoxide was achieved with the structurally characterized complex Tp^{Ph}Ni(PPh$_3$)(o-Tol), which contained a bidentate Tp^{Ph} ligand.[1264]

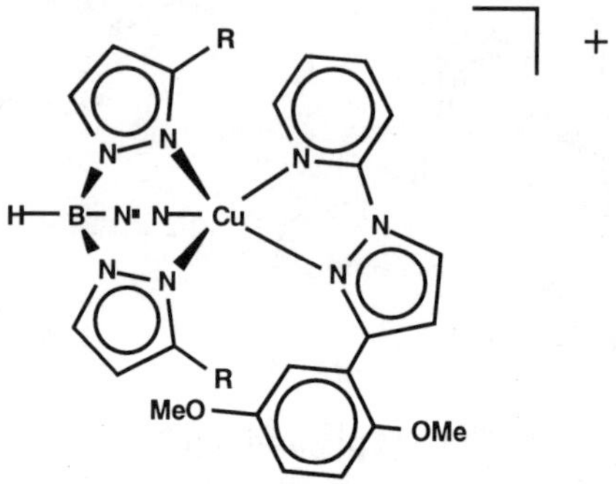

167

The Tp^{Ph} ligand was used extensively in Zn and Cd chemistry, directed in part at the preparation of catalysts, and of biological enzyme models. Thus, the synthesis of Tp^{Ph}ZnX (X = halide, NO$_3$, SR),[1240,1265,1266] was reported, and Tp^{Ph}ZnBr was structurally characterized.[1267] The two carboxylato complexes Tp^{Ph}Zn(OOCCH$_2$CN),[1268] and Tp^{Ph}Zn(OAc),[1246,1269] were studied as catalysts for the decarboxylation of cyanoacetic acid, and for malonate decarboxylation, respectively. However, the reaction of KTp^{Ph} with Zn(ClO$_4$)$_2$ led to ligand degradation.[1245] The homoleptic complex Tp$^{Ph}_2$Cd, unlike its zinc analog, was octahedral, with the phenyl rings interpenetrating between the pyrazolyl planes, as

3.2 LIGANDS Tp^R

was the heteroleptic Tp^{Ph}Cd[pzTp]. Also reported were Tp^{Ph}CdBp*, and related heteroleptic complexes,[1270] the structurally characterized Tp^{Ph}CdBp,[1092] and the mixed complexes {Tp^{Ph}Cd[HC(pz*)$_3$]}[BF$_4$] and {Tp^{Ph}Cd[HC(pz^{Ph})$_3$]}[BF$_4$].[299] The compound Tp^{Ph}Cd(OAc) was converted to a variety of cyclic ether or thioether derivatives, Tp^{Ph}Cd(OAc)L, where L was THF, dioxane, propylene oxide, cyclohexene oxide, and propylene sulfide, several of which were structurally characterized by means of X-ray crystallography. The ligand L dissociated in solution, generating a five-coordinate species, which served as a model for the initiation step in the copolymerization of epoxides with carbon dioxide, catalyzed by metal carboxylates.[1271,1272] The reaction of [Tp^{Ph}]$^-$ with indium iodide produced the air-stable, monomeric Tp^{Ph}In, the structure of which was determined by X-ray crystallography.[1273] Lanthanide complexes Tp^{Ph}$_2$M (M = Sm and Yb) have also been prepared.[1493]

The tetrakis-ligand, [pz^{o}Tp^{Ph}]Na was synthesized, characterized as the Tl salt, and converted to complexes [pz^{o}Tp^{Ph}]Mo(CO)$_2$NO, [pz^{o}Tp^{Ph}]Mo(CO)$_2$(η^3-allyl) and [pz^{o}Tp^{Ph}]Pd(η^3-methallyl).[3]

3.2.2.15   Tp^{Tol} and pz^{o}Tp^{Tol} (Tol = p-Tolyl)

The ligand Tp^{Tol}, and other *para*-substituted Tp^{Ph} ligands, behaved similarly to Tp^{Ph} in their coordination chemistry, although they protruded a bit further beyond the metal and, depending on their substituents, could increase or decrease the electron density of the Tp^x ligand. Tp^{Tol} was synthesized by the method used for preparing Tp^{Ph},[86,470] and it was used to prepare complexes of the type Tp^{Tol}MX, (M = Co, Ni, Zn, and X = NCS, NCO and N$_3$).[86] The tetrakis ligand, pz^{o}Tp^{Tol} was also reported.[470] Both, Tp^{Tol} and pz^{o}Tp^{Tol} were converted to the anions [Tp^{x}Mo(CO)$_3$]$^-$ and their reaction with various aryldiazonium cations was studied with respect to the formation of the three possible products: Tp^{x}Mo(CO)$_2$(N$_2$Ar), the aroyl species Tp^{x}Mo(CO)$_2$(η^2-COAr), and the chlorocarbyne Tp^{x}Mo(CO)$_2$($\equiv$CCl).[470] The structure of Tp^{Tol}Tl was found by X-ray crystallography to consist of head-to-head dimers in the crystal.[1274]

Structures of the monomeric Tp^{Tol}MgSH and Tp^{Tol}MgSeH complexes, which were obtained by treating Tp^{Tol}MgMe with H$_2$S and H$_2$Se, respectively, were also determined,[1275] as was that of the monomeric Tp^{Tol}MgSePh, which was prepared by the reaction of Tp^{Tol}MgMe with PhSeH. The same compound could also be obtained from the reaction of Tp^{Tol}MgMe with (PhSe)$_2$.[1276] Monofunctional zinc derivatives, Tp^{Tol}ZnX (X = Cl, Br, I, NO$_3$) were synthesized, as were the alkyl derivatives Tp^{Tol}ZnR (R = Me, Et, Bu^t, Ph), obtained in good yields either by the reaction of Tp^{Tol}ZnCl and LiR, or from ZnR$_2$ and Tp^{Tol}K.[1265] At the same time, and somewhat surprisingly, the reaction of KTp^{Tol} with Zn(ClO$_4$)$_2$ led to ligand degradation.[1245]

3.2.2.16   Tp^{An} and pz^{o}Tp^{An} (An = p-Anisyl)

This ligand was prepared by the melt method,[86, 1277] was purified as the Tl salt, and was used to synthesize Tp^{An}MX complexes (M = Co, Ni, Zn; X = NCO, NCS, N$_3$),[86] as well as the molybdenum derivatives Tp^{An}MoCl$_2$NO and Tp^{An}Mo(CO)$_2$NO, the latter structure being established by X-ray crystallography.[1277] Other molybdenum species prepared were the structurally characterized Tp^{An}Mo(NO)Cl(NH-2-MeC$_6$H$_4$) and Tp^{An}Mo(NO)Cl(OPh) complexes, and a substantial number of related analogs, which were studied by electrochemistry.[1278] The five-coordinate copper cation **167** (R = An) was investigated by UV/VIS, by EPR, and by cyclic voltammetry.[90] Tetrahedral zinc complexes Tp^{An}ZnX were obtained from the reaction of Tp^{An}K with ZnX$_2$ (X = Cl, Br, I, NO$_3$), while Tp^{An}ZnR (R = Me, Et, Bu^t, Ph) resulted from the reaction of Tp^{An}ZnCl with RLi, or the reaction of Tp^{An}K with ZnR$_2$.[1265] On the other hand, the reaction of Tp^{An}K with Zn(ClO$_4$)$_2$(H$_2$O)$_6$ led to hydrolysis of the ligand, and to the isolation of the structurally characterized dinuclear zinc complex, [pz^{An}Zn(μ-OH)]$_2$.[1245] The pz^{o}Tp^{An} ligand was converted to [pz^{o}Tp^{An}]Mo(CO)$_2$NO, the structure of which showed all anisyl substituents to be in the 3-position of the pyrazolyl ring, which was in line with the lack of observed rearrangement in any of the pz^{o}Tp^x ligands.[1254]

3.2.2.17   Tp^{oAn} (o-An = o-Anisyl) and Tp$^{Ph(oSMe)}$

The isomer of Tp^{An} (*para*-anisyl), Tp^{oAn}, was synthesized and structurally characterized as Tp^{oAn}Tl.[87] This ligand formed a very unusual cationic trinuclear complex with silver ion of composition {[Tp^{oAn}]$_2$Ag$_3$}$^+$, isolated as the perchlorate salt, in which the Tp^x ligand was interacting with the silver ions in a trinucleating (μ_3-η^1:η^1:η^1) bridging mode, as was determined by X-ray crystallography. Such arrangement involved "inversion" of the Tp^{oAn} ligand, so that the two B-H bonds were pointing at each other.[88] Structure of the complex [Tp$^{Ph(oSMe)}$Cu(H$_2$O)][PF$_6$] was determined by X-ray crystallography, which showed the SMe groups not be involved in coordination.[1560]

3.2.2.18   Tp$^{(4ClPh)}$

This ligand was prepared in the usual fashion and was characterized as the Tl salt. It was converted to Tp$^{(4ClPh)}$MX derivatives (M = Co, Zn; X = N$_3$, NCS, NCO), and to Tp$^{(4ClPh)}$Mo(CO)$_2$NO and Tp$^{(4ClPh)}$Mo(CO)$_2$(η^3-methallyl).[83] Its K and Tl salts were studied by ^{1}H and ^{13}C NMR spectroscopy.[92]

3.2.2.19   Tp$^{(4FPh)}$

The reported chemistry of this ligand was limited to the preparation of the copper cation **167** (R = p-fluorophenyl, see p. 124) which was studied by UV/VIS, by EPR, and by cyclic voltammetry.[90]

3.2 LIGANDS Tp^R

3.2.2.20 Tp^{Tn} (Tn = 2-thienyl)

This ligand turned out to be remarkably unhindered, as it readily formed stable octahedral Tp$^{Tn}_2$M complexes with first row transition metals, of which Tp$^{Tn}_2$Co was structurally characterized, but it failed to provide stable Tp^{Tn}MX species, except for Zn. The octahedral heteroleptic and paramagnetic complex Tp^{Tn}CoTp^{iPr,4Br} was obtained from the reaction of Tp^{Tn}Tl with Tp^{iPr,4Br}CoCl and it was characterized by NMR, while Tp$^{Tn}_2$Fe was studied by Mössbauer spectroscopy.[91] The five-coordinate copper cation **167** (R = 2-thienyl) was also prepared, and it was studied by UV/VIS, by EPR, and by cyclic voltammetry.[90] Lanthanide complexes Tp$^{Tn}_2$M (M = Sm and Yb) have also been prepared.[1493]

3.2.2.21 Tp^{Fn} (Fn = 2-furyl)

The only reported chemistry of this ligand was the preparation of the copper cation **167** (R = 2-furyl), the structure of which was established by X-ray crystallography, and which was studied by UV/VIS, by EPR, and by cyclic voltammetry.[90] It would be expected to be as unhindered as Tp^{Tn}.

3.2.2.22 Tp^{Ms} and Tp^{Ms*} (Ms = mesityl)

Among the various Tp^x ligands, Tp^{Ms} is unique in many ways. It was synthesized with the purpose of having a Tp^x ligand with the phenyl groups essentially orthogonal to the pyrazolyl planes, and having a minimal amount of rotational freedom. Tp^{Ms} has the smallest wedge angle among the Tp^x ligands. It is the only one where the "as prepared" ligand is predominantly (80%) the 3,3,5-isomer, Tp^{Ms*}. It is also the only one where the thallium salt of the asymmetric Tp^{Ms*} ligand, **168**, rearranges completely to the symmetric one, **169** above its melting point.

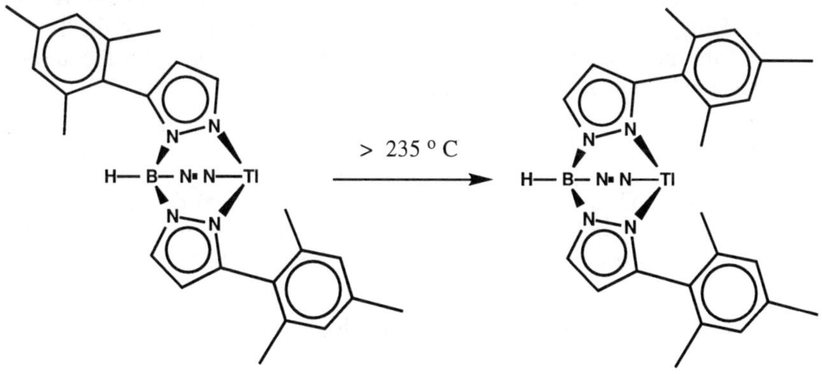

168 > 235 °C **169**

The two complexes Tp^{Ms}Mo(CO)$_2$(η^3-methallyl) and Tp^{Ms*}ZnI were structurally characterized. Tetrahedral Tp^{Ms}MX and Tp^{Ms*}MX complexes (M = Co, Zn, Cd; X = I, NCS, NCO) were readily obtainable. Also synthesized, and characterized by NMR were the complexes Tp^{Ms*}Mo(CO)$_2$(η^3-methallyl), Tp^{Ms}Rh(COD), Tp^{Ms}Rh(CO)$_2$, Tp^{Ms}Pd(η^3-methallyl), Tp^{Ms*}Rh(COD), Tp^{Ms*}Rh(CO)$_2$, and Tp^{Ms*}Pd(η^3-methallyl).[93] A surprisingly stable, monomeric copper(I) complex, Tp^{Ms}Cu(THF), was structurally characterized, and converted to the nitrosyl derivative, Tp^{Ms}CuNO, which was studied by spectroscopy, along with Tp^{Ms}CuCO. In the presence of excess NO, Tp^{Ms}CuNO disproportionated to Tp^{Ms}Cu(NO$_2$) and N$_2$O.[1279]

3.2.2.23 Tp^{Ant} and Tp^{Ant*} (Ant = 9-anthryl)

Another Tp^x ligand with a flat 3-substituent, nearly orthogonal to the pyrazolyl plane was Tp^{Ant}, isolated as Tp^{Ant}Tl. The cobalt thiocyanate derivative, Tp^{Ant}CoNCS, was also synthesized, and its structure was determined by X-ray crystallography. This revealed that about 6 % of Tp^{Ant}Tl had co-crystallized with Tp^{Ant}CoNCS. The anthryl substituents were all twisted on the average by 28° from orthogonality to the pyrazolyl plane, presumably as a result of non-bonding interactions of their terminal 2,3 and 6,7 hydrogens. The pocket, generated by the three anthryl groups, was capable of stabilizing crystallographic disorder between a vacancy, and a chain of three atoms.[89]

3.2.2.24 Tp$^{\alpha Nt}$ and Tp$^{\beta Nt}$ (Nt = naphthyl)

These two ligands, based on 3-(α-naphthyl)pyrazole and 3-(β-naphthylpyrazole), respectively,[92] were of intermediate steric hindrance between Tp^{Ph} and Tp^{Ant}. They did not form octahedral homoleptic complexes, Tp$^x{}_2$M, but yielded readily tetrahedral species, Tp^{x}MX, from which the heteroleptic complexes Tp$^{\alpha Nt}$CoTp and Tp$^{\beta Nt}$CoTp were obtained, and structurally characterized by X-ray crystallography. While the Tp$^{\alpha Nt}$ ligand enveloped the sides in this complex, making the back of the Tp ligand stick out of the pocket, the Tp$^{\beta Nt}$ ligand generated a more open on the sides, but deeper pocket, which totally engulfed the Tp ligand.[83]

3.2.2.25 Tp$^{(2,4(OMe)_2Ph)}$

The structure of the Tl salt of this ligand, Tp$^{(2,4(OMe)_2Ph)}$Tl, was determined by X-ray crystallography. The molecule was monomeric in the crystal, and there was no interaction between the *ortho*-methoxy group and the Tl atom.[83]

3.2.2.26 Tp^{Py} (Py = 2-pyridyl)

In contrast to the other 3-substituted Tp^x ligands, which could coordinate at most in tridentate fashion, the ligand Tp^{Py},[92,1280] with its additional three coordination sites was potentially hexadentate, as shown in **170**, and this hexadenticity was

3.2 LIGANDS Tp^R

demonstrated in several complexes, notably those of lanthanides and actinides, which have the best fit for this particular ligand pocket. The first reported complex where Tp^{Py} was coordinated in hexadentate fashion was the structurally characterized Tp^{Py}EuF(MeOH)$_2$.[1280] Utilizing all their nitrogen atoms, two Tp^{Py} ligands provided an icosahedral N$_{12}$ coordination environment, in the structurally characterized cations [Tp^{Py}$_2$M]$^+$ (M = Sm, U).[1281] In the complex Tp^{Py}Th(NO$_3$)$_3$ the Tp^{Py}ligand was hexadentate, and the nitrate ions were bidentate, with the molecule having C$_{3v}$ symmetry. On the other hand, in Tp^{Py}UO$_2$(OEt) the Tp^{Py} ligand was tetradentate, had

170

one arm detached, and the structure was a pentagonal bipyramid.[1282] The structures of Tp^{Py}Tl and Tp^{Py}Ag were also determined. In the former, there were three short (averaging 2.67 Å) Tl bonds to the pyrazolyl nitrogens, and three long interactions (average 3.18 Å) to the pyridyl nitrogen atoms. The structure of the cationic silver complex, [Tp^{Py}$_2$Ag$_3$]$^+$ was more complicated, consisting of a triangular cluster of Ag atoms, containing the Tp^{Py} ligands above and below the triangle, each contributing one pz^{Py} arm to each silver atom.[1283] A detailed study of Tp^{Py} coordination with trivalent lanthanide ions led to three series of complexes, two of which were cationic, and one neutral, each type being characterized by X-ray crystallography: [Tp^{Py}M(MeOH)F][PF$_6$], [Tp^{Py}M(NO$_3$)$_2$] as well as [Tp^{Py}$_2$M][BPh$_4$]. The [Tp^{Py}M(MeOH)F]$^+$ cation contained hexadentate Tp^{Py}, and eleven-coordinate lanthanide ion. [Tp^{Py}M(NO$_3$)$_2$] contained a hexadentate Tp^{Py}, while the lanthanide ion was ten-coordinate. Finally, the cation [Tp^{Py}$_2$Eu]$^+$ had the already described icosahedral geometry.[1284] In the complex [Tp^{Py}Pb(NO$_3$)](0.5Et$_2$O), the Pb^{II} ion had four strong interactions (bond lengths 2.49-2.657 Å) with the pyrazolyl nitrogen atoms and one oxygen atom from the nitrate ligand, and three weak ones (bond lengths 2.80-2.99 Å) with pyridyl nitrogens and with one nitrate oxygen atom.[87]

With first row transition metals, the structures of some Tp^{Py} complexes were more complicated. For instance, the reaction of Tp^{Py}K with Mn(OAc)$_2$ produced a tetranuclear tetracation [Tp^{Py}Mn]$^{4+}$, which was structurally characterized.[1285] With

Cu^I, the Tp^{Py} ligand afforded the complex [Tp$^{Py}_2$Cu$_3$][PF$_6$] with an intricate structure, including a triangle of copper atoms, as was determined by X-ray crystallography. This complex was studied by EPR and by electrochemistry.[1286] In the mononuclear [Tp^{Py}Cu(H$_2$O)][PF$_6$], the structure revealed a square-pyramidal copper, coordinated to two arms of the Tp^{Py} ligand and to an axial water molecule, while the uncoordinated arms interacted in chelating fashion with a proton from the axial water molecule.[1287]

3.2.2.27   Tp$^{(Py6Me)}$ [Py6Me = 2-(6-methyl)pyridyl]

Related to Tp^{Py} was the ligand Tp$^{(Py6Me)}$, **171**, differing from the former by containing additional 6-Me pyridine substituents. These substituents precluded the Tp$^{(Py\ 6Me)}$ ligand from coordinating in hexadentate fashion, and lanthanide complexes (Eu, Tb or Gd) showed tetra- and penta-coordination instead, with one or two pyridyl groups remaining detached. On the other hand, these pyridyl groups formed strong hydrogen bonds to coordinated water, as was shown by structure determinations for several such complexes. Luminescence properties of such complexes were also investigated.[1433] Structure of the cationic complex [Tp$^{(Py6Me)}$Cu(H$_2$O)][PF$_6$] was established by X-ray crystallography.[1560]

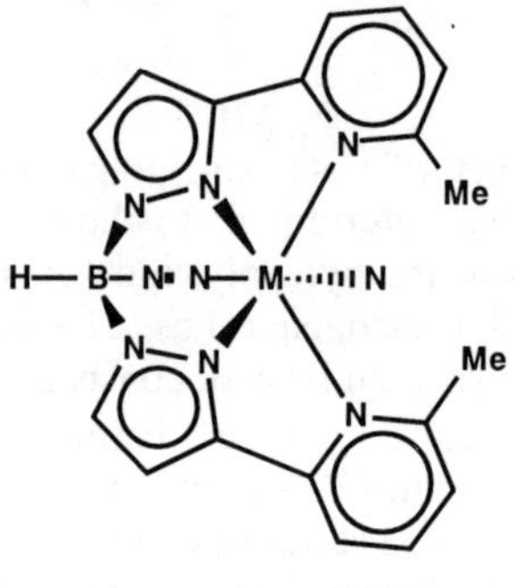

171

3.2.2.28   Tp^{CHPh_2}

This ligand, **172**, resembled Tp^{iPr}, except that instead of two methyls, the carbon attached to the 3-position contained two phenyl groups. It was prepared regiochemically pure, and gave no evidence of rearrangement. Structures of the complexes [Tp^{CHPh_2}]CoCl and [Tp^{CHPh_2}]Tl of this ligand were determined by X-ray crystallography. The six phenyl groups were not symmetrically disposed around the Co ion in the crystal, but in solution they were all NMR-equivalent. Despite the presence of six phenyl groups around the metal ion, [Tp^{CHPh_2}]CoCl reacted rapidly

3.2 LIGANDS Tp^R

with TpTl, forming the heteroleptic octahedral complex, Tp^{CHPh_2}CoTp, in which the Tp^{CHPh_2} ligand was unrearranged, as was confirmed by NMR.[83]

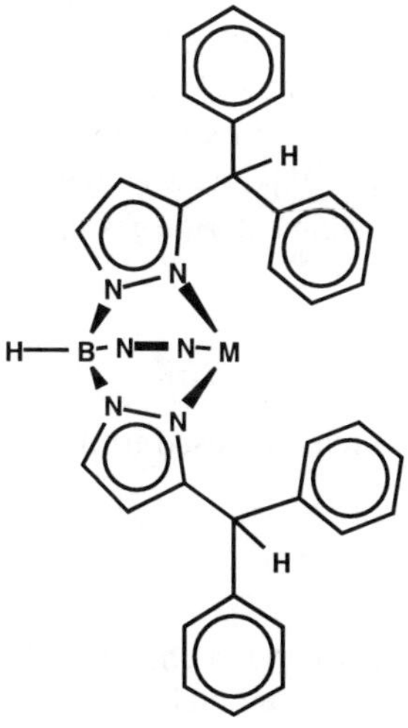

172 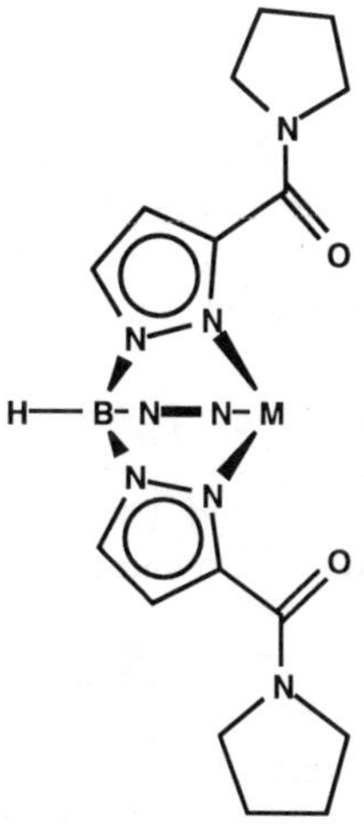   **173**

3.2.2.29   Tp$^{CO(NC_4H_8)}$

The ligand hydrotris(3-pyrrolididopyrazol-1-yl)borate, **173**, was synthesized and converted to the Tl salt, the NMR of which showed no rotation around the N-CO bond. The structure of [Tp$^{CO(NC_4H_8)}$]Tl was determined by X-ray crystallography, which showed all the polar oxygen atoms to be directed at the Tl ion, although they were not within bonding range. Unlike most Tp^{x}Tl salts, [Tp$^{CO(NC_4H_8)}$]Tl was soluble in methanol, due to the presence of the polar pyrrolidido substituents.[83]

3.2.3   4-Monosubstituted Ligands, Tp^{4R}

The 4-substituent on the pyrazolyl ring is remote from both, the coordinated metal, and from boron, being thus of little steric consequence. However, such 4-substituents may influence the electron density of the ligand through electron donation or withdrawal, although such effects have not been studied in any systematic way.

3.2.3.1   Tp^{4Me} and pz^{o}Tp^{4Me}

The first synthesis of this ligand was reported in 1996, and it was converted to a variety of Sn^{IV} derivatives of general structure Tp^{4Me}SnCl$_n$R$_{3-n}$, which were studied by Mössbauer spectroscopy. Two of these tin complexes, $Tp^{4Me}SnCl_3$, and $Tp^{4Me}SnCl_2Ph$, were structurally characterized.[99] A similar series of tin(IV) complexes was also synthesized, and studied by Mössbauer spectroscopy, employing the tetrakis-ligand, $[pz^oTp^{4Me}]^-$. The complex $[pz^oTp^{4Me}]SnCl_3$ was structurally characterized.[1288] An extended series of octahedral complexes $[Tp^{4Me}]_2M$ and $[pz^oTp^{4Me}]_2M$ (M = Mn, Fe, Co, Ni, Cu, Zn, Cd) was prepared, and studied by spectroscopy, while the structure of $[Tp^{4Me}]_2Zn$ was determined by X-ray crystallography.[1289] It was found that a reversible chlorine gas uptake and release can be achieved with the solid complex $[Tp^{4Me}]_2Ru$.[886]

3.2.3.2 Tp^{4iPr}

This is the oldest reported Tp^{4R} ligand, which was used to prepare the octahedral $Tp^{4iPr}{}_2M$ complexes of Co and Ni.[11]

3.2.3.3 Tp^{4tBu}

The Tp^{4tBu} ligand behaves in terms of its coordination chemistry like the parent Tp, however, its derivatives have higher crystallinity and melting point.[83]

3.2.3.4 Tp^{4Cl}

Octahedral complexes of this ligand, $[Tp^{4Cl}]_2M$, were synthesized for M = Mn, Fe, Co, Ni and Cu. The first four displayed unusually high thermal stability, decomposing above 430 °C, and even the copper complex was thermally stable up to 300 °C.[11] The complex $Tp^{4Cl}Mo(CO)_2(\eta^3$-allyl) has also been synthesized.[4]

3.2.3.5 Tp^{4Br}

The first reported complexes of this ligand were $[Tp^{4Br}]Mo(CO)_2(\eta^3$-allyl) and $[Tp^{4Br}]Mo(CO)_2(\eta^3$-methallyl).[17] Also reported were the octahedral $[Tp^{4Br}]_2M$ compounds (M = Mn, Fe, Co, Ni, Cu, Zn, Cd, and Pb), which were studied by spectroscopy, including ^{113}Cd NMR, and the structure of the octahedral $[Tp^{4Br}]_2Cd$ complex was established by X-ray crystallography.[100] The tetrahedral complexes $[Tp^{4Br}]Ag(PR_3)$ (R = Ph or benzyl) have also been prepared, and the former was structurally characterized.[1079] Organomercury complexes of general structure $[Tp^{4Br}]HgR$ (R = Me, Et, Pr, iPr, Ph and CN) were synthesized, and studied by NMR.[1290]

3.2.4   5-Monosubstituted Ligands, Tp^{5R}

As of now, there are no examples of 5-monosubstituted Tp^x ligands, as asymmetric pyrazoles give rise to 3-substituted Tp^x ligands or, rarely, to 3,3,5-substituted ones,

3.2 DISUBSTITUTED LIGANDS

Tp^{x*}. Only one example of rearrangement to a 5,5,3-ligand was reported for the complex [HB(5-Mepz)$_2$(3Mepz)]Ir(COD), arising from the original Tp^{Me}Ir(COD).^{979}

3.2.5 C-Disubstituted Ligands

All three types of C-disubstituted homoscorpionates are known. By far, the most commonly encountered ones are those with 3- and 5-substituents, Tp$^{R,R'}$, including ligands with either identical 3 and 5-substituents (R = R'), abbreviated as Tp^{R2}, and also those where R is different from R'. Although the key feature of the ligand is still the 3-R substituent, which defines the steric environment of the coordinated metal, the presence of a 5-substituent does provide some steric protection at the B—H end of the ligand. Furthermore, in the ligands Tp^{R2}, where the 3- and 5-substituents are identical, there is no problem with ligand rearrangement since, even if a borotropic or metallotropic rearrangement were to take place, it would be degenerate, and the rearranged ligand would not be different from the original one, and still contain the same 3- and 5-substituents. As a rule, when the two substituents differ in size, the larger one winds up exlusively (or predominantly, if the size difference is only modest) in the 3-position, while the smaller one becomes the 5-substituent. Ligands Tp^{tBu,Me}, Tp^{tBu,Tn} and even Tp^{tBu,iPr} are good examples of this generalization, being obtained as single isomers.

The 3,4-disubstituted Tp^x ligands are of two types: those derived from a 3-substituted pyrazole, which was additionally substituted in the 4-position either by halogenation, which is very facile with pyrazoles, or by the introduction of an alkyl group (usually *tert*-butyl), which is more difficult. The other type contains a fused benzo-, or naphtho-ring at the 3,4-positions of the pyrazole, that is, the Tp^x ligands are hydrotris(indazol-2-yl)borates. The known examples are limited to ligands derived from indazoles which contain either a 7-alkyl substituent, or a 6,7-fused benzo ring (plus additional substitutents elsewhere).

There is a fair number of examples of 4,5-disubstituted Tp^x ligands. They all are derived from indazole itself, and from substituted indazoles containing any combination of 3-, 4-, 5-, and 6- alkyl or aryl substituents, but excluding those with a 7-substituent. The reason for this "abnormal" regiochemistry of indazolylborates seems to lie in the higher electron density on the 2-N in indazole, which overrides the small steric effect of the C-H from the fused benzo ring, and facilitates the 2-N bonding to boron. Only when a 7-substituent is present, does the steric effect overcome the electronic one. This sub-class of homoscorpionate ligands provides a unique example, where the Tp^x ligand contains a protective pocket around boron, but at the same time the access to the coordinated metal is as unhindered as with the parent Tp.

3.2.5.1   3,4-Disubstituted Ligands

3.2.5.1.1 $Tp^{iPr,4Br}$

This particular variant of Tp^{iPr} was a convenient ligand, because its derivatives were easier to crystallize, as compared with Tp^{iPr}. Unlike pz^oTp^{iPr}, the $pz^oTp^{iPr,4Br}$ analog remains unknown.[5] Tetrahedral $Tp^{iPr,4Br}MX$ complexes, such as the structurally characterized $Tp^{iPr,4Br}CoCl$,[1291] were very convenient sources for the preparation of numerous heteroleptic $Tp^{iPr,4Br}MTp^x$ complexes. These were exemplified by the structurally characterized $Tp^{iPr,4Br}CoTp$, $Tp^{iPr,4Br}CoBp^{Ph}$, $Tp^{iPr,4Br}CoTp^{Ph}$ with a κ^2 Tp^{Ph} ligand plus an agostic B—H—Co bond, $Tp^{iPr,4Br}Co[Ph_2Bp]$ containing a five-coordinate cobalt,[1292] and by $Tp^{iPr,4Br}NiTp*$.[1293] The dinuclear nickel complex $[Tp^{iPr,4Br}NiNCS]_2$ had five-coordinate nickel, bridged by two NCS ligands (see **142**, p. 104).[86] On the other hand, the analogous cobalt species was monomeric.[5]

A stable hydrido bis(dihydrogen) complex $Tp^{iPr,4Br}RuH(H_2)_2$, prepared via hydrogenation of $Tp^{iPr,4Br}RuH(CO)$, has also been reported in conjunction with a study of ruthenium polyhydrogen complexes,[866,868] and the imidotitanium complex, $Tp^{iPr,Br}Ti(=NBu^t)Cl(py-4-Bu^t)$, has also been reported.[350] The three rhodium derivatives, $Tp^{iPr,4Br}Rh(CO)_2$ $Tp^{iPr,4Br}Rh(COD)$ and $Tp^{iPr,4Br}Rh(NBD)$, were synthesized as part of a κ^2-κ^3 isomerism study in such species.[114] Rather interesting heteroleptic dinuclear complexes of the type $Tp^{iPr,4Br}Ni[(\mu-pz)_3IrCp*]$, $Tp^{iPr,4Br}Ni[(\mu-pz)_3Rh(\eta^6-p\text{-cymene})]$ and $Tp^{iPr,4Br}Co[(\mu-pz)_2(\mu-Cl)IrCp*]$ were synthesized and several of these structures were determined by X-ray crystallography. These complexes contained what may be regarded as metal-based scorpionate analog ligands, $[ArM(pz)_3]^-$.[305] The five-coordinate heteroleptic complex, **174**, was synthesized and structurally characterized.[1294]

174

3.2 3,4-DISUBSTITUTED LIGANDS

3.2.5.1.2   Tp^{tBu,4Br}

This derivative of Tp^{tBu} had very similar coordination chemistry to that of Tp^{tBu}, but a comparison of their Tl salts showed it to have somewhat smaller ^{205}Tl-^{13}C coupling, suggestive of a looser structure than in Tp^{tBu}.[83]

3.2.5.1.3   Tp^{Cy,4Br}

This analog of Tp^{Cy} formed readily Tp^{Cy,4Br}MX derivatives, and these could be converted to heteroleptic paramagnetic Tp^{Cy,4Br}CoTp^x complexes, which had well-defined, confirmatory NMR spectra. The structure of Tp^{Cy,4Br}Tl was determined by X-ray crystallography.[83]

3.2.5.1.4   Tp^{iPr,4tBu}

The interesting feature of this ligand was that the t-butyl group in the 4-position prevented the 3-isopropyl group from rotating its methyl groups away from the coordinated metal, making Tp^{iPr,4tBu,} in effect, the equivalent of Tp^{tBu}, but probably without free rotation. The structure of its tetrahedral cobalt(II) complex, [Tp^{iPr,4tBu}]CoNCS, was determined by X-ray crystallography.[83]

3.2.5.1.5   Tp^a (175)

The ligand Tp^a, with the systematic name (Hydrotris(2H-benz[g]-4,5-dihydroindazol-2-ylborate), **175**, was prepared by tethering the 3-phenyl group with a -CH$_2$CH$_2$-

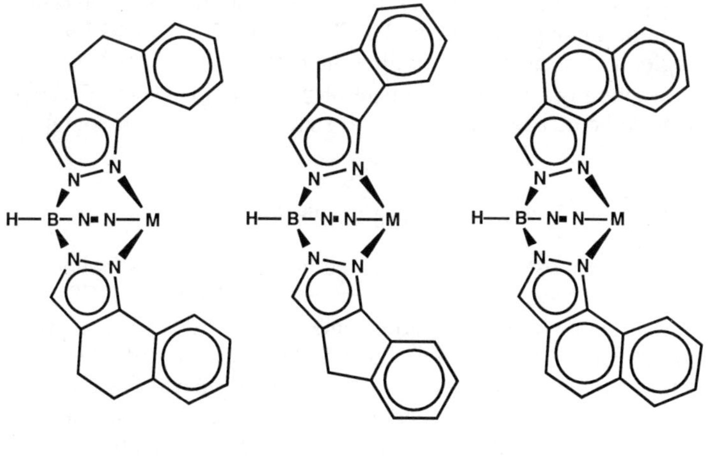

175   **176**   **177**

link to the 4-position of pyrazole. Octahedral $Tp^a{}_2M$ (M = Co, Fe, Zn) complexes, the heteroleptic Tp^aCoTp^{Np}, plus rhodium complexes $Tp^aRh(COD)$, and the structurally characterized $Tp^aRh(CO)_2$ were synthesized.[104]

3.2.5.1.6 Tp^b (176)

This ligand, hydrotris(1,4-dihydroindeno[1,2-c]pyrazol-1-yl)borate, **176**, had the phenyl groups coplanar with the pyrazolyl plane. However, the short methylene tethers pulled the phenyl groups away from the metal, making the ligand sterically unhindered and forming unrearranged homoleptic and heteroleptic complexes.[104]

3.2.5.1.7 Tp^{a*} (Hydrotris(2H-benz[g]indazol-2-yl)borate)

The ligand Tp^{a*}, **177**, was similar to **175**, but the change from an ethylene to ethyne tether resulted in coplanarity of all three rings. Octahedral $Tp^{a*}{}_2M$, and heteroleptic $Tp^{a*}MTp^{Np}$ complexes were formed easily, but the large cone angle led to a κ^2 structure in $Tp^{a*}Mo(CO)_2(\eta^3$-methallyl), along with an agostic B—H—Mo bond. A κ^3 structure would have entailed prohibitive non-bonding interactions of the third Tp^{a*} ligand arm with the η^3-methallyl ligand.[101,103] Complexes $Tp^{a*}CoTp^{Np}$, $Tp^{a*}Rh(COD)$ and $Tp^{a*}Rh(CO)_2$ were also reported, and the first one was characterized by NMR.

3.2.5.1.8 $Tp^{3Bo,7Me}$

This was one of the two reported indazole-based homoscorpionate ligands, containing boron bonded to the less-hindered nitrogen atom. It is related to structure **178**. $Tp^{3Bo,7Me}$ was characterized as the Tl salt, which was found to have the highest ^{205}Tl-^{13}C coupling among the known Tp^xTl compounds, J = 416 Hz. Despite its large cone angle, it was capable of forming the octahedral $[Tp^{3Bo,7Me}]_2Co$ complex, and also the structurally characterized heteroleptic one, $[Tp^{3Bo,7Me}]CoTp^{Np}$, which was determined to be octahedral in the solid state by X-ray crystallography, and in solution by NMR.[101]

3.2.5.1.9 $Tp^{3Bo,7tBu}$ and $Tp^{(3Bo,7tBu)*}$

This was the most sterically hindered homoscorpionate ligand prepared so far, as can be seen from structure **178**. $Tp^{3Bo,7tBu}$ reacted with $Co(NCS)_2$ only under forcing conditions, and with ligand rearrangement to $Tp^{3Bo,7tBu*}$, which produced the structurally characterized $[Tp^{(3Bo,7tBu)*}]CoNCS$. The cobalt-based ligand cone angle derived from that complex (calculated from the unrearranged part of the ligand) was an awesome 315°.[101,102] As was mentioned before, indazoles containing substituents in positions other than 7, produce ligands of the type **179**, in which boron is bonded to the more hindered nitrogen.

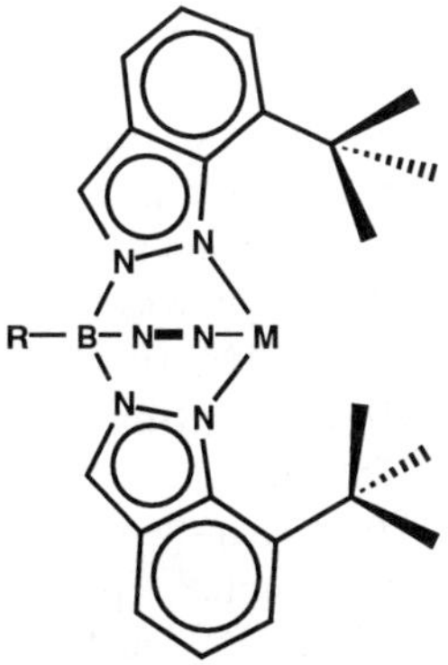

3.2 DISUBSTITUTED LIGANDS Tp^{R2}

178 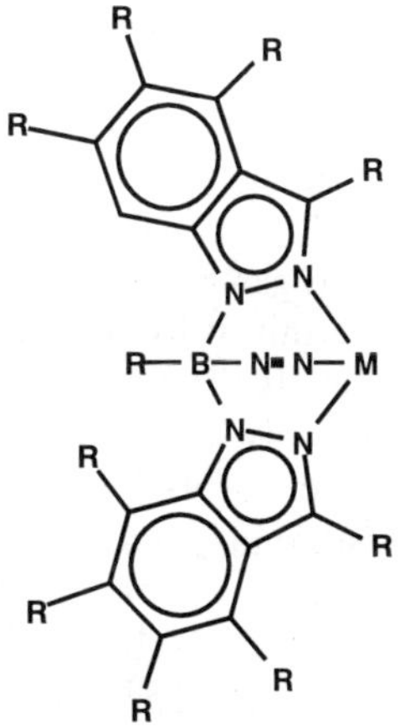 **179**

3.2.5.1.10 $Tp^{Bn,4Ph}$

This ligand was characterized by NMR as the Tl salt, and also as the heteroleptic complex $[Tp^{Bn,4Ph}]CoTp$.[83]

3.2.5.1.11 $Tp^{Pr,4Et}$ and $Tp^{(3,4-(CH_2)_n)}$ (n = 3, 4, 6 and 10)

These very lipophilic ligands were isolated as Tl^I salts, the NMR of which indicated an isomer mixture for the $Tp^{(3,4-(CH_2)_n)}$ ligands, but not for $Tp^{Pr,4Et}$, which was structurally characterized as the octahedral $[Tp^{Pr,4Et}]_2Co$ complex.[83]

3.2.5.2a 3,5-Disubstituted Ligands of Type Tp^{R2}

3.2.5.2a.1 Tp^{Et2}

The only reported complex of this ligand was $Tp^{Et2}Mo(CO)_2NO$, obtained from nitrosation of the anion $[Tp^{Et2}Mo(CO)_3]^-$.[19]

3.2.5.2a.2 Tp^{iPr2}

The Tp^{iPr2} ligand has been used quite extensively, mainly with the first row transition metals, and with zinc, much of the work being aimed at the modelling of various metalloenzymes.[1295] Dinuclear complexes, $[Tp^{iPr2}M(\mu-OH)]_2$, have been converted to the carbonato-bridged derivatives, $[Tp^{iPr2}M]_2(\mu-CO_3)$, in which the Fe,

Co, Ni, and Cu complexes contained a bis-bidentate carbonate ligand, while in the Zn complex, the carbonate ion was bidentate to one Zn ion, and monodentate to the other.[1296] The [Tp^{iPr2}Mn(μ-OH)]$_2$ complex was oxidatively converted to the (μ-O)$_2$ moiety,[1297] but in the presence of excess H$_2$O$_2$ a monomeric side-on peroxo species, Tp^{iPr2}Mn(O$_2$)(Hpz^{iPr2}) was isolated, exhibiting thermochromism, according to the formation of a hydrogen bond between the Hpz^{iPr2} proton and the peroxo oxygen. Its structure was determined by X-ray crystallography,[1298] as was that of the complex Tp^{iPr}Mn(OOCPh)(Hpz^{iPr2}).[1299] The reaction of [Tp^{iPr2}Mn(μ-OH)]$_2$ with Hpz^{iPr2} yielded the structurally characterized [Tp^{iPr2}Mn]$_2$(μ-OH)(μ-pz^{iPr2}).[1300] Acetic acid converted [Tp^{iPr2}Mn(μ-OH)]$_2$ to [Tp^{iPr2}Mn]$_2$(μ-OH)(μ-OAc) which was oxidized, yielding two isolable species: Tp^{iPr2}Mn^{III}(μ-O)$_2$(μ-OAc)Mn^{IV}Tp^{iPr2} and [Tp^{iPr2}Mn^{III}]$_2$(μ-O)(μ-OAc).[1301] The trinuclear complex Tp^{iPr}Mn^{III}(μ-OH)(μ-OAc)$_2$Mn^{II}(μ-OH)(μ-OAc)$_2$Mn^{III}Tp^{iPr2} was also reported.[1302] The tertiary hydrogen of the 3-isopropyl group in [Tp^{iPr2}Mn(μ-OH)]$_2$ was sensitive to oxidation by the Mn-coordinated oxygen, and gave rise to the structurally characterized **180**. It was also possible to isolate a novel Zn(OAc) derivative containing the Tp^{iPr2} ligand with one

180

isopropyl group per ligand hydroxylated at the tertiary position.[1303,1304] Similar results were obtained starting with a thiolato complex, Tp^{iPr2}MnSR.[1305] The asymmetric complex, Tp^{iPr2}Mn(μ-O$_2$CPh)$_3$Mn(Hpz^{iPr2})$_2$ has also been reported.[1306]

A chemistry similar to that above has also been demonstrated with iron. Mononuclear complexes of structure Tp^{iPr2}Fe(OAr) were prepared by treating Tp^{iPr2}FeCl with [OAr]$^-$, and they were converted by oxygen to Tp^{iPr2}Fe(OAr)$_2$ species,[1307] which reacted with *m*-chloroperbenzoic acid producing catecholato derivatives through hydroxylation of the phenolate *ortho*-position.[1308] Extradiol oxygenation of such catecholato complexes was studied,[1309] and rather persuasive evidence was presented for the reversible formation of a peroxo adduct of

3.2 DISUBSTITUTED LIGANDS Tp^{R2}

$Tp^{iPr2}Fe(O_2CPh)(NCMe)$.[1310] Binuclear iron complexes of structure $[Tp^{iPr2}Fe(\mu-OH)]_2$, convertible to $[Tp^{iPr2}Fe]_2(\mu-OH)(\mu-O_2CPh)$,[1311] and the structurally characterized $[Tp^{iPr2}Fe]_2(\mu-O_2)(\mu-O_2CCH_2Ph)_2$, obtained by the reaction of oxygen with $Tp^{iPr2}Fe(O_2CR)$, were also reported.[1312] Protonolysis of $Tp^{iPr2}RuH(H_2)_2$ with trifluoromethanesulfonic acid yielded the aqua complex $[Tp^{iPr2}Ru(H_2O)_2(THF)]^+$, in which the THF could be replaced by a variety of N- and P-donor ligands.[1313]

Oxidation of $[Tp^{iPr2}Co(\mu-OH)]_2$ with H_2O_2 yielded a dinuclear $\mu-\eta^2:\eta^2$-peroxo complex, $[Tp^{iPr2}Co]_2(\mu-\eta^2:\eta^2-O_2)$, which on decomposition formed the Co analog of **90** (see p. 67).[1314] A very detailed study of this reaction, including the isolation and structural characterization of the partially oxygenated species leading, ultimately, to the fully oxygenated complex $HB[3-(CMe_2OH)-5-Pr^ipz]_3CoOH$, has been conducted.[1315] Complexes $Tp^{iPr2}MCl$ (M = Fe, Co, Ni) reacted with $CH_2=CHCH_2MgCl$, to yield the respective $Tp^{iPr2}M(allyl)$ derivatives. However, in the of Ni and Co derivatives the allyl substituent was η^3, and the structures were square-pyramidal, while with Fe it was η^1, and the structure was tetrahedral, as was that of $Tp^{iPr2}Fe(CH_2\text{-}Tol)$.[1316] The structure of the complex $Tp^{iPr2}Ru(H)(COD)$ showed a κ^2 Tp^{iPr2} ligand, in which the unattached pz^{iPr2} arm was hovering above the Ru atom, which was in a square pyramidal configuration.[122] The ligand denticity equilibria in the complex $Tp^{iPr2}Ir(COD)$ were studied by NMR.[979] Dehydrative condensation of the palladium complex $\kappa^2\text{-}Tp^{iPr2}Pd(py)(OH)$ with HOOZ yielded a series of hydroperoxo, *tert*-butylperoxo and μ-peroxo species, of which $Tp^{iPr2}Pd(py)(O_2Bu^t)$ and $[Tp^{iPr2}Pd(py)]_2(\mu-O_2)$ were structurally characterized.[1317] It was suggested on the basis of the structures and IR spectra of $Tp^{iPr2}Rh(COD)$ and $Tp^{iPr2}Rh(NBD)$ that the BH stretch frequency can be used to differentiate between κ^2 (≈ 2470 cm^{-1}) and κ^3 (> 2470 cm^{-1}) bonding of the Tp^x ligand.[1318]

The $Tp^{iPr2}Cu$ chemistry resembled that of Mn and Fe. For instance, $[Tp^{iPr2}Cu(\mu-OH)]_2$ was converted to $Tp^{iPr2}Cu(O_2CPh-3-Cl)$,[44] to $[Tp^{iPr2}Cu]_2(\mu-\eta^2:\eta^2-O_2)$,[1319] studied in detail by spectroscopy,[1320] to $Tp^{iPr2}Cu(p\text{-nitroacetanilido})$ involving N,O-chelation,[1321] to the carbonato-bridged $[Tp^{iPr2}Cu]_2(\mu-CO_3)$,[45] and to thiolato derivatives $Tp^{iPr2}CuSBu^t$,[1322] and $Tp^{iPr2}CuSC_6F_5$.[1323] Such thiolate complexes were used to model Resonance Raman spectra of the blue copper proteins.[1324] An unusual C-S bond cleavage in the thiolate complex $Tp^{iPr2}CuSCPh_3$ gave rise to the dinuclear complex $[Tp^{iPr2}Cu]_2(\mu-\eta^2:\eta^2-S_2)$, **181**.[1074] The reaction of $[Tp^{iPr2}Cu]_2(\mu-\eta^2:\eta^2-O_2)$ with CO or with PPh_3 produced the respective $Tp^{iPr2}CuL$ species, with loss of oxygen,[107,1325] while the reaction with NaN_3 yielded $[Tp^{iPr2}Cu]_2(\mu-OH)(\mu-N_3)$.[1326] The structures of most of the above complexes were determined, and their electronic states subjected to extended Hückel MOP calculations.[107] Two different alkylperoxo complexes $Tp^{iPr2}Cu(OOBu^t)$ and $Tp^{iPr2}Cu(OOCMe_2Ph)$ were prepared, and the latter one was structurally characterized.[1327] The monomeric $Tp^{iPr2}ZnOH$ was found to cleave phosphate esters, and the structure of $[Tp^{iPr2}Zn]_2(\mu-O_3PAr)$ was determined by X-ray crystallography.[1328] It also reacted rapidly with CO_2, forming the asymmetrically bridged [Tp^{iPr2}Zn]$_2$(μ-CO_3).[1329] Tp^{iPr2}ZnX complexes were compared with those of other Tp^x ligands,[1246] and the structure of Tp^{iPr2}CdI was determined by X-ray crystallography.[1247]

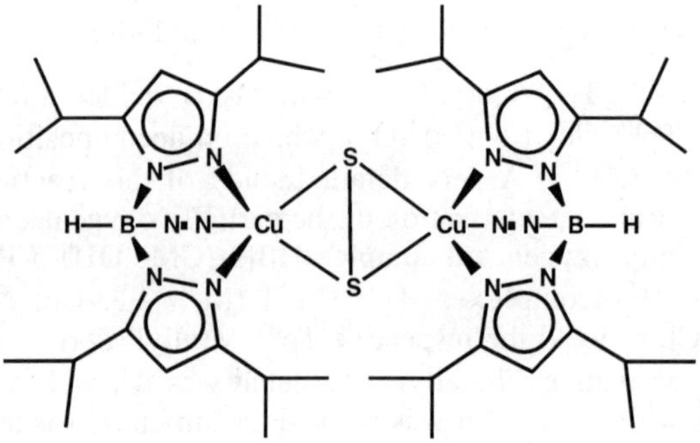

181

3.2.5.2a.3 Tp^{tBu2}

Tp^{tBu2} contained, unlike the other Tp^x ligands with 3-Bu^t substituents, a second Bu^t group in the 5-position, which introduced considerable strain around the boron atom, resulting in a twisted conformation of this ligand. In addition to the K salt, the Cs, and Tl salts, as well as the Tp^{tBu2}MI (M = Zn, Cd) complexes have been prepared, and the structures of all were determined by X-ray crystallography.[108] The reaction of [Tp^{tBu2}]$^-$ with "GaI" produced the structurally characterized monovalent gallium complex Tp^{tBu2}Ga, as well as its GaI$_3$ adduct, Tp^{tBu2}Ga—GaI$_3$.[1330] Tp^{tBu2}Ga could be oxidized with Se or Te to yield the corresponding Tp^{tBu2}Ga=E (E = Se, Te) complexes, the structures of both being established by X-ray crystallography.[1331] In a very similar reaction the sulfur analog Tp^{tBu2}Ga=S was obtained.[1332] By contrast, the reaction of Tp^{tBu2}In with sulfur yielded the tetrasulfido derivative Tp^{tBu2}In(η^2-S$_4$).[1333] The highly twisted Tp^{tBu2}In was also prepared, and in comparing its structure to that of the previously reported Tp^{tBu}In,[1249] the structure of the latter was corrected.[1334] Tp^{tBu2}In could be oxidized with Se (but not with Te) to yield Tp^{tBu2}In=Se, the first structurally characterized example of an indium complex with a valence multiple bond.[1335]

3.2.5.2a.4 Tp$^{(CF_3)2}$

Unlike the other Tp^{R2} ligands containing electron-donating substituents, the ligand Tp$^{(CF_3)2}$ contains six strongly electron-withdrawing trifluoromethyl substituents, which are sterically resembling methyl groups. It was used mainly to study Cu and

3.2 LIGANDS Tp^{R2}

Ag complexes. The Tp$^{(CF_3)2}$ ligand was synthesized in standard fashion,[94,116] and was structurally characterized as the Tl salt.[117] It formed complexes such as Tp$^{(CF_3)2}$AgPPh$_3$, Tp$^{(CF_3)2}$CuPPh$_3$, and the salt [Et$_4$N][Tp$^{(CF_3)2}$], the structures of which, and of related species, were determined by X-ray crystallography,[94] as well as the carbonyls Tp$^{(CF_3)2}$CuCO,[1336] and Tp$^{(CF_3)2}$Mn(CO)$_3$. These complexes, which were characterized by high CO stretching frequency, were studied by cyclic voltammetry.[95] The structures of Tp$^{(CF_3)2}$Cu(CNBu^t),[1337] of the potassium salt Tp$^{(CF_3)2}$K,[116] of the silver complexes Tp$^{(CF_3)2}$AgCO and Tp$^{(CF_3)2}$Ag(CNBu^t),[1339] as well as of the silver(I) and indium(I) derivatives Tp$^{(CF_3)2}$Ag(THF) and Tp$^{(CF_3)2}$In,[1340] were determined by X-ray crystallography. In a rather detailed and thorough study of Tp$^{(CF_3)2}$AgL complexes, those with L = THF, toluene, CO, H$_2$C=CH$_2$, HC≡CH, CNBu^t, and Bu^{t}CN were synthesized and structurally characterized.[1341] The gold complex Tp$^{(CF_3)2}$Au(CO) was found to have a tetrahedral structure, but Tp$^{(CF_3)2}$Au(CNBu^t) was highly asymmetric, with one short, and two long Au—N bonds.[1342]

3.2.5.2a.5 Tp^{Ph2}

This ligand was reported at the same time as Tp^{iPr2}, although it has never quite achieved the popularity of the former, being somewhat less user-friendly. Copper(I) derivatives thereof, Tp^{Ph2}CuCO and Tp^{Ph2}CuPPh$_3$ were synthesized and characterized.[1295] Treatment of the copper(I) acetone adduct Tp^{Ph2}Cu(OCMe$_2$) with dioxygen, gave rise to [Tp^{Ph2}Cu]$_2$(O$_2$),[107] which was studied by spectroscopy.[1320] The copper(I) dimer, [Tp^{Ph2}Cu]$_2$, was synthesized and its structure was determined by X-ray crystallography. There was a linear arrangement of N—Cu—N units, each Cu being coordinated to nitrogens from two different ligands, while the third pz^{Ph2} arm remained uncoordinated. By contrast, in Tp^{Ph2}Cu(Hpz^{Ph2}) the ligand was κ^3 and copper was in a tetrahedral environment. These complexes were studied by electrochemistry.[1238]

The reaction of NO with Tp^{Ph2}Cu(Hpz^{Ph2}) produced Tp^{Ph2}Cu(NO), which was viewed as a possible model of nitrite reductase,[1239] and some Tp^{Ph2}ZnX complexes have also been reported.[1265] Silver complexes stabilized by phosphines, having the general structure Tp^{Ph2}Ag(PR$_3$) (R = Ph, o-tolyl, m-tolyl, p-tolyl, benzyl or cyclohexyl) were prepared by the reaction of a mixture of KTp^{Ph2}, Ag(O$_3$SF$_3$) and PR$_3$ in tetrahydrofuran. Related complexes, Tp^{Ph2}Ag(PMePh$_2$) and Tp^{Ph2}Ag(NCR) were also synthesized, and the structures of Tp^{Ph2}Ag(PPh$_3$) and of Tp^{Ph2}Ag(PBz$_3$) were determined by X-ray crystallography, showing in each case a tetrahedrally coordinated Ag ion.[1343] Zinc derivatives Tp^{Ph2}ZnX (X = Cl, Br, I) have also been reported.[1265] Numerous Tp^{Ph2}HgR complexes (R = Me, Et, Pr, iPr, Ph, Tol, Bz, and Fc) were synthesized, and the structure of Tp^{Ph2}HgEt was found to be the first instance of the HgR moiety being tetrahedrally coordinated to a tripodal N-donor ligand.[1290]

3.2.5.2b 3,5-Disubstituted Ligands of Type Tp$^{R,R'}$

3.2.5.2b.1 Tp^{iPr,Me} and Tp^{iPr,Me*}

Unlike Tp^{iPr} which was obtained regiochemically pure, the "as prepared" ligand from KBH$_4$ and 3-isopropyl-5-methylpyrazole, was already a roughly 7:3 mixture of Tp^{iPr,Me} and Tp$^{(iPr,Me)*}$. The derived LMX complexes, and the LTl salt could be converted to pure Tp^{iPr,Me} derivatives through recrystallization, but the L$_2$M species were those of Tp$^{(iPr,Me)*}$. Molybdenum complexes LMo(CO)$_2$NO, LMoI$_2$NO and LMo(OR)$_2$NO were also prepared, and the regiochemically pure Tp^{iPr,Me}MoNO(OEt)$_2$ was structurally characterized.[106] If was found that when the L ligands in [Tp^{iPr,Me}/Tp^{iPr,Me*}]MoNO(L$_2$) were large (OPr^i, OBu^t, NHPh, NHTol) the mixed complex rearranged to pure Tp^{iPr,Me}MoNO(L$_2$).[1344] These rearrangements were studied in detail, and the structure of Tp^{iPr,Me}MoNO(OPr^i)$_2$ was also determined.[1345] A convenient method called "inverse recrystallization" was effective in converting the mixed LCoI complex to pure Tp^{iPr,Me}CoI.[1346] This cobalt complex was used in the activation of dioxygen, through its reduction in the presence of N$_2$ or CO to yield [Tp^{iPr,Me}Co]$_2$N$_2$ or Tp^{iPr,Me}CoCO, respectively. The latter could be converted to Tp^{iPr,Me}Co(O$_2$) by treatment with oxygen, while the former yielded Tp^{iPr,Me}Co-O$_2$-CoTp^{iPr,Me}. Decomposition of these complexes led to [Tp^{iPr,Me}Co]$_2$(μ-OH)$_2$, and to dehydrogenation of one isopropyl group per ligand to isopropenyl.[1347] The complex Tp^{iPr,Me}Co(O$_2$) could be converted to the dimeric, structurally characterized species Tp^{iPr,Me}Co(μ-O$_2$)$_2$CoTp^{iPr,Me}, with each Co five-coordinate, and linked via two -O–O- bridges.[36] The diamagnetic complex Tp^{iPr,Me}Co(CO)$_2$ was prepared, and structurally characterized. It was in equilibrium with CO + Tp^{iPr,Me}Co(CO), and the pertinent thermochemical parameters were determined.[1348]

3.2.5.2b.2 Tp^{tBu,Me}

This ligand was formed as a single isomer, and no rearrangement to Tp$^{(tBu,Me)*}$ has ever been observed. Its tetrahedral derivatives were somewhat more stable than those of Tp^{tBu}, and the ^{205}Tl-^{13}C coupling in Tp^{tBu,Me}Tl was larger than that of Tp^{tBu}Tl, suggesting a tighter "bite".[83] The first structurally characterized complex was Tp^{tBu,Me}Co(O$_2$) containing a side-on superoxo ligand,[35] followed by the monomeric Tp^{tBu,Me}NiNCS (the complex Tp^{iPr,4Br}NiNCS was a dimer).[86] The structure of [Tp^{tBu,Me}CoCO]$_2$Mg(THF)$_4$ has been determined,[1253] as were the structures of Tp^{tBu,Me}CuCl, of [Tp^{tBu,Me}Cu]$_2$ and of Tp^{tBu,Me}Tl. A comparison of the last three with their Tp^{tBu} analogs suggested an increase in structural rigidity due to the 5-Me substituents.[1349] The Cr^{II} derivatives, Tp^{tBu,Me}CrCl and Tp^{tBu,Me}CrCl(Hpz^{tBu,Me}) have been structurally characterized, and the former could be converted to Tp^{tBu,Me}CrR derivatives (R = Et, Ph, CH$_2$SiMe$_3$), all of which possessed the cis-divacant octahedral structure, with the R substituent deviating by about 30° from the B·····Cr

3.2 LIGANDS Tp$^{R,R'}$

characterized chromium(IV) complex, **182**, thought to have arisen through an intramolecular rearrangement of the precursor κ^2-Tp^{tBu,Me}Cr(O$_2$)Ph.[1351]

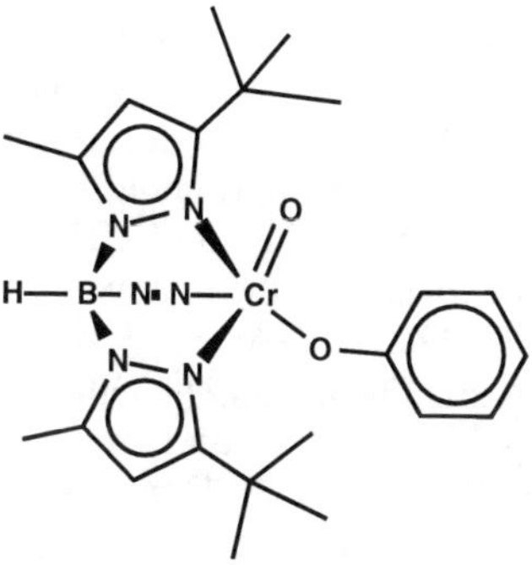

182

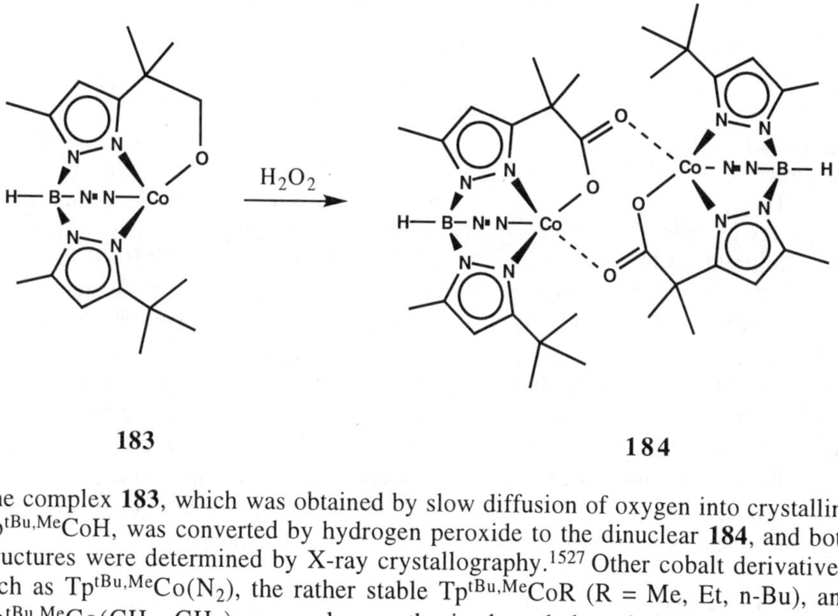

183　　　**184**

The complex **183**, which was obtained by slow diffusion of oxygen into crystalline Tp^{tBu,Me}CoH, was converted by hydrogen peroxide to the dinuclear **184**, and both structures were determined by X-ray crystallography.[1527] Other cobalt derivatives, such as Tp^{tBu,Me}Co(N$_2$), the rather stable Tp^{tBu,Me}CoR (R = Me, Et, n-Bu), and Tp^{tBu,Me}Co(CH$_2$=CH$_2$), were also synthesized, and the ethylene complex was structurally characterized.[1529]

The Tp^{tBu,Me} ligand was used extensively in the studies of zinc, particularly in trying to model the enzyme carbonic anhydrase. This required the availability of a stable, monomeric $Tp^x ZnOH$ complex, and the structurally characterized $Tp^{tBu,Me}ZnOH$ met this criterion.[1352,1353] Its reaction with CO_2 produced very rapidly and reversibly the bicarbonate complex, $Tp^{tBu,Me}Zn(OCO_2H)$, which was then slowly transformed into the dinuclear carbonato complex, $[Tp^{tBu,Me}Zn]_2(\mu-\eta^1,\eta^1-CO_3)$.[1329,1354] $Tp^{tBu,Me}ZnOH$ could also be readily converted to the alkylcarbonato derivative, $Tp^{tBu,Me}ZnOC(O)OR$, which was decarboxylated under high vacuum, producing $Tp^{tBu,Me}ZnOR$.[1355] $Tp^{tBu,Me}ZnOR$ could be prepared much more cleanly by the reaction of the hydride $Tp^{tBu,Me}ZnH$ with ROH. The resulting alkoxide readily equilibrated in the presence of water with the hydroxide. Upon reaction with aromatic aldehydes, the complex $Tp^{tBu,Me}ZnOCH_2Ar$ was formed, along with an aldehyde or ketone, derived from the original OR ligand.[1526] Other reported zinc complexes had the general structure $Tp^{tBu,Me}Zn(L)$, and included $Tp^{tBu,Me}ZnOAc$,[1246] $Tp^{tBu,Me}ZnF$,[109] $Tp^{tBu,Me}Zn(OR)$ and $Tp^{tBu,Me}Zn(OAr)$,[1356] and also those where L was either cysteine or histidine,[40] a drug molecule,[39] or a sulfonate ligand.[1357] Various phosphate esters were cleaved by $Tp^{tBu,Me}ZnOH$,[1338] and this complex was also used to cleave diphosphates, sulfonatophosphates and disulfonates.[1358] Tp^x ligand degradation was found to occur in the reactions with $Zn(ClO_4)_2$.[1245]

$Tp^{tBu,Me}Cd(NO_3)$ was synthesized, along with several other $Tp^xCd(NO_3)$ complexes, and the relevance of the nitro ligand denticity to carbonic anhydrase activity was discussed.[1247] The structures of $Tp^{tBu,Me}CuCO$ and $[Tp^{tBu,Me}Cu]_2$ were determined, with the dimer containing "inverted" ligands, and B—H—Cu bonds.[1359] The structure of $Tp^{tBu,Me}SnCl$ was a distorted trigonal bipyramid with the chloride in an axial site, and the lone pair in the equatorial position.[1248] Reported lanthanide complexes included $Tp^{tBu,Me}YbI(L)$ (L = THF or lutidine),[1360] the iodide in which could be replaced to produce $Tp^{tBu,Me}Yb[N(SiMe_3)_2]$, $Tp^{tBu,Me}Yb[CH(SiMe_3)_2]$, $Tp^{tBu,Me}Yb(CH_2SiMe_3)$, and $Tp^{tBu,Me}Yb[\mu-HBEt_3)(THF)$, the first example of a complex with a coordinated $[HBEt_3]^-$ ligand.[41] The structures of $[Tp^{tBu,Me}]_2Yb$, and of its Sm analog, contained one tridentate and one bidentate ligand, along with an agostic B—H—M bond. Exchange processes in these complexes were studied by NMR.[42] Isocarbonyl complexes of divalent samarium and ytterbium were obtained in good yield by the reaction of $Tp^{tBu,Me}LnI(THF)$ with the $[(C_5H_4Me)Mo(CO)_3]^-$ anion. The air- and moisture-sensitive products were structurally characterized, and found to be dimers, $[Tp^{tBu,Me}Ln(THF)(\mu-CO)_2Mo(\eta^5-C_5H_4Me)(CO)_3]_2$, in which two of the carbonyls per Mo were O-bonded to the lanthanide ion.[1362]

3.2.5.2b.3 $Tp^{tBu,iPr}$

The synthesis and NMR study of the Tl salt of this particular variant of the Tp^{tBu} ligand have been reported,[92] and so was $Tp^{tBu,iPr}Cu(O_2)$, containing side-on bond oxygen molecule,[1363] the copper(I) dimer $[Tp^{tBu,iPr}Cu]_2$, and the copper carbonyl $Tp^{tBu,iPr}CuCO$.[1359] Complexes of structure $Tp^{tBu,iPr}Fe(L)$, where L was OAc, OH, and O(O)COPh, some of them with an additional coordinated $Hpz^{tBu,iPr}$ molecule,

3.2 LIGANDS Tp$^{R,R'}$

were synthesized and several of them were structurally characterized. The benzoylformato complex, Tp^{tBu,iPr}Fe[O(O)COPh] contained five-coordinate iron, including chelation through the vicinal oxygen atoms of the benzoylformato ligand.[1364] Several tetrahedral cobalt complexes, including Tp^{tBu,iPr}CoOH, Tp^{tBu,iPr}Co(OAc) and Tp^{tBu,iPr}Co(OOCCMe$_2$Ph) have also been prepared. The structure of the last one was established by X-ray crystallography.[1315] Another reported, and structurally characterized, complex of this ligand was Tp^{tBu,iPr}Mn(OOCMe$_2$Ph), which was obtained by the reaction of Tp^{tBu,iPr}MnOH with cumyl hydroperoxide.[1365]

3.2.5.2b.4   Tp^{tBu,Tn} (Tn = 2-thienyl)

This ligand was another variant of Tp^{tBu}. It was characterized as the Tl salt, and its coordination chemistry resembled that of Tp^{tBu,Me}. Complexes Tp^{tBu,Tn}MX (M = Co, Ni, Zn; X= Cl, Br, NCS, NCO) were prepared and studied by NMR.[83]

3.2.5.2b.5   Tp^{CF_3,Me}

This ligand was first reported in the reaction of converting the ethylene complex Tp^{CF_3,Me}Ir(CO)(CH$_2$=CH$_2$), in which the Tp^x ligand was κ^2, into the octahedral hydrido vinyl complex Tp^{CF_3,Me}Ir(CO)(H)(CH=CH$_2$).[1366] The stability of ethylene versus hydridovinyl complexes in Rh and Ir systems was the subject of theoretical studies.[964] The Tp^{CF_3,Me} ligand was also used, along with other Tp^x analogs, in a study of κ^2-κ^3 isomerism in Tp^{x}Rh(diene) systems.[114] Among various other Rh complexes, that of Tp^{CF_3,Me}Rh(CO)$_2$ was structurally characterized.[1367] The ligand denticity equilibria in the complex Tp^{CF_3,Me}Ir(COD) were studied by NMR.[979] The copper(I) complex Tp^{CF_3Me}Cu(MeCN) was structurally characterized, and it was converted to the unstable nitrosyl complex Tp^{CF_3,Me}CuNO, which reacted with additional NO to produce the copper(II) derivative, Tp^{CF_3,Me}CuNO$_2$, containing a symmetrically bonded nitrite ion. The kinetics of this reaction were studied, and the CO complex Tp^{CF_3,Me}CuCO was synthesized.[1279]

3.2.5.2b.6   Tp^{CF_3,Tn}

The synthesis of this ligand, and the structure of its Tl salt, Tp^{CF_3,Tn}Tl have been reported.[115] The structure of Tp^{CF_3,Tn}Rh(CO)$_2$ was determined by X-ray crystallography, and it was similar to that of Tp^{CF_3,Me}Rh(CO)$_2$.[979]

3.2.5.2b.7   Tp^{Ph,Me} and Tp^{Ph,Tn}

This ligand is the prototype of a larger family of its Tp^{Ar,Me} variants. It was obtained as a single regioisomer, and structurally characterized as the Tp^{Ph,Me}ZnI complex.[104]

Other $Tp^{Ph,Me}ZnL$ complexes (L = OR, OAr, cysteine, histidine) have also been prepared,[40,1356] as was $Tp^{Ph,Me}Rh(COD)$.[114] The κ^2-κ^3 equilibria in the complex $Tp^{Ph,Me}Ir(COD)$ were studied by NMR.[979] Complexes $Tp^{Ph,Me}ZnSR$ (R = Et, CH_2Ph) were obtained by the reaction of $Tp^{Ph,Me}ZnOH$ with the corresponding thiols. Their conversion to the appropriate methyl thioethers was regarded as modelling comparable chemistry catalyzed by cobalamine independent methionine synthase.[1089] Various types of phosphate esters were cleaved by $Tp^{Ph,Me}ZnOH$.[1338] Nickel complexes, $Tp^{Ph,Me}Ni(O$-ethylcysteinato$)$ and $Tp^{Ph,Me}NiCl$ were synthesized as part of a study of five-coordinate nickel-cysteinate centers.[1534] The reported complexes of the $Tp^{Ph,Tn}$ ligand were $Tp^{Ph,Tn}NiNO_3$ and $Tp^{Ph,Tn}ZnOAc$.[1564]

3.2.5.2b.8 $Tp^{Tol,Me}$ (Tol = *para*-tolyl)

A zinc complex of this ligand, $Tp^{Tol,Me}ZnF$, and the dinuclear malonate derivative $[Tp^{Tol,Me}Zn(OOC)]_2CH_2$ have been reported, the former complex being structurally characterized.[109]

3.2.5.2b.9 $Tp^{Cum,Me}$ and $Tp^{Cum,Me*}$
(Cum = 4-isopropylphenyl)

This ligand, which has a deeper pocket than $Tp^{Tol,Me}$, was first introduced by Vahrenkamp in the complex $Tp^{Cum,Me}ZnOH$, which was effective in cleaving stoichiometrically some esters and amides,[110,1368] various types of phosphate esters,[1338] as well as diphosphates, sulfonatophosphates and disulfonates.[1358] During a detailed study of the rich chemistry of $Tp^{Cum,Me}ZnOH$, the presence of a small amount of the isomeric ligand, $Tp^{Cum,Me*}$ was discovered, and some of its derivatives were isolated. The reaction with H_2S afforded $Tp^{Cum,Me}ZnSH$ and $Tp^{Cum,Me*}ZnSH$, and the structures of both were determined, as were those of $[Tp^{Cum,Me*}Zn]_2O$ and $[Tp^{Cum,Me}Zn]_2S$. The Zn–O–Zn sequence was linear, while Zn–S–Zn was bent. Other derivatives of structure $Tp^{Cum,Me}ZnL$ were those with L = OCOOR, SC(S)OEt, SC(O)NHPh, the dinuclear sulfito complex $[Tp^{Cum,Me}Zn]_2[\mu$-$O_2SO]$,[1355,1369] as well as those with L = alkoxide or aryloxide,[1356] SR (R = Ph, CH_2Ph, CH_2CH_2Ph),[1370] sulfonates,[1357] cysteinates and histidinates,[40] drug molecules,[39] nucleobases, their natural precursors, nucleosides and nucleotides.[1371] Alkyl carbonate derivatives, $Tp^{Cum,Me}ZnOC(O)OR$ could be prepared from $Tp^{Cum,Me}ZnOH$ and dialkylcarbonates, or from alcohol and CO_2.[1355] In the unusual dinuclear monocation, $\{[Tp^{Cum,Me}ZnOH]_2H\}^+$, the two oxygens were bridged asymmetrically by a proton, which could be regarded as a mode of stabilization of the Zn-(H_2O) species.[1372]

Despite the bulk of the $Tp^{Cum,Me}$ ligand, the octahedral complex $[Tp^{Cum,Me}]_2Co$ was prepared without ligand rearrangement. It reacted with 3,5-di-*tert*-butylcatechol to yield the 3,5-di-*tert*-butyl-1,2-semiquinonate (3,5-DBSQ) complex, $Tp^{Cum,Me}Co(3,5$-DBSQ$)$. Similar $Tp^{Cum,Me}M(3,5$-DBSQ$)$ complexes were also

3.2 LIGANDS Tp$^{R,R'}$

prepared for M = Cu and Zn, by the reaction of Tp^{Cum,Me}M(OH) with 3,5-di-*tert*-butylcatechol. They were structurally characterized, and their redox properties were explored.[1373] In a closely related study, the reaction of (2,3,9,10,16,17,23,24-octahydroxyphthalocyaninate)nickel(II) (= NiPc(OH)$_8$) with Tp^{Cum,Me}M(OH) gave rise to NiPc(OH)$_4$(OZnTp^{Cum,Me})$_4$, which upon addition of base was converted to the fully chelated species, NiPc(O$_2$ZnTp^{Cum,Me})$_4$. Similar chemistry was observed between Tp^{Cum,Me}Zn(OH) and 5,6-dihydroxyphthalimide.[1374] The unusual trinuclear copper complex, **185**, contained square planar copper, with one long Cu-N bond per Tp^{Cum,Me} ligand, and it could be regarded as a complex of Cu(OMe)$_2$ with two identical Tp^{Cum,Me}CuOMe entities. The structure of the sulfanyl derivative, Tp^{Cum,Me}Cu[O(MeS)C$_6$H$_4$] has also been reported.[1375]

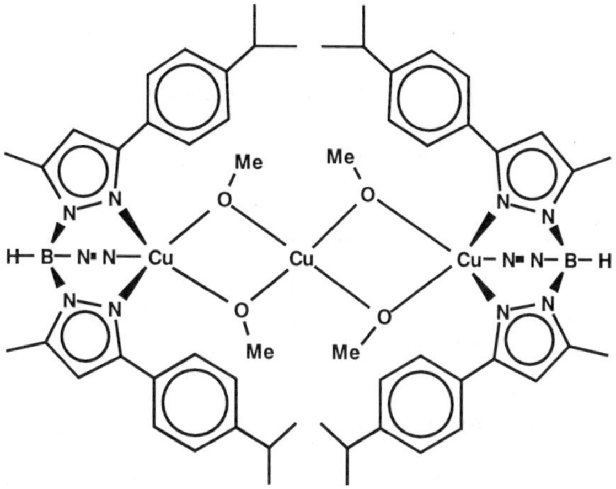

185

3.2.5.2b.10 Tp^{tBuPh,Me}

One of the reported, and structurally characterized, complexes of this ligand (tBuPh = 4-Bu^t-phenyl) was Tp^{tBuPh,Me}MgOH, which was prepared from Tp^{tBuPh,Me}MgMe by treatment with one equivalent of water. Despite the bulk of this ligand, this complex was not a monomer, but rather a hydroxide-bridged dimer.[111] Other complexes of this ligand which were prepared and structurally characterized were Tp^{tBuPh,Me}Cu(NCMe), Tp^{tBuPh,Me}Cu(CO), Tp^{tBuPh,Me}Cu(PPh$_3$), and Tp^{tBuPh,Me}Cu(PBu^t)$_3$. In the first three complexes copper was tetrahedrally coordinated with a κ^3 Tp^{tBuPh,Me} ligand, but in the last one the Tp^{tBuPh,Me} ligand was κ^2, and the copper ion was in a trigonal environment. An effort was made to gain insight into the structure of the pocket around the metal by calculating and comparing the areas of triangles formed by connecting the midpoints of the 3-phenyl rings.[1538]

3.2.5.2b.11 $Tp^{3Py,Me}$

The preparation of this ligand followed standard procedures,[112] and it was converted to several zinc complexes, $Tp^{3Py,Me}ZnX$.[1376] The structure of a dimeric cationic complex, containing two $[Tp^{3Py,Me}Zn]$ moieties linked by a $[H_3O_2]^-$ bridging entity was determined by X-ray crystallography. One of the 3-pyridyl nitrogens from a neighboring Tp^x ligand coordinated to each zinc ion, making it five-coordinate.[1372] $Tp^{3Py,Me}ZnOH$ was found to cleave several types of phosphate esters.[1338]

3.2.5.2b.12 $Tp^{3Pic,Me}$ (Pic = 4-methylpyridyl)

The chemistry of $Tp^{3Pic,Me}$ resembled that of $Tp^{Py,Me}$, and its dinuclear cationic complex $\{[Tp^{3Pic,Me}Zn]_2[H_3O_2]\}^+$ had a similar structure to that of the $Tp^{Py,Me}$ analog.[1372] Its hydroxy derivative, $Tp^{Pic,Me}ZnOH$, was found to be effective in cleaving phosphate esters,[1338] sulfonatophosphates and disulfonates.[1358]

3.2.5.2b.13 $Tp^{Bn,Me}$ and $Tp^{Tn,Me}$

The $Tp^{Bn,Me}$ ligand (Bn = benzyl) was characterized as the Tl salt, and also by the NMR spectrum of its homoleptic Co^{II} complex $[Tp^{Bn,Me}]_2Co$.[83] A Cu(II) complex of the $Tp^{Tn,Me}$ ligand (Tn = 2-thienyl), $Tp^{Tn,Me}CuCl$, was reported.[1564]

3.2.5.3 4,5-Disubstituted Ligands

The only examples of 4,5-disubstituted Tp^x ligands were those derived from indazoles containing a variety of alkyl or aryl substituents in the 3-, 4-, 5-, and 6-, but not in the 7-position.

3.2.5.3.1 Tp^{4Bo} and pz^oTb^{4Bo}

The synthesis of Tp^{4Bo} (= hydrotris(indazol-2-yl)borate) and of pz^oTp^{4Bo}, along with a number of their complexes, has been reported some time ago, although none of the reported compounds had been structurally characterized, and no basis for the particular structure assignements was given; moreover, some of the claims could not be verified.[1377,1378] A more careful synthesis of the tetrakis(indazolyl)borato anion led to a tentative structural assignment of this ligand, on the basis of analogy with the benzotriazolylborato analog, as pz^oTp^{4Bo} (rather pz^oTp^{3Bo}).[119,160] The first structural characterization by X-ray crystallography of the Tp^{4Bo} ligand was in its molybdenum complex, $Tp^{4Bo}Mo(CO)_2(\eta^3$-methallyl).[118] Apart from X-ray crystallography, a reliable way of resolving the Tp^{4Bo} versus Tp^{3Bo} issue was on the basis of the NMR spectra of paramagnetic Co^{II} homoleptic Tp^{Bo}_2Co, or heteroleptic $Tp^{Bo}CoTp^{Np}$

3.2 4,5-DISUBSTITUTED LIGANDS

complexes, where the remaining 3- or 5-protons of the pyrazolyl core could be readily distinguished, appearing at about -100 ppm, and at +70 ppm, respectively.[101] Complexes $Tp^{4Bo}{}_2MMe_3$ (M = Pd, Pt) were synthesized, and SCF calculations have indicated that Tp has a larger intrinsic "bite" than the isoelectronic $HC(pz)_3$.[1379]

3.2.5.3.2 $Tp^{4Bo,5NO_2}$, $pz^oTp^{4Bo,5NO_2}$

This ligand and some of its complexes have been reported, but no satisfactory structure proof has been provided.[1380] Similarly, several SnR_3 derivatives (R = Me, Bu, Ph) of the ligand $pz^oTp^{4Bo,5NO_2}$ have also been reported.[1381]

3.2.5.3.3 $Tp^{4Bo,5NH_2}$

This ligand and some of its complexes have been reported, but no satisfactory structure proof has been provided.[1382]

3.2.5.3.4 $Tp^{4Bo,5Me}$

In a definitive paper, representative examples of the Tp^{3Bo} and Tp^{4Bo} families of ligands were described. Their most convenient characterization rested on the ^1H NMR of their octahedral Co^{II} homoleptic, or heteroleptic (with Tp^{Np} as the other ligand) complexes, as each proton appeared in a unique, and easily distinguishable region of the NMR spectrum. The $Tp^{4Bo,5Me}$ ligand was characterized as the Tl salt, and it was also converted to the octahedral complexes $[Tp^{4Bo,5Me}]_2Co$, $Tp^{4Bo,5Me}CoTp^{Np}$ and $Tp^{4Bo,5Me}Mo(CO)_2(\eta^3$-methallyl). Compounds $[Tp^{4Bo,5Me}]_2M$ (M = Fe, Zn) were also synthesized.[101]

3.2.5.3.5 $Tp^{4Bo,5Et}$

The $Tp^{4Bo,5Et}$ was characterized as the Tl salt, and structurally identified through the NMR of its $[Tp^{4Bo,5Et}]_2Co$ and $[Tp^{4Bo,5Et}]CoTp^{Np}$ complexes. The compound $[Tp^{4Bo,5Et}]_2Fe$ was also synthesized.[101]

3.2.5.3.6 $Tp^{4Bo,5tBu}$

The $Tp^{4Bo,5tBu}$ ligand was characterized as the Tl salt, and its regiochemistry was elucidated through the NMR of its $[Tp^{4Bo,5tBu}]CoTp^{Np}$ complex. The compound $[Tp^{4Bo,5tBu}]Mo(CO)_2(\eta^3$-methallyl) was also synthesized.[101]

3.2.5.3.7 $Tp^{4Bo,5Ph}$

The $Tp^{4Bo,5Ph}$ ligand was characterized as the Tl salt, and its regiochemistry was established through the NMR of the $[Tp^{4Bo,5Ph}]CoTp^{Np}$ complex.[101]

3.2.5.3.8 Tp^{4Bo,4,6Me2}

The Tp^{4Bo,4,6Me2} ligand was characterized as the Tl salt, and its regiochemistry was elucidated through the NMR spectra of the cobalt(II) complexes [Tp^{4Bo,4,6Me2}]$_2$Co and [Tp^{4Bo,4,6Me2}]CoTp^{Np}. Other derivatives of this ligand, such as [Tp^{4Bo,4,6Me2}]$_2$Fe and [Tp^{4Bo,4,6Me2}]Mo(CO)$_2$(η^3-methallyl) were also reported.[101]

3.2.6 Trisubstituted Ligands

There is quite a variety of known 3,4,5-trisubstituted Tp^x ligands, but only two of them, Tp^{Me3} and Tp^{Br3}, had three identical R groups. Some trisubstituted ligands were derivatives of Tp*, where either a halogen, or an alkyl group has been introduced into the 4-position prior to synthesizing the Tp^x ligand, and in some cases halogenation of Tp* took place upon halogenation of the coordinated metal. In general, the 4-substituents did not introduce major deviations from the coordination chemistry of the ligand with the same 3- and 5-substituents. Since many of the trisubstituted Tp^x ligands contained 3- and 5-methyl substituents, and Tp^{Me2} is also denoted as Tp*, we shall abbreviate Tp^{Me2,4x} ligands as Tp*^{x}. Other ligands where all three substituents are identical, which will be Tp^{R3}.

3.2.6.1 Tp^{Me3} = Tp*^{Me}

The ligand synthesis, and its octahedral complexes [Tp^{Me3}]$_2$M (M = Mn, Fe, Co, Ni and Zn), were reported in 1967,[11] the complex [Tp^{Me3}]$_2$Fe was studied by Mössbauer spectroscopy,[14,801,819,821] and a study of the temperature and pressure dependence of the electronic spin states of [Tp^{Me3}]$_2$Fe (and of related Fe and Co complexes) was carried out.[826] Dinuclear complexes of Ni^{III} and Co^{III} were prepared by oxidation with H$_2$O$_2$ of the hydroxy-bridged precursors, [Tp^{Me3}M(OH)]$_2$ to [Tp^{Me3}M(O)]$_2$. Both were of low stability, but the structure of the Ni^{III} complex could be established by X-ray crystallography.[1383] The rhodium complex Tp^{Me3}Rh(COD) was prepared and studied in the context of ligand κ^2-κ^3 equilibria.[114] The κ^2-κ^3 solution equilibria of the complex Tp^{Me3}Ir(COD), and of numerous related Tp^{x}Ir(COD) complexes were studied by NMR.[979] Various complexes, such as Tp^{Me3}HgSEt,[1223] Tp^{Me3}MoCl$_2$NO and Tp^{Me3}Mo(CO)$_2$NO have been reported,[1384] as also were [Tp^{Me3}Mo(CO)$_3$]$^-$ and Tp^{Me3}Mo(CO)(η^2-COMe).[455] A number of tin(IV) derivatives Tp^{Me3}SnCl$_n$Me$_{3-n}$ was synthesized, and the structures of Tp^{Me3}SnCl$_2$Me and of Tp^{Me3}K were determined by X-ray crystallography,[1385] but attempts to recrystallize Tp^{Me3}SnClMe$_2$ led to decomposition of the ligand.[1386]

3.2.6.2 Tp*^{Et}

This ligand was converted to two complexes: Tp*^{Et}Mo(CO)$_2$NO and Tp*^{Et}MoCl$_2$NO, which were characterized by NMR.[1384]

3.2 3,4,5-TRISUBSTITUTED LIGANDS

3.2.6.3 Tp*^{Bu}

The early compounds derived from Tp*^{Bu} were the octahedral [Tp*^{Bu}]$_2$M (M = Co, Zn) complexes.[11] More recently, Tp*^{Bu}Mo(CO)$_2$NO and Tp*^{Bu}MoCl$_2$NO were reported, and the structure of the former was established by X-ray crystallography.[1384]

3.2.6.4 Tp*^{Am} (Am = n-pentyl)

This ligand was converted to two molybdenum complexes: Tp*^{Am}Mo(CO)$_2$NO and Tp*^{Am}MoCl$_2$NO, which were characterized by NMR.[1384]

3.2.6.5 Tp*^{Bn} (Bn = benzyl)

A variety of compounds was prepared from this ligand, all derived from the starting anion [Tp*^{Bn}M(CO)$_3$]$^-$ (M = Mo, W). These included the structurally characterized complex Tp*^{Bn}Mo(CO)$_2$NO, which exhibited an "inverted bowl" structure. This complex, and its W analog, were converted to derivatives Tp*^{Bn}MNO(X)(Y) (X = Y = Cl, I or OC$_6$H$_4$-p-Me; X = Cl, Y = OMe or OC$_6$H$_4$-p-Me), which were characterized by spectroscopy.[121]

3.2.6.6 Tp*^{Cl}

Inadvertent formation of the Tp*^{Cl} ligand occurred during chlorination of Tp*Mo(CO)$_2$NO,[120] but the direct synthesis of this ligand was easy.[114] Niobium complexes of Tp*^{Cl} containing, in addition, the thematic PhC≡CMe ligand, and halogen or alkyl groups on niobium, such as Tp*^{Cl}NbCl(CH$_2$SiMe$_3$)(PhC≡CMe), Tp*^{Cl}NbCl(Et)(PhC≡CMe), Tp*^{Cl}NbMe$_2$(PhC≡CMe), Tp*^{Cl}NbCl$_2$(PhC≡CMe), and Tp*^{Cl}NbCl(Bn)(PhC≡CMe), were synthesized, and used as catalysts in ethylene polymerization.[407] Tp*^{Cl}$_2$Fe was prepared and studied by Mössbauer spectroscopy.[821] The rhodium complex Tp*^{Cl}Rh(COD) was investigated in the context of ligand κ^2-κ^3 equilibria,[114] while complexes Tp*^{Cl}Cu(PPh$_3$) and Tp*^{Cl}Cu[P(Tol)$_3$] were studied by spectroscopy and by NMR, and the structure of Tp*^{Cl}Cu(PPh$_3$) was determined by X-ray crystallography.[1057,1058] Also reported were Tp*^{Cl}HgCN,[1100] and the anion [Tp*^{Cl}Mo(CO)$_3$]$^-$, which was used in a detailed study of the formation of η^2-aroyl-, and η^1-halogenocarbyne derivatives.[470] The κ^2-κ^3 equilibria in the complex Tp*^{Cl}Ir(COD) were studied by NMR, and its structure was determined by X-ray crystallography.[979] Treatment of Tp*^{Cl}Rh(CO)$_2$ with one equivalent PMePh$_2$ produced square planar Tp*^{Cl}Rh(CO)(PMePh$_2$), which was structurally characterized as κ^2. However, when an excess of the phosphine PMePh$_2$ was used, the exclusive product was Tp*^{Cl}Rh(CO)(PMePh$_2$)$_2$, **186**, in which the Tp*^{Cl}ligand was bonded κ^1, a rare mode of bonding for homoscorpionate ligands, and there was also an agostic B—H—Rh bond present.[1540]

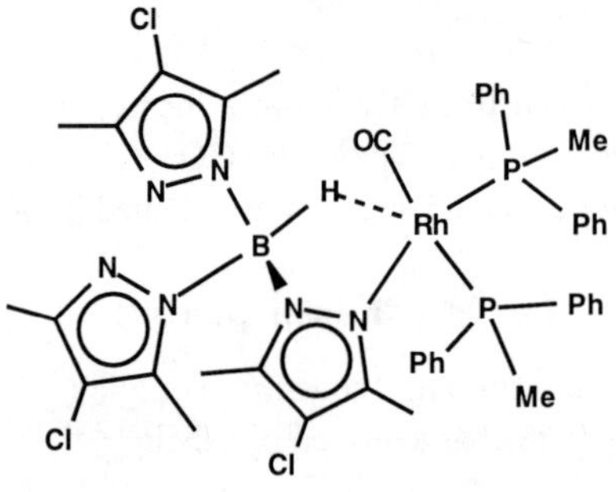

[Structure diagram of compound **186**]

186

3.2.6.7 Tp*Br

This ligand was prepared inadvertently during bromination of the coordinated metal coordinated to Tp*, but it was also synthesized on purpose, and the κ^2-κ^3 equilibria in the complex Tp*^{Br}Ir(COD) were studied by NMR.[979]

3.2.6.8 Tp^{iPr2,Br}

A conventional reaction of Hpz^{iPr2,Br} with NaBH$_4$ produced the above ligand, which was converted to the structurally characterized complex Tp^{iPr2,Br}Ru(H)(COD). In contrast to its analog derived from Tp^{iPr2}, it also contained an agostic B—H—Ru bond, in addition to the bidentate Tp^{iPr2,Br} ligand, thus placing Ru in a somewhat distorted octahedral environment,[122] similar to that found in Tp^{iPr,4Br}Ru(H)(COD).[868]

3.2.6.9 Tp^{Br3}

Heating of 3,4,5-tribromopyrazole with KBH$_4$ produced readily the Tp^{Br3} ligand, and it was characterized as the Tl salt. Structures of the tetrahedral, though slightly distorted, [Tp^{Br3}]Tl, and of the octahedral [Tp^{Br3}]$_2$Co, were determined by X-ray crystallography. Although [Tp^{Br3}]$_2$Co contained eighteen bromine atoms in the molecule, it was exceedingly stable, and the bromines therein were unreactive towards nucleophiles.[83]

3.2.6.10 Tp^{Ph,Me,Ph}

The "as prepared" ligand hydrotris(3,4-diphenyl-5-methylpyrazol-1-yl)borate was a mixture, containing some of its isomer Tp$^{(Ph,Me,Ph)*}$. The pure ligand could be obtained by means of recrystallization of the K salt, and it was converted to the tetrahedral cobalt complex [Tp^{Ph,Me,Ph}]CoI, which was structurally characterized. While the homoleptic compound [Tp^{Ph,Me,Ph}]$_2$Co could not be prepared, the mixed

3.2 3,4,5-TRISUBSTITUTED LIGANDS

octahedral complex, [Tp^{Ph,Me,Ph}]Co[Tp$^{(Ph,Me,Ph)*}$] was synthesized, and its structure was determined by X-ray crystallography.[123]

3.2.6.11   Tp^{4Bo,3Me}

This ligand was prepared as a single isomer from 3-methylindazole, and it readily formed octahedral complexes [Tp^{4Bo,3Me}]$_2$M, as well as heteroleptic ones, such as Tp^{4Bo,3Me}CoTp^{Np}, the NMR specrum of which indicated over-all C$_{3v}$ symmetry of the latter molecule. The structure of the rhodium complex, Tp^{4Bo,3Me}Rh(COD), showed a κ^2 ligand, with the third arm in a non-interactive mode with respect to Rh.[101]

3.2.6.12   Tp^{a*,3Me} (= 187) and Tp$^{(a*,3Me)*}$

This ligand, **187**, the systematic name of which is hydrotris(3-methyl-2*H*-benz[g]indazol-2-yl)borate, was similar to the ligand Tp^{a*} (see structure **177**, p. 135),

187

except for the additional 3-methyl groups. This structural change already made a difference in the regiospecificity observed during the synthesis of this ligand. Whereas Tp^{a*} was produced as the only isomer, Tp^{a*,3Me} was obtained as a mixture of the symmetric and asymmetric isomers in about 4:6 ratio, although the symmetric isomer could be obtained pure through recrystallization. The flatness of the three-ring system permitted the formation of homoleptic octahedral [Tp^{a*,3Me}]$_2$Co, and heteroleptic [Tp^{a*,3Me}]$_2$CoTp^{Np} complexes, which were studied by NMR, and the structure of the latter was determined by X-ray crystallography. Rhodium complexes [Tp^{a*,3Me}]$_2$Rh(CO)$_2$ and [Tp^{a*,3Me}]$_2$Rh(COD) were also prepared. The asymmetric isomer, Tp$^{(a*,3Me)*}$, was not investigated.[101]

Chapter 4

Heteroscorpionates, RR'Bp^x

4.1 General Considerations

This chapter covers heteroscorpionates, i.e. ligands of general structure **188** and **189**. The difference between these two, otherwise identical, structures is that in **188** there is

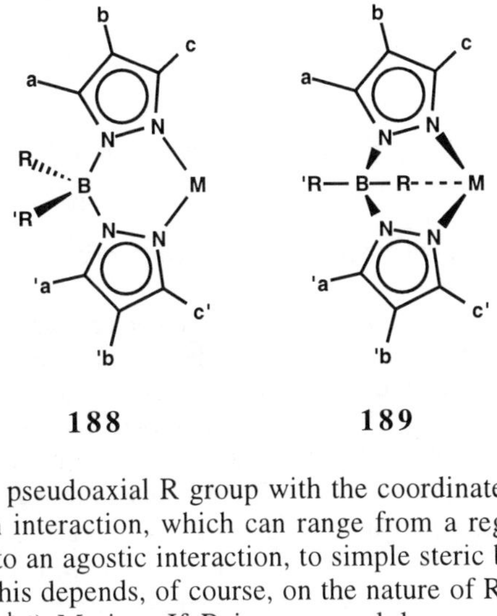

188   **189**

no interaction of the pseudoaxial R group with the coordinated metal ion, while in **189** there is such an interaction, which can range from a regular bond, through a long-distance bond, to an agostic interaction, to simple steric blocking of the access to the metal. All of this depends, of course, on the nature of R, and the depth of the boat in the B(μ-pz^{a,b,c})$_2$M ring. If R is a pyrazolyl group different from those forming the B(μ-pz^{a,b,c})$_2$M ring in **189** we have a variant of a Tp^x ligand. In this chapter we are excluding Tp^{x*} ligands, resulting from a rearrangement of the Tp^x ligand, or formed in the course of Tp^x ligand synthesis, since they were covered in Chapter 3. We will include in Chapter 4 ligands where R is a pyrazole unrelated to those forming the B(μ-pz^{a,b,c})$_2$M ring. It should be noted that the ligand Me$_2$NTp acted as a heteroscorpionate, [pzB(μ-pz)$_2$(μ-NMe$_2$)]M, coordinating through only two of the three pz groups, and the dimethylamino group, in the molybdenum complex [Me$_2$NTp]Mo(CO)$_2$(η^3-allyl). The third pz group was not coordinated to Mo, and did not exchange with the coordinated pz groups.[130]

Considering, that heteroscorpionates possess the same options of containing different 3-, 4-, and 5-substituents as the Tp^x ligands, but at the same time having two groups on boron that can be varied, as compared with only one in homoscorpionates, it follows that a much greater variety of scorpionate ligands should be accessible in the heteroscoropionate sub-area. Yet until now, most (over 90%) research dealing with scorpionate ligands centered on homoscorpionate derivatives. Clearly, there are still many novel structures in this particular sub-area, with specific coordination abilities, to be designed, synthesized, and used.

The known types of heteroscorpionate ligands are:

1. H$_2$B(pz^x)$_2$ (= Bp^x), including H$_2$B(pz^x)(pz^y) (= Bp^{x*}), where pz^y is the same pyrazolyl group as pz^x, but bonded through the other nitrogen atom; the latter ligand was usually obtained through rearrangement of H$_2$B(pz^x)$_2$.

2. R$_2$B(pz^x)$_2$ (= R$_2$Bp^x) where R = alkyl (including ring structures), aryl, or halogen.

3. R(R'Z)B(pz^x)$_2$ (= R(R'Z)Bp^x), where R = H, alkyl, aryl, and Z is a heteroatom (O, S, NR').

4. H$_2$B(pz^x)(pz^z) where pz^x and pz^z are different pyrazolyl groups.

The first heteroscorpionates reported included already many of the important ligand variants.[1] In 1967, the parent Bp ligand, and its complexes with Mn, Fe, Co, Ni, Cu and Zn were synthesized,[9] followed by carbon- and boron-substituted analogs of Bp, including complexes (Co, Ni, Cu and Zn) of the ligands Bp* and Bp^{Me3}, Et$_2$Bp, Bu$_2$Bp and Ph$_2$Bp.[11] At the same time, pyrazaboles, which are neutral heterocycles of structure R$_2$B(μ-pz^x)$_2$BR'$_2$, and which may be regarded as [R'$_2$B]$^+$ complexes of the [R$_2$B(pz^x)]$^-$ ligand, have also been reported along with the essentials of their chemistry.[1,10,12] The Bp$_2$M complexes (M = Mn, Fe, Co, Ni, Cu) were studied by spectroscopy,[13] and the molybdenum complexes Bp*Mo(CO)$_2$(η^3-CH$_2$CRCH$_2$) and Et$_2$BpMo(CO)$_2$(η^3-CH$_2$CRCH$_2$), were synthesized.[21] Both of them had nominally 16-electron structures. To remedy this electron deficiency, the former complexes formed an agostic B—H—Mo bond, as was shown by an X-ray structure determination of Bp*Mo(CO)$_2$(η^3-allyl),[23] while the latter readily added donor ligands (Hpz, pyridine), and in their absence formed an agostic C—H—Mo bond through one of the methylene hydrogens of the pseudoaxial ethyl group, as was suggested by low-

4.2 LIGANDS Bp^x

frequency C—H stretches in the IR, and by a shift of the pseudoaxial methylene protons into the "hydridic" range.[22] The presence of this agostic bonding was confirmed by X-ray crystallographic structure determination of Et$_2$BpMo(CO)$_2$(η^3-CH$_2$CPhCH$_2$),[24] and the dynamic processes of making and breaking such agostic bonds were later elucidated.[25]

4.2 Specific Ligands

4.2.1 Bp^x Ligands

A key feature of the Bp^x ligands is the presence of the BH$_2$ grouping which, on the one hand, renders the Bp^x ligands more hydrolytically labile than their Tp^x counterparts but, on the other hand, permits elaboration of the Bp^x ligand through the addition of various unsaturated systems to the B—H bond. This way of creating novel ligands has been explored only to a rather limited extent. Another theme in the Bp^x ligand system is the formation of agostic B—H—M bonds, which has been found to occur with many metals. Finally, the residual reducing power of the BH$_2$ grouping makes the Bp^x complexes of easily reducible cations, such as silver(I) and palladium(II), unstable, and can also be used in some instances for organic reductions.

4.2.1.1 Bp

This is the simplest of all scorpionate ligands, and it is basically bidentate, forming the typical boat-shaped H$_2$B(μ-pz)$_2$M ring, in which the pseudoaxial B—H may form an agostic bond to the metal in some instances. The Bp$_2$M complexes of first row transition metals are usually square planar or tetrahedral, although some octahedral anionic [Bp$_3$M]$^-$ species of low stability have also been isolated. The preparation of the Bp ligand, and of some of its complexes has been described in detail, and it was noted that the Bp$_2$Mn and Bp$_2$Fe complexes were air-sensitive.[9,74] The spectra of these complexes were determined,[13] and some were later reassigned.[1387] The structure of the free acid, BpH, was determined by X-ray crystallography.[1388]

A study of solvent extraction of divalent transition metals with the Bp ligand showed it to be an efficient process, but Mg^{2+}, Ca^{2+}, Sr^{2+}, and Ba^{2+} were not extracted, and Be^{2+} was only slightly so.[1389,1390] Several Bp$_2$M complexes were investigated as reducing agents for cyclohexanone and cyclohexenone.[1391] Although the reaction of [Bp]$^-$ with divalent transition metal ions usually yielded the neutral Bp$_2$M complexes (M = Cr, Mn, Fe, Co, Ni, Cu, Zn), and the tris-complexes were only formed with excess Bp ligand in acetonitrile,[1392] it was possible to isolate the anionic tris complexes, [Bp$_3$M][Et$_4$N] of Fe, Co and Ni.[1393]

The heteroleptic titanium(III) complex, $BpTiCp_2$, was synthesized,[344] as was $BpZr(Cp)Cl_2$, which contained an agostic B—H—Zr bond.[361] The complex $[Bp_3V]K$ was also reported.[1394] The reaction of BpK with Nb_2Cl_{10} yielded the anion $[BpNbCl_5]^-$, along with other species.[392] A seven-coordinate tantalum complex Bp_2TaMe_3 was synthesized from $TaCl_2Me_3$ and BpK.[1395,1396]

The structure of the square planar Bp_2Cr was determined by means of X-ray crystallography.[1394] Five-coordinate anionic species $[Bp_2MX]^-$ (M = Cr, Mn; X = halides or pseudohalides) were prepared, and the square pyramidal structure of $[Bp_2MnCl][AsPh_4]$ was established by X-ray crystallography. Oxidation of $[Bp_2CrNCS]^-$ yielded the structurally characterized dinuclear complex $[(Bp_2Cr)_2(\mu\text{-}OEt)_3(NCS)_2]^-$.[1397,1398] The complexes $BpMo(\eta^2\text{-}C(O)R)(\eta^2\text{-}CNBu^t)(CO)(PMe_3)$ were thermally isomerized to the more thermodynamically stable η^2-iminoacyl-carbonyl derivatives, $BpMo(\eta^2\text{-}RN{=}CBu^t)(CO)_2(PMe_3)$.[1399,1400] A brief review of the structural and reactivity aspects of molybdenum acyl complexes derived from the Bp and Bp* ligands was published.[1401] The neopentylidyne compound $BpMo(\equiv CBu^t)(CO)(PX_3)_2$ has also been reported,[1402] as were the various clusters $[(Bp)Co_2W(\mu_3\text{-}CC_6H_4Me\text{-}4)(CO)_9]$, and also $[(Bp)Rh_2W(\mu_3\text{-}CMe)(\mu\text{-}CO)(CO)_3)(\eta^5\text{-indenyl})]$, $[(Bp)FeW(\mu\text{-}CC_6H_4Me\text{-}4)(CO)_6]$, $[(Bp)PtW(\mu\text{-}CC_6H_4Me\text{-}4)(CO)_3(COD)]$ as well as $[Bp_2MW_2(\mu\text{-}CR)_2(CO)_6$ (M = Ni or Pt).[1403] The biochemical distribution of a Bp derivative of ^{99m}Tc with unspecified composition was used in radiochemical analysis.[1404]

Tetranuclear iron complexes which contained a nonplanar $[Fe_4O_2]^{8+}$ core, $[Bp_2Fe_4O_2(O_2CR)_7]^-$ (R = Me or Ph), have been synthesized, and were studied by Mössbauer and Raman spectroscopy. The structure of the compound with R = Ph has been determined by X-ray crystallography.[1405] Bp-based acetyl complexes of both, Fe and Ru, such as $BpM(PMe_3)_2(CO)(COMe)$, were reported.[806] Although the reaction of $RuCl_2(DMSO)$ with the Bp ligand caused its degradation,[1406] $BpRu(PPh_3)_2$ could be prepared and it was characterized by NMR.[1407] A large number of octahedral complexes of the BpRu moiety were prepared. They included $BpRuH(CO)(PPh_3)_2$, $BpRuH(CS)(PPh_3)_2$, the two sigma-phenyl complexes $BpRuPh(CO)(PPh_3)_2$ and $BpRuPh(CS)(PPh_3)_2$, vinylic derivatives such as $BpRu(CR{=}CR'H)(CO)(PPh_3)_2$, as well as $BpRu(C{\equiv}CR)(CO)(PPh_3)_2$, and also $BpRu[C(S)CPh{=}CPhH](CO)(PPh_3)_2$. The structure of the vinyl complex, $BpRu(CH{=}CH_2)(CO)(PPh_3)_2$, was determined by X-ray crystallography.[1408]

A single-crystal EPR study was carried out on the structurally characterized tetrahedral Bp_2Co. The molecule contained two different dihedral angles, of 121° and 114°.[1409] The perfluoroalkyl complex $BpCoCp(i\text{-}C_3F_7)$ was obtained as a mixture of two stereoisomers, which could be separated by chromatography.[802] Rhodium compounds $BpRh(COD)$, and $BpRh(CO)_2$ were also synthesized.[934,935]

4.2 LIGANDS Bp^x

The structure of Bp$_2$Ni was established as square planar,[1410] and the charge density in this complex was determined through X - X$_{HO'}$ multipolar analysis, as well as from ab initio calculations.[1411] Numerous BpPd(PMe$_3$)(R) complexes have been reported, including BpPd(CH$_2$CMe$_2$Ph)(PMe$_3$) and BpPd(CH$_2$SiMe$_3$)(PMe$_3$),[1412] as were some Bp derivatives of various cyclopalladated systems,[1021] including the bimetallic palladium complex BpPd(COFc)(PPh$_3$).[1012] The platinum species BpPt(PEt$_3$)Cl has also been synthesized,[1041] and BpPtMe$_3$ was found to contain an agostic B—H—Pt bond.[1042]

Copper(I) complexes of structure BpCuL (L = phosphines, phosphites, arsines, etc.) were synthesized,[1413] as was BpAg(PPh$_3$).[1414] The structure of the anionic dinuclear complex [BpCu(μ-pz)$_2$(μ-Cl)CuBp]$^-$ was determined by X-ray crystallography, and its single-crystal EPR spectra were studied.[1415] The complex BpCu was prepared from BpK and CuI, and it was converted to derivatives of general structure BpCuL (L = bipy, dppe, (py)$_2$, (PPh$_3$)$_2$, and PCy$_3$). BpCu was an effective catalyst for the cyclopropanation of olefins with diazoacetic ester in homogeneous and heterogeneous systems.[1416]

The compounds of structure H$_2$B(μ-pz)$_2$BR$_2$, called "pyrazaboles" comprise a separate category of neutral heterocycles,[1] which may be regarded as Bp chelates of the boronium cation [BR$_2$]$^+$. In fact, a number of asymmetric pyrazaboles has been

$$[Bp]^- + R_2BX \rightarrow H_2B(\mu\text{-pz})_2BR_2 + X^- \qquad (4.1)$$

prepared in this fashion.[1417] This class of compounds has its own chemistry, and it will be discussed in Section 4.3.

The reported gallium complexes of Bp included Bp$_2$GaCl, Bp$_3$Ga, Bp$_2$GaMe, and BpGaMe$_2$, the latter being converted by acetic acid to BpGaMe(OAc), while Bp$_3$Ga yielded the structurally characterized Bp$_2$Ga(OAc).[1418] The structure of Bp$_2$GaCl was found to contain no agostic bonds,[1419] and some structural details thereof were corrected.[1420] The indium(III) complex Bp$_3$In was structurally characterized,[1421] as were BpInMeCl and BpInMe$_2$, among a number of other related In^{III} complexes which contained the [BpIn] moiety,[1423] as well as the heteroleptic complex BpInTp*(Cl).[1109] Tin(IV) complexes, Bp$_2$SnR$_2$ were also reported.[1422]

Both, Bp$_2$Sn and BpSnCl have been synthesized, and the structure of the latter showed it to be dinuclear, with long-range halide interactions between the two tin ions, which were in a pyramidal environment.[1119,1120] The complex BpSnClMe$_2$ was five-coordinate, and its structure was a distorted trigonal bipyramid,[1124] while that of BpSn(CH$_2$CH$_2$COOMe)$_2$Cl contained six-coordinate tin, including a bond to the ester carbonyl.[1388] Numerous other tin complexes of structure BpSnClR$_2$ and

Bp_2SnR_2 (R = Me, Bu, Ph) have been prepared. It has been reported that halogenation of the BH_2 portion of Bp occurs without ligand degradation, leading to $X_2BpSnPh_2$ complexes (X = Cl, Br, I).[1422] The lead(II) heteroleptic complex $BpPb[CpCo(P(O)OEt)_2]_3$,[1122] and a homoleptic one, Bp_2Pb,[1134] have also been reported.

Luminescence studies were carried out on Bp_3Tb at different temperatures, in the solid phase, and in various solvents.[1424] The complex Tp_2SmBp was structurally characterized, and was found to contain an agostic B—H—Sm bond.[1150] Numerous actinide Bp complexes have also been reported. They included a variety of uranium complexes, such as Bp_4U and Bp_3UCl_2,[1163,1165] $BpUCl_2(py)$,[1189] $BpU(Cp)Cl_2$, $Bp_2U(Cp)Cl_2[P(O)Ph_3]$, and $BpU(Cp)Cl[P(O)Ph_3]$,[1168] as well as $Bp_3U(THF)$, which contained agostic B—H—U bonds.[1425] Other actinide complexes reported were the thorium compounds $[Bp_5Th]K$,[1166] and $BpTh(Cp)Cl_2$,[1168] and the neptunium derivatives Bp_2NpCl_2, Bp_4Np, $Bp_2Np(MeC_5H_4)_2$ and Bp_2NpCp.[1170]

4.2.1.2   Bp* (= Bp^{Me2})

A series of first row transition metal complexes $Bp*_2M$ was synthesized, and they were found to resemble their Bp_2M counterparts, except that the $Bp*_2Mn$ and $Bp*_2Fe$ complexes were no longer air-sensitive, presumably due to the screening effect of the 3-Me groups.[11] The thermochemistry of this series of complexes was studied by TGA and DTA,[1426] as was the use of the Bp* ligand, among others, in the extraction of divalent first row transition metals, including structural characterization of the complexes $Bp*_2Co$ and $Bp*_2Ni$. It was found that the distance between the M-bonded nitrogen atoms, called the bite size of the ligand, slightly decreased with increasing number of methyl groups.[1390]

The [Bp*]⁻ anion readily displaced two CO molecules from $Mo(CO)_6$, producing the anion $[Bp*Mo(CO)_4]^-$, which reacted with allylic halides yielding $Bp*Mo(CO)_2(\eta^3-CH_2CRCH_2)$ complexes, with R = H, Me, Ph or Br.[21,22] These, formally 16-electron, complexes contained agostic B—H—Mo bonds, as was shown for $Bp*Mo(CO)_2(\eta^3-allyl)$ by X-ray crystallography.[23] In $Bp*Mo(CO)_2(\eta^3-C_7H_7)$ there was also a similar agostic B—H—Mo bond, as was shown by spectroscopy and by X-ray crystallography, despite the possibility of achieving an 18-electron configuration by means of $\eta^5-C_7H_7$ bonding.[1427,1428] The compound $Bp*Mo(\eta^2-C(O)R)(\eta^2-CNBu^t)(CO)(PMe_3)$ could be thermally isomerized to the more thermodynamically stable derivative, which was an η^2-iminoacyl-carbonyl complex $Bp*Mo(\eta^2-(RN=CBu^t)(CO)_2(PMe_3)$.[1399,1400] On the other hand, when the complex $Bp*Mo(CO)_2(\eta^2-C(O)Me)(PMe_3)$ was heated, an intramolecular hydroboration occurred, forming **190**, the structure of which was determined by X-ray crystallography. A very similar reaction took place with the iminoacyl analog,

4.2 LIGANDS Bp^X

Bp*Mo(CO)$_2$(η^2-MeC=NBu^t)(PMe$_3$).[1429] The Mo^{IV} complex, Bp*MoO(S$_2$PR$_2$), has also been reported.[1430]

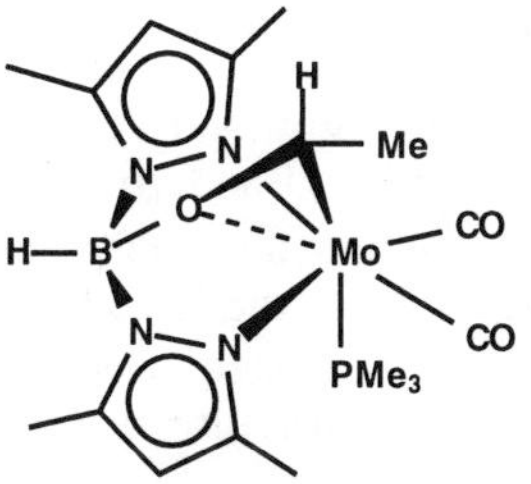

190

The rhodium(I) complex, Bp*Rh(COD), was readily synthesized,[934] and it was found to contain no agostic B—H—Rh bonds.[937] Also prepared were the complexes Bp*M(CO)$_2$ and Bp*M(CO)(PR$_3$) for the metals Rh and Ir.[1432] The photochemistry of Bp*Rh(CO)$_2$ was investigated by ultrafast IR spectroscopy. This compound, unlike its Tp* analog, did not activate C—H bonds.[954]

A variety of Bp*Pd(PMe$_3$)(R) complexes has been reported, and the structure of Bp*Pd(PMe$_3$)(CH$_2$SiMe$_3$) was determined by X-ray crystallography.[1412] The agostic B—H—Pt bond, present in the complex Bp*PtMe$_3$ was found to be easily broken by phosphines, and by CO, with the formation of Bp*PtMe$_3$L species.[1042] A short (2.06 Å) agostic B—H—Ru bond was found in the structure of [Bp*Ru(H)(η^4-C$_8$H$_{12}$)],[1434] and a B—H—Ta bond in the structure of Bp*TaMe$_3$Cl.[1435] An unusual tricopper (I,II,I) complex of the Bp* ligand has been structurally characterized, in which all three coppers were linked through sulfide bridges as shown in structure **191**. This complex was thought to be of some relationship to the proposed mixed valence dicopper electron transfer sites in the enzymes nitrous oxide reductase, or cytochrome c oxidase.[1436]

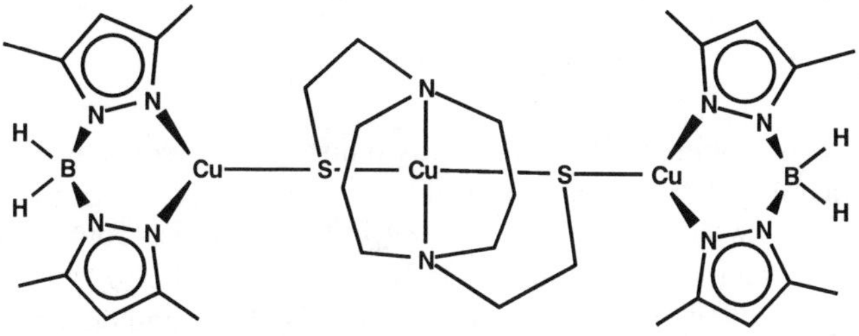

191

Several gallium(III) and indium(III) complexes with the Bp* ligand have also been synthesized. They included the structurally characterized compounds Bp*GaMeCl, Bp*GaMe$_2$, Bp*InMeCl, Bp*InMe$_2$ and Bp*$_2$InMe.[1437] The heteroleptic indium complex Bp*InTp*(Cl), has also been reported.[1109]

Bp* complexes of U^{III}, Ce^{III}, Sm^{III} and Yb^{III} were synthesized, and structures of the isomorphous compounds Bp*$_3$U and Bp*$_3$Sm were determined by X-ray crystallography. In each of these complexes the Bp* ligands were tridentate, by way of a B—H—M agostic bond, so that the metal ions were in a formally nine-coordinate environment, and the six nitrogens were arranged in a trigonal prismatic geometry.[1438] An identical type of structure was also found in Bp*$_3$Y.[1439] The complex Bp*UCl$_2$ has also been reported.[1189]

4.2.1.3   [H$_2$B(pz)(pz*)]$^-$

In contrast to Bp^{x*} ligands, where the two pyrazolyl groups are identical, even though bonded through different nitrogen atoms to boron, the ligand [H$_2$B(pz)(pz*)]$^-$ was the first purposely synthesized Bp^x ligand containing two different pyrazolyl groups.[1440] The synthesis proceeded by preparing (Hpz*)BH$_3$, its conversion to (Hpz*)BH$_2$I by treatment with iodine, followed by the reaction with another pyrazole to yield the boronium iodide [H$_2$B(Hpz*)(Hpz^x)][I] containing two different pyrazoles bonded to boron. The reaction of this salt with NaH produced [H$_2$B(pz*)(pz^x)]Na.[143] The structure of the square planar nickel complex derived from this ligand, [H$_2$B(pz)(pz*)]$_2$Ni, was determined by X-ray crystallography, and it was found to contain the (pz) and (pz*) rings in a *trans*-arrangement.[1390]

4.2.1.4   [H$_2$B(pz*)(pz^{Ph2})]$^-$

This ligand was synthesized by the method described in 4.2.1.3.[143]

4.2.1.5   [H$_2$B(pz*)(pz^{4Me})]$^-$

This ligand was synthesized by the method described in 4.2.1.3.[143]

4.2.1.6   [H$_2$B(pz)(pz^{tBu2})]$^-$

By the use of LiBH$_4$ with a 1:1 mixture of two different pyrazoles it was possible to obtain this asymmetric ligand, and its structure was confirmed by X-ray crystallography of two complexes: [H$_2$B(pz)(pz^{tBu2})]Tl, which contained an agostic B—H—Tl bond, and the zinc complex [H$_2$B(pz)(pz^{tBu2})]ZnI(Hpz^{tBu,iPr}), prepared by the addition of Hpz^{tBu,iPr} to [H$_2$B(pz)(pz^{tBu2})]ZnI.[144] This method was much more convenient for the preparation of Bp^x ligands containing two different pyrazolyl groups, than that described in 4.2.1.3.

4.2 LIGANDS Bp^x

4.2.1.7 [H$_2$B(pz*)(pz^{tBu2})]$^-$

This ligand, which was prepared by the method of 4.2.1.6, formed the zinc complex [H$_2$B(pz*)(pz^{tBu2})]ZnI, which readily added Hpz^{tBu,iPr} to produce the adduct [H$_2$B(pz*)(pz^{tBu2})]ZnI(Hpz^{tBu,iPr}), the structure of which was established by X-ray crystallography.[144]

4.2.1.8 [H$_2$B(pz^{trip})(pz^{tBu2})]$^-$

Again, the method of 4.2.1.6 was used to prepare this ligand, which was structurally characterized as the zinc complex [H$_2$B(pz^{trip})(pz^{tBu2})]ZnI. The zinc ion was in a tetrahedral environment, due to the presence of an agostic B—H—Zn bond.[144]

4.2.1.9 Bp^{Me}

This ligand was synthesized,[132] converted to a variety of first row transition metal complexes, and its enthalpy of chelation with divalent metals was determined.[1390] The tin(II) complexes Bp^{Me}$_2$Sn and Bp^{Me}SnCl have also been synthesized, and were characterized by NMR.[1120]

4.2.1.10 Bp^{iPr}

Complexes Bp^{iPr}$_2$M of the first row divalent metals Co, Ni, Cu and Zn were synthesized from the "as prepared" solution of Bp^{iPr}K, and were characterized by analysis and spectroscopy.[5]

4.2.1.11 Bp^{tBu}

The first reported compound of this ligand was the thallium salt, Bp^{tBu}Tl.[3] Several three-coordinate monoalkyl zinc derivatives Bp^{tBu}ZnR (R = Me, Et, Bu^t) were prepared, of which Bp^{tBu}ZnBu^t was structurally characterized. They reacted with HOAc yielding Bp^{tBu}Zn(κ^2-OAc), and with water they formed the structurally characterized cyclic trimer [Bp^{tBu}Zn(μ-OH)]$_3$. The B—H part of the ligand reacted with formaldehyde, acetaldehyde or acetone, yielding complexes of new ligands, [(R'O)Bp^{tBu}]ZnR, where R' was Me, Et and Pr^i, respectively.[144,1088] A number of nickel complexes Bp^{tBu}Ni(Ar)(PMe$_3$)$_2$ (Ar = Ph, Tol, An, Ph-p-NMe$_2$), containing a very rare example of a monodentate Bp^{tBu} ligand, has been reported.[1005] A structure similar to **191** was obtained by using the dimer [Bp^{tBu}Cu]$_2$ instead of [Bp*Cu]$_2$ in the same reaction.[1436] Treatment of Bp^{tBu}K with either AlMe$_3$ or with Me$_2$AlCl produced Bp^{tBu}AlMe$_2$, which on heating rearranged to Bp^{tBu*}AlMe$_2$ (i.e. [H$_2$B(3-Bu^{t}pz)(5-Bu^{t}pz)]AlMe$_2$.[1104] The rather unusual dinuclear complex, assigned structure **192**, has also been reported.[1004]

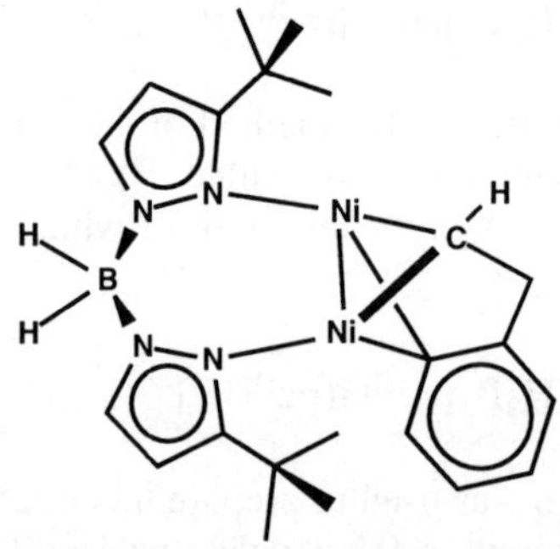

192

4.2.1.12 Bp^Trip

This ligand, containing a very bulky tripticyl group, has been structurally characterized as the thallium(I) salt, Bp^{Trip}Tl, which contained an agostic B—H—Tl bond. No other chemistry of this ligand has been reported.[97]

4.2.1.13 Bp^Ph

This ligand was isolated as the thallium(I) salt, and it was converted to several Bp^{Ph}$_2$M complexes (M = Co, Ni, Zn), as well as to the molybdenum derivatives Bp^{Ph}Mo(CO)$_2$(η^3-CH$_2$CRCH$_2$) (R = H, Me).[3]

4.2.1.14 Bp^Fc

The Tl salt of the Bp^{Fc} ligand (Fc = ferrocenyl) has been prepared and characterized, but no coordination chemistry thereof has been reported.[133]

4.2.1.15 Bp^Et2

The Bp^{Et2} ligand, somewhat more hindered than Bp*, was converted to [Bp^{Et2}]$_2$Co and [Bp^{Et2}]$_2$Ni, as well as to [Bp^{Et2}Mo(CO)$_4$]$^-$ which yielded, upon treatment with allylic halides, Bp^{Et2}Mo(CO)$_2$(η^3-CH$_2$CRCH$_2$) complexes (R = H, Me, Ph), which were characterized by NMR.[22]

4.2.1.16 Bp^(CF3)2

This ligand was synthesized,[116, 117] and the structures its Cu and Zn complexes and of the K salt were determined by X-ray crystallography. [Bp$^{(CF_3)2}$]$_2$Cu had a square planar structure, while that of zinc was tetrahedral, and quite similar to that of Tp*$_2$Zn, the structure of which was also determined.[134] In the dimeric complex

4.2 LIGANDS Bp^x

{[Bp$^{(CF_3)2}$]Cu(CNBu^t)}$_2$ the pseudoxial B—H of each ligand was coordinated to the Cu ion of its partner molecule.[1337] The ruthenium coordination chemistry of this ligand was exemplified by the structurally characterized [Bp$^{(CF_3)2}$]RuH(COD), which contained an agostic B—H—Ru bond. This molecule was converted to [Bp$^{(CF_3)2}$]RuH(PR$_3$)$_2$, to the benzene complex [Bp$^{(CF_3)2}$]RuH(C$_6$H$_6$), in which there was no agostic B—H—Ru bond, and to [Bp$^{(CF_3)2}$]RuH(H$_2$)(PCy$_3$)$_2$ in which the [Bp$^{(CF_3)2}$] ligand coordinated in the rare, monodentate, mode.[1441]

4.2.1.17   Bp^{Ph2}

The synthesis of this ligand was reported, and it was converted to the molybdenum complex Bp^{Ph2}Mo(CO)$_2$(η^3-allyl).[22]

4.2.1.18   Bp^{tBu,iPr}

The organozinc complex of this ligand, [Bp^{tBu,iPr}]ZnMe, was converted by treatment with paraformaldehyde to [(MeO)Bp^{tBu,iPr}]ZnMe, which involved insertion of CH$_2$O into the pseudoaxial B—H bond.[146] In an analogous reaction, Ph$_2$C=S was inserted into [Bp^{tBu,iPr}]ZnI to yield [(Ph$_2$CHS)Bp]ZnI, modelling the binding of histidine and cysteine residues at the active sites of several enzymes.[1442] Surprisingly, the cobalt complex of this ligand, [Bp^{tBu,iPr}]$_2$Co, turned out to be square planar, and containing very short (1.95 Å) agostic B—H—Co bonds being, when counting the agostic bonds, octahedral.[1443]

4.2.1.19   Bp^{Br3}

Although tribromopyrazole is a moderately strong acid, and its anion could act as a good leaving group, the Bp^{Br3} ligand was stable, and produced complexes typical of other Bp^x ligands. Thus, it readily formed Bp^{Br3}$_2$M complexes with first row transition metals, and the structure of its Bp^{Br3}Mo(CO)$_2$(η^3-CH$_2$CMeCH$_2$) complex, which was determined by X-ray crystallography,[83] resembled closely that of the Bp* analog.[23]

4.2.1.20   Bp^{Py}

A potentially tetradentate ligand (just as Tp^{Py} is potentially hexadentate), Bp^{Py} has formed a variety of lanthanide complexes with the general structure [Bp^{Py}$_2$LnX]$^{n+}$ (X = H$_2$O, DMF or NO$_3$). In the complexes [Bp^{Py}$_2$Eu(DMF)]$^+$, [Bp^{Py}$_2$Tb(NO$_3$)], and [Bp^{Py}$_2$Tb(H$_2$O)][Bp^{Py}] there were two tetradentate Bp^{Py} ligands, plus an ancillary ligand completing the coordination sphere. All the above complexes were studied by chemiluminescence.[147] Also prepared were the structurally characterized compounds Bp^{Py}Tl and Bp^{Py}$_2$Pb. The Bp^{Py}Tl structure revealed molecular stacking, with two long bonds from the pyridyl nitrogen atoms to Tl, while in Bp^{Py}$_2$Pb all eight nitrogen atoms were directed at the metal ion, with variable degrees of interaction.[1444]

4.2.1.21 Bp[2,4(OMe)2Ph]

The Bp[2,4(OMe)2Ph] ligand was, just like Tp[Py], also potentially tetradentate, except that unlike Bp[Py], it would involve N,N,O,O rather than N,N,N,N ligation, and the second chelating ring would be six- rather than five-membered. The Bp[2,4(OMe)2Ph] ligand was found to coordinate in N,N,O tridentate fashion, as was determined in the structure of the octahedral nickel complex, [Bp[2,4(OMe)2Ph]]$_2$Ni.[83]

4.2.1.22 Bp[Bipy] [Bipy = 6'-(2,2'-bipyridyl]

This potentially hexadentate ligand, Bp[Bipy], **193**, has been synthesized, and the structure of its K salt was determined by X-ray crystallography. It contained a dinuclear double-helical complex [Bp[Bipy]K]$_2$ with hexa-coordinate potassium ions.[1445] Other complexes of this ligand included those of Mn[II], Cu[II], Zn[II], Tl[I], Gd[III], Ce[III] and Er[III]. The first row ions produced cationic complexes of composition [Bp[Bipy]$_2$M]$_2$[BF$_4$], which were dinuclear. The structure of the complex [Bp[Bipy]Cu]$_2$[BF$_4$]$_2$ revealed a double helical ligand array, which also included stacking

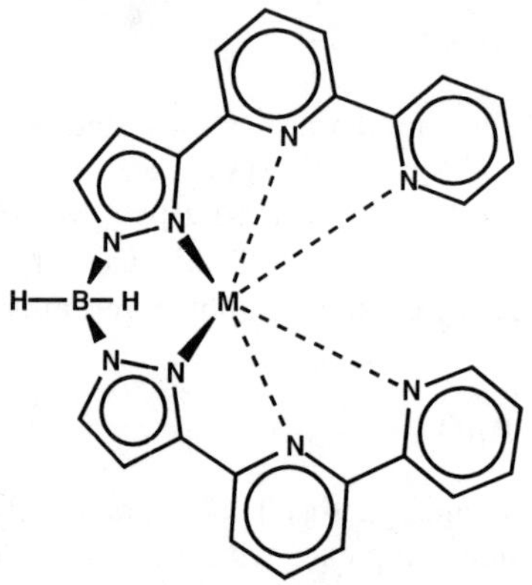

193

between ligands, while Bp[Bipy]Gd(NO$_3$)$_2$ showed a hexadentate Bp[Bipy] ligand with two κ^2 nitrate ions. In Bp[Bipy]Tl the metal was three-coordinate.[1446]

4.2 LIGANDS R_2Bp^x

4.2.1.23 Bp^{4Bo}, $Bp^{4Bo,5NO_2}$ and $Bp^{4Bo,5NO_2}$

The synthesis of the above ligands: Bp^{4Bo},[141] $Bp^{4Bo,5NO_2}$,[142] and $Bp^{4Bo,5NH_2}$,[1382] as well as of some of their metal complexes has been reported, but none of these compounds have been structurally characterized.

4.2.2 R_2Bp^x Ligands

Ligands of structure R_2Bp^x, where R substituents are alkyl or aryl groups differ from the Bp^x ligands in having no reducing power of the BH_2 grouping, and therefore can be readily used with reduction-prone metal ions, such as silver(I) or palladium(II). In contrast to the relatively easy formation of agostic B—H—M bonds with Bp^x ligands, the formation of agostic B—C—H—M bonds does not occur very frequently, and these bonds are easily broken by donor ligands. At the same time, the R groups may shield effectively the coordinated metal from other ligands as, for instance, in the case of Et_2Bp_2Ni. On the other hand, no agostic interaction has ever been found with the Ph_2Bp ligand. The reason for this is the mutually orthogonal disposition of the two phenyl groups, with the psudoaxial one being distant from the coordinated metal. There are only two reports of R_2Bp^x ligands, where R is a halogen.

4.2.2.1 Me_2Bp

The simplest R_2Bp ligand, Me_2Bp, was converted to $[Me_2Bp]_2M$ complexes for M = Ni, Cu and Zn, which were studied by spectroscopy and by NMR.[135] From the dynamic NMR spectrum of the square planar $[Me_2Bp]_2Ni$ complex the E_a for the boat inversion was found to be 67 ± 7 kcal.[1447] The Me_2Bp ligand was also converted to the structurally characterized $[Me_2Bp]GaMe_2$ complex.[1448]

4.2.2.2 Et_2Bp

The synthesis of the $[Et_2Bp]^-$ ligand from BEt_3, $[pz]^-$ and Hpz, and its conversion of $[Et_2Bp]_2M$ complexes (M = Co, Ni, Cu, Zn) was reported in 1967. The presence of the pseudoaxial ethyl groups in $[Et_2Bp]_2Ni$ prevented the addition of donor ligands (NH_3, py) through steric blocking of the space above and below the plane of the molecule, and the proximity of the one methylene group per ligand was expressed by a considerable shift of the CH_2 protons in the NMR spectrum,[11] even though there was no agostic B—C—H—Ni bonding.[1449] The $[Et_2Bp]_2Cr$ complex had a similar structure and was remarkably air-stable.[1450] In tetrahedral complexes, however, the B-ethyl groups did not prevent the $[Et_2Bp]_2Mn$ and $[Et_2Bp]_2Fe$ complexes from being just as air-sensitive as their simple Bp analogs.[11]

The [Et$_2$BpMo(CO)$_4$]$^-$ anion, prepared from Mo(CO)$_6$ and [Et$_2$Bp]$^-$, was readily converted to Et$_2$BpMo(CO)$_2$(η^3-CH$_2$CRCH$_2$) (R = H, Ph) complexes, as well as to related species Et$_2$BpMo(CO)$_2$(η^3-cyclohexenyl), and Et$_2$BpW(CO)$_2$(η^3-CH$_2$CHCH$_2$)(Hpz).[22] The nominally 16-electron complexes Et$_2$BpMo(CO)$_2$(η^3-CH$_2$CRCH$_2$) readily added donor ligands (Hpz, pyridine), and in their absence formed an agostic C—H—Mo bond through one the methylene hydrogens of the pseudoaxial ethyl group, as was suggested by low-frequency C—H stretches in the IR, and by a shift of the pseudoaxial methylene protons into the "hydridic" range. These spectroscopic features disappeared upon the addition of donor ligands.[22] The presence of this agostic bonding was confirmed by X-ray crystallographic structure determination of Et$_2$BpMo(CO)$_2$(η^3-CH$_2$CPhCH$_2$),[24] and of its (η^3-allyl) analog, **194**.[1451] The Hpz adduct of the latter complex, Et$_2$BpMo(CO)$_2$(η^3-allyl)(Hpz), **195**, was structurally characterized as devoid of the agostic bond, and with the B(μ-pz)$_2$Mo

194 **195**

(-N=N- represents μ-pz)

ring in a very unusual chair conformation.[1452] The various dynamic processes of making and breaking such agostic bonds were studied, and the pertinent energy barriers were calculated.[25] In the structurally characterized complex Et$_2$BpMo(CO)$_2$(η^3-C$_7$H$_7$), agostic C—H—Mo bonding was found to override a possible pentahapto-C$_7$H$_7$ structure, as a means of achieving the 18-electron configuration.[1453] The reaction of [Et$_2$Bp]$^-$ with Mo$_2$(OAc)$_4$ readily yielded the expected [Et$_2$Bp]$_2$Mo$_2$(OAc)$_2$ complex,[426] but, in addition, also the totally unexpected [Et$_2$Bp]$_2$Mo$_2$[Et$_2$B(pz)(OH)]$_2$, containing the ligand Et$_2$B(pz)(OH), which must have arisen through partial hydrolysis of [Et$_2$Bp]$^-$.[1454]

Rhodium(I) derivatives, Et$_2$BpRh(CO)$_2$ and Et$_2$BpRh(CNR)$_2$, were synthesized and converted to rhodium(III) species by oxidative addition of I$_2$, MeI or HgCl$_2$.[1455] Complexes [Et$_2$Bp]$_2$Pd and Et$_2$BpPd(η^3-allyl) were also reported.[1456] Stable Et$_2$BpPtMe(L) complexes (L = PR$_3$, CNR, PhC≡CPh, PhC≡CMe) were prepared, but with negatively substituted acetylenes, such as F$_3$CC≡CCF$_3$ or ROOCC≡CCOOR, insertion into the Pt-Me bond took place.[1457,1458] The structure of Et$_2$BpPtMe(PhC≡CMe) showed *cis*-bending away of the acetylenic substituents

4.2 LIGANDS R_2Bp^x

from Pt by 18 to 21°.[1459] The thermally sensitive tin complexes [Et$_2$Bp]$_2$SnR$_2$ (R = Me, Et, Bu) were synthesized and studied by NMR.[1123] Aluminum and gallium derivatives, [Et$_2$Bp]AlEt$_2$ and [Et$_2$Bp]GaEt$_2$, have also been reported.[1106]

An interesting assortment of cationic complexes, [Et$_2$B(μ-pz^x)$_2$CR$_2$]$^+$, has been prepared, and isolated as [PF$_6$]$^-$ or [B$_{12}$H$_{12}$]$^{2-}$ salts. The bridging pyrazoles included pz and pz*, while the R groups were H, Me, pz, tetramethylene, (H,Me), and (H,pz). They can be regarded as arising from the reaction of [Et$_2$Bp]$^-$ with R$_2$CX$_2$, resulting in the replacement of two X$^-$ anions, which would make them Et$_2$Bp complexes of a carbon moiety. In actuality, their mode of preparation involved the reverse reaction, namely that of a geminal polypyrazolylmethane with Et$_2$BX, where X was CF$_3$COO or OTs.[187]

$$R_2C(pz^x)_2 + Et_2BX \rightarrow [Et_2B(\mu\text{-}pz^x)_2CR_2]^+ + X^- \quad (4.2)$$

4.2.2.3   Et$_2$Bp^{Fc}

This ligand has been prepared as the K salt, and it was converted to the complex [Et$_2$Bp^{Fc}]Pd(η^3-allyl).[133]

4.2.2.4   Pr$_2$Bp

Reacting Me$_2$NBPr$_2$ with pyrazole plus Kpz produced the Pr$_2$Bp ligand, which was isolated as the K salt.[1456]

4.2.2.5   Bu$_2$Bp

Although there is a close similarity between the [Bu$_2$Bp]$^-$ and [Et$_2$Bp]$^-$ ligands, including the shift of the pseudoaxial methylene protons in the NMR of the [Bu$_2$Bp]$_2$Ni complex,[11] not much has been done with this ligand, because of its high lipophilicity, and difficulty of obtaining crystalline complexes. Nevertheless, the palladium(II) and platinum(II) complexes [Bu$_2$Bp]MCl(PEt$_3$) have been synthesized and characterized by NMR.[1013]

4.2.2.6   (BBN)Bp

A special case of an R$_2$Bp ligand was (BBN)Bp, **196**, prepared easily from 1,5-borabicyclononane, [pz]$^-$ and Hpz, in which the R$_2$ portion of the ligand formed a rigid cage structure.[137] Because of this, only one tertiary hydrogen was directed at the coordinated metal ion, and it was shown by X-ray crystallography that the complex [(BBN)Bp]$_2$Co contained two agostic C—H—Co bonds.[1460] This molecule has been studied in detail by spectroscopy.[1461] Other homoleptic and heteroleptic complexes

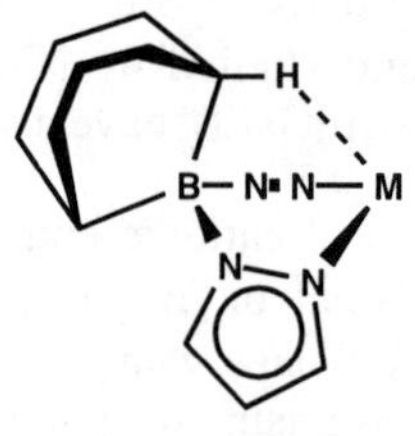

196

containing the (BBN)Bp ligand have also been synthesized. They included (BBN)BpTl, [(BBN)Bp]$_2$M (M = Ni, Zn), [(BBN)Bp]Pd(η^3-CH$_2$CRCH$_2$) (R = H, Me), and [(BBN)Bp]M(CO)$_2$(η^3-CH$_2$CRCH$_2$) (M = Mo, W; R = H, Me, Ph). While in homoleptic complexes it was possible for the agostic interaction to occur, in sterically hindered heteroleptic ones this did not happen because of the bulk of the (BBN)Bp ligand, as was demonstrated in the case of (BBN)BpCoTp^{iPr,4Br}. In that complex, the cobalt ion was only five-coordinate, the tertiary apical C—H being pushed away by the three 3-isopropyl groups of the Tp^{iPr,4Br} ligand.[137] The rhodium(I) complex (BBN)BpRh(CO)$_2$, and its COD and NBD analogs have also been synthesized, and characterized by multidimensional NMR spectroscopy. The structure of (BBN)BpRh(COD) indicated the presence of a weak agostic C—H—Rh bond, which may have been responsible for the lack of inversion of the B(μ-pz)$_2$Rh ring.[138] The free acid (BBN)Bp has also been reported.[1463]

4.2.2.7 (BBN)Bp^{Me}

This analog of (BBN)Bp has been synthesized, and converted to the complexes [(BBN)Bp^{Me}]Rh(CO)$_2$, [(BBN)Bp^{Me}]Rh(COD) and [(BBN)Bp^{Me}]Rh(NBD), which were studied by multidimensional NMR spectroscopy.[138] The free acid [(BBN)Bp^{Me}]H has also been reported.[1463] The reaction of K(BBN)Bp^{Me} with Fe(CO)$_2$Me(I)(PMe$_3$)$_2$ resulted in partial degradation of the ligand, yielding the compound **197**

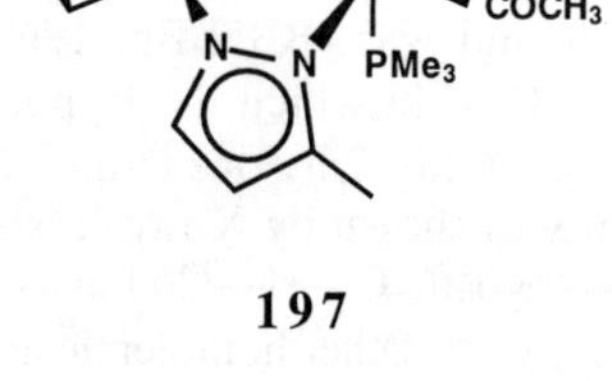

197

4.2 LIGANDS R_2Bp^x

4.2.2.8 (BBN)Bp^{Ph}

This ligand is only known as the free acid, [(BBN)Bp^{Ph}]H.[1463]

4.2.2.9 (BBN)Bp*

This ligand is only known as the free acid, [(BBN)Bp*]H.[1463]

4.2.2.10 (BBN)Bp^{Ph,Me}

This ligand is only known as the free acid, [(BBN)Bp^{Ph,Me}]H.[1463]

4.2.2.11 (BBN)Bp^{Fc}

A more complicated analog of (BBN)Bp, the (BBN)Bp^{Fc} ligand, was prepared as the K salt, and it was converted to [(BBN)Bp^{Fc}]Pd(η^3-allyl).[133]

4.2.2.12 Ph$_2$Bp

Unlike the aliphatic [R$_2$Bp]$^-$ ligands, the aromatic [Ph$_2$Bp]$^-$, which was available either from NaBPh$_4$ and Hpz, or from H$_3$NBPh$_3$, [pz]$^-$ and Hpz, was very stable itself, and formed numerous highly crystalline derivatives.[11] A major distinguishing feature of this ligand was, that none of its complexes have ever shown an agostic C—H—M interaction. The reason for this is the relationship of the two phenyl groups, in which the pseudoequatorial one is bisecting the angle between the pyrazolyl planes in the B(μ-pz)$_2$M ring, while the pseudoaxial phenyl group is orthogonal to it. This has been found in simple complexes such as [Ph$_2$Bp]$_2$Ni,[1464] and complicated ones, such as the 16-electron species Ph$_2$B(pz)$_2$Mo(CO)$_2$(η^3-CH$_2$CMeCH$_2$).[26] The 16-electron complex Ph$_2$BpMo(CO)$_2$(η^3-C$_7$H$_7$), and its Hpz adduct Ph$_2$BpMo(CO)$_2$(η^3-C$_7$H$_7$)(Hpz) were prepared, and the structure of the latter complex was established by X-ray crystallography.[1213]

Rhodium(I) complexes, such as [Ph$_2$Bp]Rh(LL) (LL = COD, NBD, (CO)$_2$) were studied by NMR, and compared to their (BBN)Bp analogs. Unlike the latter, they did display inversion of the B(μ-pz)$_2$Rh ring.[138] A heteroleptic complex containing both, carbollide and scorpionate ligands, the anion [*closo*-3(Ph$_2$Bp)-3,1,2-NiC$_2$B$_9$H$_{11}$]$^-$ was prepared, and oxidized by FeCl$_3$ to the neutral species [*closo*-3(Ph$_2$Bp)-3,1,2-NiC$_2$B$_9$H$_{11}$].[931] Stable [MePt^{II}] complexes of general structure [Ph$_2$Bp]PtMe(L), where L was PR$_3$, RNC, or a variety of acetylenes, were synthesized,[1457] as were the copper(I) species [Ph$_2$Bp]Cu(L), (L = various phosphines and arsines).[1414] It was noted, that while [Ph$_2$Bp]Ag was a stable molecule, the corresponding Bp analog rapidly reduced the silver(I) ion.[1078] Also reported were tin(IV) complexes, [Ph$_2$Bp]SnCl(R$_2$) (R = Bu, Ph), [Ph$_2$Bp]SnR$_3$, and [Ph$_2$Bp]$_2$SnR$_2$ (R = Bu, Ph),[1422] and tin(II) compounds Ph$_2$BpSnCl and [Ph$_2$Bp]$_2$Sn.[1120] [Ph$_2$Bp]$_2$SiCl$_2$, the only example of a structurally characterized silicon scorpionate complex, was prepared from Ph$_2$BpK and SiCl$_4$, and its structure was determined by X-ray crystallography.[1466] An actinide derivative of the [Ph$_2$Bp] ligand, [Ph$_2$Bp]$_2$UCl$_2$, has also been reported.[1163]

4.2.2.13   Ph$_2$Bp^{pm}

This ligand, having the *cis*- structure, **198**, was synthesized, and characterized as the Tl salt. It was converted to [Ph$_2$Bp^{pm}]CuI, which was used in a study of the cyclopropanation of styrene with ethyl diazoacetate.[1257]

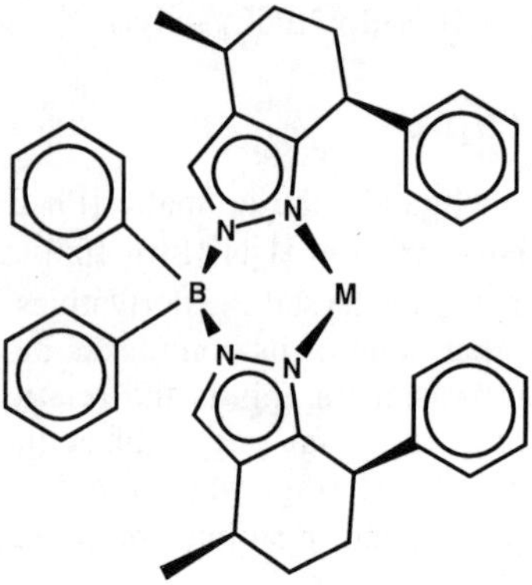

198

4.2.2.14   F$_2$Bp*

Only a single report exists about the preparation of the F$_2$Bp* ligand, and its conversion to the square planar [F$_2$Bp*]$_2$Ni and tetrahedral [F$_2$Bp*]$_2$Co complexes.[11]

4.2.2.15   X$_2$Bp   (X = Cl, Br, I)

Although the ligands themselves have not been prepared, it has been reported that halogenation with chlorine, bromine or iodine, respectively, converted the BH$_2$

4.3 LIGANDS R(R'Z)Bp^X

moiety in Bp$_2$SnR$_2$ complexes to BX$_2$, thus yielding [X$_2$Bp]$_2$SnR$_2$ species, which were characterized by analysis, and by spectroscopy.[1422]

4.2.2.16   Ph(Me)Bp^{Me}

This is the only heteroscorpionate ligand containing one alkyl and one aryl group bonded to boron. It was synthesized by first converting MeB(OPr^i)$_2$ to Ph(Me)B(OPr^i), which was followed by its reaction with [pz^{Me}]$^-$ + Hpz^{Me}. Rhodium(I) complexes, [Ph(Me)Bp^{Me}]Rh(LL) (LL = (CO)$_2$, NBD and COD) were prepared, and it was shown by NMR studies that the COD and NBD complexes exist in solution in single isomeric forms, having the phenyl group in the pseudoaxial position.[77]

4.2.2.17   Ligands of Structure 199

Although no complexes have been prepared from the ligand having structure **199** in its "free acid" form, which was obtained by the exothermal addition of pyrazole to 1,3-dimethyl-1,3-diaza-2-boracyclopentane, such complexes should be accessible either via deprotonation of **199**, followed by the reaction with a metal cation, or even via a direct reaction of this free acid with an appropriate metal salt.[1467]

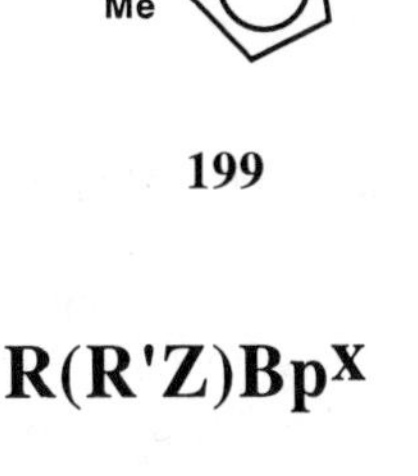

199

4.3   Ligands R(R'Z)Bp^X

4.3.1   General Considerations

In this section will be covered Bp^x ligands which are definitely tridentate, due to the presence of a heteroatom Z which can, and does, coordinate to the metal, in addition anion [R(R'Z)Bp^x]$^-$, and were then used for coordination purposes, others were prepared within an existing complex, at times by accident. The [R(R'Z)Bp^x]$^-$ ligands will also include those where R'Z is a pyrazolyl group, provided it is different from pz^x, and is not its regioisomer. It will not include Tp^{x*} ligands, arising from a rearrangement of a Tp^x ligand, as those were dealt with in Chapter 3.

4.3.2 Specific Ligands

4.3.2.1 (p-TolS)Bp*

The first ligand of this category was prepared from KBp* and p-TolSH, and it was converted to several copper(II) and cobalt(II) complexes of structure (p-TolS)Bp*Cu(L), **200**, where L was O-ethylcysteinato, p-nitrophenylthiolato, and pentafluorophenylthiolato. Their spectroscopic properties were compared with those of the N$_2$S$_2$ active site in poplar plastocyanin.[139]

200   **201**

4.3.2.2 (MeBnS)Bp* (MeBn = p-methylbenzyl)

This ligand was not prepared as such, but was obtained when the complex [W(≡C-p-Tol)Br(CO)$_4$ was treated sequentially with sulfur, and then with KBp*. The resulting product was assigned the structure [(MeBnS)Bp*]W(CO)$_2$(S$_2$C-p-Tol).[140]

4.3.2.3 (Ph$_2$CHO)Bp^{tBu,iPr}

This particular ligand was obtained by the insertion of benzophenone into the pseudoaxial B—H bond of the zinc complex [Bp^{tBu,iPr}]ZnI to yield the fully characterized compound [(Ph$_2$CHO)Bp^{tBu,iPr}]ZnI, **201**.[1442]

4.3 LIGANDS R(R'Z)Bp^X

4.3.2.4 (Ph$_2$CHS)Bp^{tBu,iPr}

An *in situ* synthesis of this ligand involved insertion of Ph$_2$C=S into the pseudoaxial B—H bond in [Bp^{tBu,iPr}]ZnI to yield [(Ph$_2$CHS)Bp]ZnI, and this complex modelled the binding of histidine and cysteine residues at the active sites of several enzymes.[1442]

4.3.2.5 (*i*-PrO)Bp*

A molybdenum complex of this ligand, (Pr^{i}O)Bp*Mo(CO)$_2$NO, was obtained during one particular synthesis of Tp*Mo(CO)$_2$NO, and its genesis remains obscure. Nonetheless, its structure was confirmed by X-ray crystallography.[1468]

4.3.2.6 (MeO)Bp^{tBu}

The three-coordinate alkylzinc derivative Bp^{tBu}ZnEt was found to react readily with paraformaldehyde, yielding the tetrahedral zinc complex (MeO)Bp^{tBu}ZnEt, arising from insertion of the carbonyl group into the pseudoaxial B—H bond.[1088]

4.3.2.7 (EtO)Bp^{tBu}

This ligand was formed *in situ* during the reaction of acetaldehyde with Bp^{tBu}ZnEt yielding the tetrahedrally coordinated zinc complex (EtO)Bp^{tBu}ZnEt, containing a bridging (μ-OEt) group.[1088]

4.3.2.8 (Pr^{i}O)Bp^{tBu}

Treatment of Bp^{tBu}ZnEt with acetone produced the complex (Pr^{i}O)Bp^{tBu}ZnEt, containing an isopropoxy bridge between boron and zinc, another example of *in situ* ligand formation.[1088]

4.3.2.9 (MeO)Bp^{tBu,iPr}

The reaction of Bp^{tBu,iPr}ZnMe with formaldehyde resulted in a net insertion of CH$_2$O into the pseudoaxial B—H bond, and formation of the structurally characterized tetrahedral zinc complex [(MeO)Bp^{tBu,iPr}]ZnMe, containing a coordinated B—OMe group.[145]

4.3.2.10 (HCOO)Bp^{tBu,iPr}

Another conversion of Bp^{tBu,iPr} to a N,N,O ligand was achieved by the reaction of CO$_2$ with [Bp^{tBu,iPr}]ZnCl. This resulted in insertion of CO$_2$ into the B—H bond, and the structure of the resulting complex, **202**, was confirmed by X-ray crystallography.[1469]

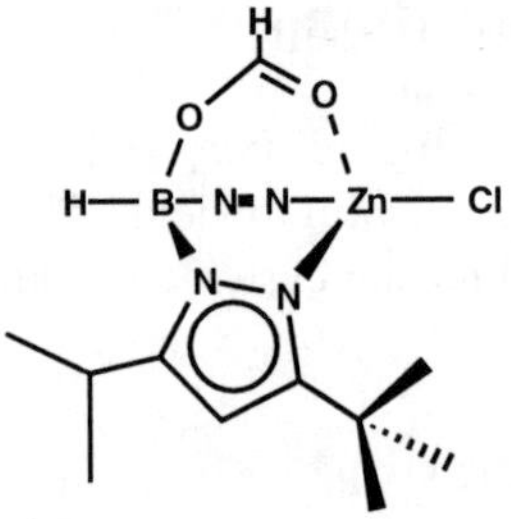

202

4.3.2.11 CpBp*

Thermolysis of Tp*$_2$SmCp resulted in the loss of Hpz*, and formation of the structurally characterized species [CpBp*]SmTp*, in which the Cp was σ-bonded to the pseudoaxial position on boron, at the same time retaining the η^5-bonding to Sm. The unusual CpBp* ligand, capable of being dinegative, and occupying a total of seven coordination sites, was not independently isolated.[1470]

4.3.2.12 (pz^{4CN})Bp

Again, the ligand was not prepared directly, but it was obtained by heating the nickel complex Bp$_2$Ni with Hpz^{4CN}, which led to the formation of the octahedral species [(pz^{4CN})Bp]$_2$Ni which is, in effect, a Tp$_2$Ni complex containing, in addition, two 4-cyano groups, presumably *trans* to each other. These cyano groups were not hydrolyzed by hydrochloric acid, but were hydrolyzed by alkali, and the resulting dicarboxylic acid was shown to be capable of various standard organic transformations without destruction of the Tp$_2$Ni core.[52,83]

4.3.2.13 (pz*)Bp

This ligand was prepared by heating BpK with Hpz*, and it was converted directly to the Fe^{II} complex [(pz*)Bp]$_2$Fe which was studied by optical spectroscopy and magnetic susceptibility measurments.[825]

4.3.2.14 (pz)Bp*

This ligand was prepared by heating Bp*K with Hpz, and it was converted directly to the Fe^{II} complex [(pz)Bp*]$_2$Fe which was studied by optical spectroscopy and magnetic susceptibility measurments.[825]

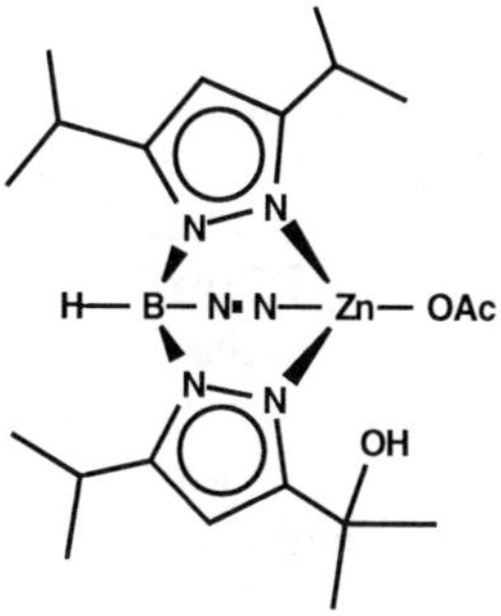

4.4 PYRAZABOLES

203

4.3.2.15 [3-(CMe$_2$OH)-5-iPrpz]Bp^{iPr2}

This unusual heteroscorpionate ligand was prepared from the manganese complex [Tp^{iPr2}Mn]$_2$(μ-OH)$_2$ via a two-stage oxidation which led to oxygenation of one tertiary isopropyl position per ligand, and it was isolated as the structurally characterized zinc complex {[3-(CMe$_2$OH)-5-iPrpz]Bp^{iPr2}}Zn(OAc), **203**.[1304]

4.3.2.16 (RO)Bp* (R = Me, Et)

Although these ligands were not prepared directly, their rhenium tetrahydride complexes of structure (RO)Bp*ReH$_4$(PPh$_3$) have been obtained in rather low yields by the reaction of ReOCl$_3$(PPh$_3$)$_2$ in the appropriate alcohol with a large excess of NaBp*, and the structure of the complex with R = Et was established by X-ray crystallography.[1565]

4.3.2.17 (RO)$_2$Bp (R = Me, Et)

When the above reaction of ReOCl$_3$(PPh$_3$)$_2$ was carried out with excess NaBp, instead of NaBp*, the reaction took a different course, as both boron-bonded hydrogens were replaced by alkoxy groups, and the products were rhenium dihydrides, (RO)$_2$BpReH$_2$(PPh$_3$)$_2$, of which the one with R = Me was structurally characterized.[1565]

4.4 Pyrazaboles (= $R_2B(\mu\text{-pz}^x)_2BR'_2$)

The pyrazaboles are a special class of boron-nitrogen heterocycles, and their relationship to scorpionate ligands rests on several considerations. They are co-products of scorpionate ligand synthesis through the reaction of RBX_2 with Hpz^x, when both, the scorpionate free acid, $RB(pz^x)_3H$, and the pyrazabole, $R(pz^x)B(\mu\text{-pz}^x)_2BR(pz^x)$, are formed. They are also the thermal decomposition products of the scorpionate free acids, $[H_nB(pz^x)_{4-n}]H$, along with the pyrazole Hpz^x. Furthermore, they may be regarded as boronium scorpionate complexes, $[R_2Bp][BR'_2]$, and indeed, they are accessible by the synthetic route:

$$[R_2Bp^x]^- + R'_2BX \rightarrow R_2B(\mu\text{-pz}^x)_2BR_2' + X^- \qquad (4.3)$$

Lastly, some of the pyrazaboles containing two exocyclic boron-bonded pyrazolyl groups as, for instance, $Et_2B(\mu\text{-pz})_2B(pz)_2$, are also good chelating agents through these exocyclic pyrazolyl groups. They coordinate readily to metal ions, but differ from Bp ligands in being neutral, rather than anionic. For this reason they give rise to cationic metal complexes.

The two main types of pyrazabole structures, and their numbering system are shown below:

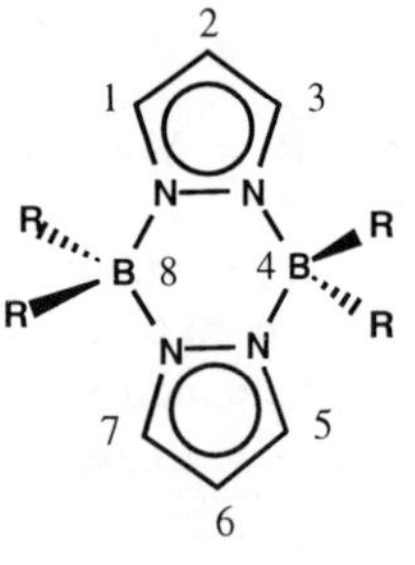

204 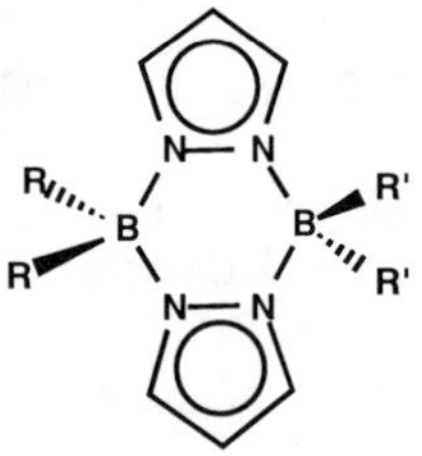 **205**

The main distinction between the structures **204** and **205** is that the symmetrical **204** is obtained through dimerization of the 1,3-dipole, $R_2B(pz)$, which is an energetically favored process, according to CNDO and ab initio/IGLO/NMR studies.[1471,1472] Pyrazaboles of type **205**, which have different 4- and 8-substituents are synthesized from two different boron precursors: $[R_2Bp^x]^-$ and R'_2BX, as shown in Eq. (4.3).

4.4.1 Symmetrical pyrazaboles

The parent unsubstituted pyrazabole, **204** R = H, was prepared most readily by the reaction of the Me_3NBH_3 complex with pyrazole, as shown in Eq. (4.4).

$$2\ Me_3NBH_3 + 2\ Hpz \rightarrow H_2B(\mu\text{-}pz)_2BH_2 + 2\ NMe_3 \uparrow + 2\ H_2 \uparrow \qquad (4.4)$$

Borane complexes with pyridine or THF could also be used instead of Me_3NBH_3. This reaction also produced C-substituted pyrazaboles, when C-substituted pyrazoles were substituted for Hpz. For instance, the reaction of 4-X-substituted pyrazoles yielded 2,6-disubstituted pyrazaboles (X = Cl, Br, Me, CN, NO_2, $CF(CF_3)_2$). From Hpz^{R2} (R = Me, Ph, CF_3) the 1,3,5,7-tetrasubstituted pyrazaboles were obtained, and from Hpz^{R3} (R = Me, Br) the appropriate 1,2,3,5,6,7-hexasubstituted pyrazaboles. In the case of $Hpz^{(CF_3)_2}$ it was necessary to use the (THF)BH_3 complex in order to synthesize the pyrazabole, since the reaction with Me_3NBH_3 produced the compound $Me_3NBH_2[pz^{(CF_3)_2}]$ which was so stable that it could be distilled unchanged, and could not be converted to the pyrazabole.[1,10] When a 3-monosubstituted pyrazole, such as 3-methylpyrazole was used, a mixture of 1,5- and 1,7-dimethyl isomers was obtained.[79]

Similarly, heating pyrazoles with trialkyl or triarylboranes generated the 4,4,8,8-tetraalkylpyrazaboles, or their tetraaryl analogs.

$$2\ Hpz^x + 2\ BR_3 \rightarrow R_2B(\mu\text{-}pz^x)_2BR_2 + 2\ RH \uparrow \qquad (4.5)$$

The 4,4,8,8-tetraethylpyrazaboles included those with 2,6-substituents (Cl, Br, CN, NO_2, and $CF(CF_3)_2$), as well as the 1,3,5,7-tetramethyl, and the 1,2,3,5,6,7-hexasubstituted (Br, Me) derivatives.[1,10] 1,5-Diferrocenyl pyrazabole and its 4,4,8,8-tetraethyl analog were also reported.[133] The formation of 4,4,8,8-tetraethylpyrazabole from BEt_3 and Hpz was catalyzed by pivalic acid, or by $Bu^tCOOBEt_2$.[1473] Heating of pyrazole with a various $B(SR)_3$ compounds gave rise to the appropriate 4,4,8,8-tetrakis(alkylmercapto)pyrazaboles, and the same could be done with the BH_2 group in 4,4-dialkylpyrazaboles, forming 4,4-dialkyl-8-8-bis(alkylmercapto)pyrazaboles.[1474] The replacement of boron-bonded SR groups with F by means of boron trifluoride, and of boron-bonded Cl with Grignard reagents or with alcohols, has also been reported.[1475]

The boron-bonded hydrogens in pyrazaboles were readily replaced with halogens (Cl, Br, I) to yield the 4,4,8,8-tetrahalo derivatives,[12] although partial halogenation could also be achieved.[1476] Boron trihalides could also be used for this purpose.[1477] Boron-fluorinated pyrazaboles were prepared by the reaction of 4,8-unsubstituted pyrazaboles with Et_2OBF_3, or with $MeOHBF_3$, which yielded the separable 4,4-difluorinated and 4,4,8,8-tetrafluorinated derivatives. The BCl_2 moiety in pyrazaboles could be converted to $B(OAc)_2$, $B(O_2CCF_3)$, or $B(OMe)_2$. Thus, combining the direct synthesis of pyrazaboles with halogenation, and subsequent other transformations, a very large variety of substituted pyrazaboles was prepared.[1478] Some active hydrogen compounds, such as pyrocatechol or pyrazole, reacted with elimination of hydrogen and formation of the 4,4,8,8-tetrasubstituted derivatives **206** and **207**. When only two equivalents of pyrazole were used, the 4,8-

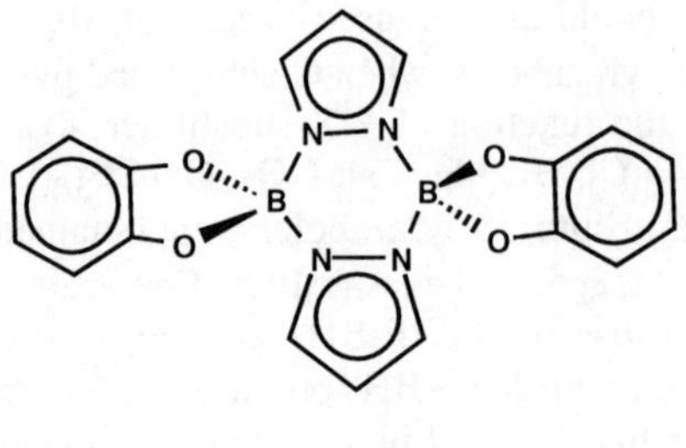

206 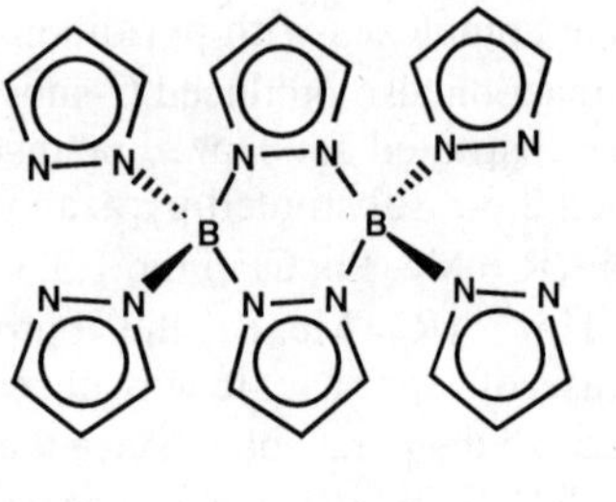   **207**

dipyrazolylpyrazabole was obtained as a mixture of *cis* and *trans* isomers. The reaction of RBX_2 compounds with pyrazole produced the scorpionate free acid, $RB(pz)_3H$, and the pyrazabole, $R(pz)B(\mu-pz)_2BR(pz)$.[12] The reaction of pyrazabole with diverse other pyrazoles was also studied.[1479] Several pyrazaboles containing an

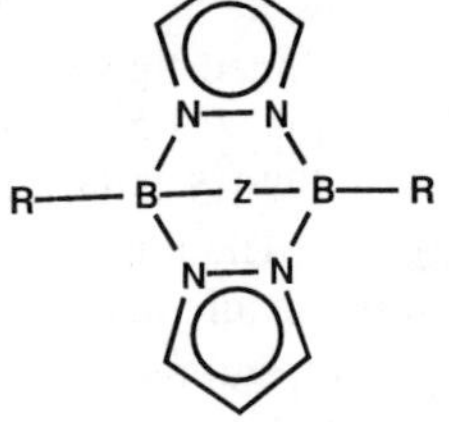

208

additional bridge, Z, between the boron atoms, with a general structure **208** were synthesized. The 4—8 bridge, Z, could be monoatomic, diatomic or triatomic. A monoatomic bridge example was the structurally characterized complex with Z = Se.[1480] Diatomic bridges were exemplified by the structurally characterized complexes **208** with Z = —S—S—,[1481] and —Se—Se—.[1480] The reaction of pyrazole with tri-B-organylboroxins produced pyrazaboles **208**, with a triatomic bridge Z = (O—BR—O).[1101,1482] The triply bridged (**208**, Z = R_2SN_2) pyrazabole was also reported,[1483] as was the rather unusual, but structurally characterized complex, where

4.4 PYRAZABOLES - UNSYMMETRICAL

Z was 1,1'-ferrocenyl. Other types of pyrazaboles with the 1,1'-ferrocenyl bridge, and containing substituents on the pz ring were also reported.[1484,1485] The structures of several of these pyrazaboles were determined by X-ray crystallography, and they were also studied by electrochemistry.[1486]

The stability of the pyrazabole ring was demonstrated by performing a number of typical organic reactions on variously 2,6-disubstituted 4,4,8,8-tetraethylpyrazaboles, without destruction of the tricyclic pyrazabole core. The diverse 2,6- substituents included CN, COOH, COONa, Br, Li, CHO, NO_2, NH_2 and $N(COMe)_2$.[12] A particularly telling example was the nitration with fuming nitric acid of 4,4,8,8-tetrafluoropyrazabole, which resulted in clean nitration of the 2,6-positions, with their subsequent reduction to the 2,6-diamino derivative.[1462] The structures of the parent pyrazabole, of the planar 4,4,8,8-tetrabromopyrazabole,[1477] and of other pyrazaboles, exemplified by 4,4'-bis(pyrazol-1-yl)pyrazabole, 1,3,5,7-tetramethylpyrazabole, and 4,4,8,8,-tetrakis(pyrazol-1-yl)pyrazabole,[1488] and also by 4,8-dichloro-4,8-diethylpyrazabole, 4,4-dichloro-8,8-diphenylpyrazabole, 4,4,8,8-tetrafluoropyrazabole,[1478] $[EtB]_2(\mu$-Pz$)_2(\mu$-OBRO),[1489] 4,8-diethyl-4,8-bis(pyrazol-1-yl)pyrazabole,[1490] and also by 2,6-dibromo-4,4,8,8,-tetraethylpyrazabole,[1487] were established by X-ray crystallography.

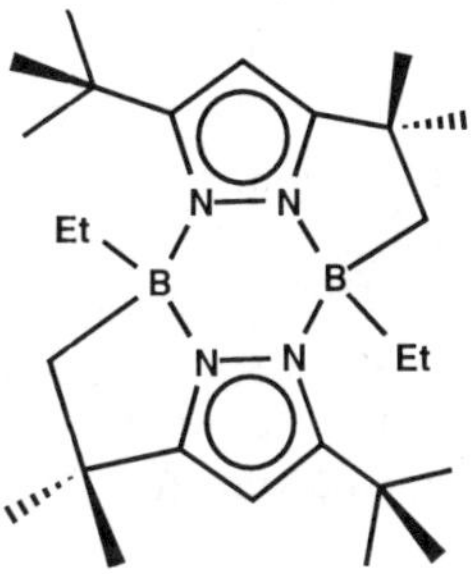

209

Pyrazaboles could not be obtained from R_2BH and Hpz^x components in those instances where both reactants contained very bulky substituents. Thus, the reaction of 1,5-borabicyclononane (BBN) with Hpz^{tBu2} yielded a monomeric pyrazolylborane, (BBN)-pz^{tBu2}, and the same happened with a number of other very hindered pyrazoles. Only with Hpz, Hpz^{4Br} and Hpz^{Me} were the appropriate pyrazaboles (BBN)(μ-pz^x)$_2$(BBN) obtained.[1491] In a more detailed study of this reaction it was shown that Et_2Bpz^{tBu2} can be first cracked to lose ethylene, producing the pyrazabole Et(H)B(μ-pz^{tBu2})$_2$B(H)Et, which was then converted at higher temperatures to the complicated pyrazabole **209**.[1492]

4.4.2 Unsymmetrical pyrazaboles

These pyrazaboles are defined as those where the 4- and 8-substituents are nonidentical. They are typically of structure $R_2B(\mu-pz^x)_2BR'_2$, and can be synthesized by the reaction of a Bp, Tp or pz°Tp ligand with an R'_2BX compound, in which X is a good leaving group, which can be a halide, an aryl- or alkylsulfonate, and even trifluoroacetate,[1417,1532] as was shown in equation (4.3). Considering the wealth of scorpionate ligands available, this reaction permits the synthesis of a wide variety of asymmetric pyrazaboles. On the scorpionate side the ligands Bp, Bp*, Et_2Bp, Ph_2Bp, and pzTp were used, while on the R_2BX side $Et_2B(OTs)$, $Bu_2B(OTs)$ were typical reagents, and Me_3BH_2I was employed as a source of the $[BH_2]^+$ fragment.[136] The BH_2 reactivity in pyrazaboles of structure $H_2B(\mu-pz^x)_2BR_2$ was the same as that of the symmetrical pyrazaboles, further expanding the synthetic scope of this area.

The ability of pyrazaboles containing geminal 4,4- and 4,4,8,8-pyrazolyl groups to have their own coordination chemistry was demonstrated by the synthesis of various types of spiro-cationic species. These may be exemplified by the cation **210** and the dication **211**. The cation **210** (R = Et) could be prepared in two ways: by the reaction of $pzTpPd(\eta^3$-allyl) with Et_2BX or, alternatively, by the reaction of 4,4-diethyl-8,8-bis(pyrazol-1-yl)pyrazabole with the palladiumm complex $[ClPd(\eta^3$-allyl)]$_2$. Treatment of pzTp with two equivalents of Et_2BX, produced the spiro cation $[Et_2B(\mu-pz)_2B(\mu-pz)_2BEt_2]^+$.[311] Finally, the dication **211** could be prepared starting with 4,4,8,8-tetrakis(pyrazolyl)pyrazabole and reacting it with $ZnCl_2$, Et_2BX, or $[ClPd(\eta^3$-allyl)]$_2$, respectively, which produced versions of **211** with $MX_2 = ZnCl_2$, Et_2B, or $Pd(\eta^3$-allyl), respectively.[1493-1495] Various boron-unsubstituted pyrazaboles, containing C-substituents, such as Me, Cl, CN, and Br, located at different positions of the pyrazolyl ring,[1496] and also those containing boron tetrasubstituted with ethyl groups, have been studied by mass spectrometry.[1946,1047]

210   **211**

Chapter 5
Applications of Scorpionate Ligands

5.1 General Considerations

This chapter presents examples of scorpionate ligands being studied for a variety of different applications. These include catalysis of polymerization, carbene/nitrene transfer, oxidation, the construction of models duplicating, or approximating, the spectral properties or the activity of various enzymes, metal extraction, metal deposition, and other uses. Some examples of studies using a series of Tp^x ligands to explore a particular phenomenon, plus a number of miscellaneous items have also been included in this chapter.

5.2 Catalysis

5.2.1 Polymerization and Oligomerization

The carbynes $Tp^*W(Cl)_2{\equiv}CBu^t$ and $Tp^*W(Br)_2{\equiv}CPh$ were converted by alumina to $Tp^*W(O)Cl(=CHBu^t)$, and $Tp^*W(O)Br(=CHPh)$, respectively, while the reaction of $Tp^*W(O)Br(=CHPh)$ with aniline yielded $Tp^*W(=NPh)Br(=CHPh)$. The latter compound, when combined with one equivalent of $AlCl_3$, catalyzed the acyclic diene metathesis (ADMET) oligomerization of 1,9-decadiene, producing a mixture of 1,9,17-octadecatriene, 1,9,17,25-hexadodecatetraene, and ethylene in about 30 % conversion. On the other hand, the complex $Tp^*W(=NPh)Br(=CHPh)$ was an efficient ring-opening metathesis polymerization (ROMP) catalyst, with $AlCl_3$ as a cocatalyst. While by itself, $Tp^*W(O)Cl(=CHBu^t)$ did not catalyze ring-opening polymerization of cyclooctene, in the presence of $AlCl_3$ a rapid reaction ensued, producing polyoctenamer in over 90 % yield.[679] Similar activity was shown by Mo complexes $TpMo(=CHR)(=NAr)(OTf)$ and $TpMo(=CHR)(=NAr)(Me)$.[671]

An air- and moisture-stable ring-opening metathesis polymerization (ROMP) catalyst was prepared on the basis of the air-stable, moisture-stable tungsten(VI) precursor, Tp*W(=CHBu^t)(O)Cl. Addition of AlCl$_3$ to this complex in cyclooctene at 25 °C, resulted in rapid ring opening polymerization of cyclooctene, producing high molecular weight polyoctenomer within 15 min. The catalyst retained its activity within the solidified polymer, as addition of more monomer led to further polymerization. The catalyst activity was identical in air, or in an inert atmosphere. Similar polymerization of norbornene could also be effected.[742]

The compound TpMo(=CHCMe$_2$Ph)(=NAr)(OTf), where Ar was 2,6-bis(isopropyl)phenyl, was by itself inert toward the metathesis of neat cyclooctene or 1,9-decadiene, and it did not polymerize norbornylene. However, in the presence of AlCl$_3$, it quantitatively catalyzed the ring-opening metathesis polymerization of cyclooctene, at a rate comparable to that of its tungsten analog. The complex TpMo(=CHCMe$_2$Ph)(=NAr)(Me) in the presence of AlCl$_3$ polymerized quantitatively norbornylene to a high polymer. However, neither of the above complexes, even in the presence of AlCl$_3$, would polymerize 1,9-decadiene. Only low yields of a dimer were obtained.[671] Related to the above molybdenum-based catalyst precursors were those based on tungsten, as exemplified by TpW(=CHBu^t)(=NPh)(CH$_2$Bu^t), TpW(=CHCMe$_2$Ph)(=NPh)(CH$_2$CMe$_2$Ph), as well as by TpW(=CHBu^t)[=N(2,6-(Pr^i)$_2$C$_6$H$_3$](CH$_2$Bu^t). In the presence of AlCl$_3$, these compounds catalyzed the ring-opening polymerization of cyclooctene.[728]

Dimerization of terminal alkynes was found to be catalyzed by the neutral complexes of Ru^{II}, such as TpRuCl(PPh$_3$)$_2$, TpRuCl(PPh$_3$)(py), and TpRuH(PPh$_3$)$_2$. The acetylenes studied were of structure RC≡CH (R = Ph, SiMe$_3$, n-Bu and t-Bu), and they yielded mixtures of 1,4- and 1,2-substituted butenynes, with the conversion and selectivity strongly dependent on the alkyne substituent. Phenylacetylene produced only head-to-head dimers, predominantly *trans*, while Me$_3$SiC≡CH yielded only the *cis* head-to-head dimer, plus some of the head-to-tail dimer. An almost random mixture of all three product isomers was obtained from Bu^{n}C≡CH, while Bu^{t}C≡CH led to the exclusive formation of the head-to-head *cis* isomer.[903] TpRu(=CHPh)Cl(PCy$_3$) catalyzed ring-closing olefin metathesis in the presence of either HCl, CuCl, or AlCl$_3$.[893]

The homopolymerization of phenylacetylene derivatives containing a variety of *para*-substituents (H, Me, Cl, CN, COOMe, NO$_2$) was studied in detail, employing Rh^I derivatives of sterically hindered homoscorpionates Tp^{R2}Rh(COD) (R = Me, Ph, Pr^i). The resulting polymers had a head-to-tail, *cis*-transoidal structure. The reaction was strongly dependent on the Tp substituents, with the larger ones giving better yields, and at higher rates (Pr^i > Ph > Me) . Reducing the bulk of the Rh complex by replacing COD with NBD decreased the yield, and replacement of COD with (CH$_2$=CH$_2$)$_2$ gave no product at all. No product was also obtained when

5.2 CATALYSIS

CpRh(COD) was used. The complexes [Bp*]Rh(COD) and [Bp$^{(CF_3)_2}$]Rh(COD) were also active, implying that the active species involved κ^2 coordination of the Tp^x ligand. While the polymerization was tolerant of a wide range of phenyl substituents, the *ortho* substituted 2,6-dimethyl-4-Bu^t-phenylacetylene, and aliphatic acetylenes RC≡CH (R = Bu^t, SiMe$_3$ and COOMe) were inactive.[969]

The hindered complex Tp^{tBu}MgOEt acted in methylene chloride at 22 °C as a catalyst precursor for the fast, stereoselective ring-opening polymerization of L,L-dilactide, which was by an acyl cleavage (rather than alkyl cleavage) mechanism, to

212

yield isotactic poly(L,L-lactide), **212**, with PDI = 1.2. No syndiotactic linkages were observed by NMR. This polymerization was thought to involve a single polymer chain per Mg center.[1234]

Ring-opening metathesis polymerization of norbornene was catalyzed by the complex TpRu(=C=CHPh)(Cl)(PPh$_3$). The yield of polymer was 76 % after 24 hrs at 80 °C, and 99% after 72 hrs. By contrast, replacement of Tp with Cp* gave only 29% after 24 hrs, and replacement of Tp with Cp gave no polymer at all. The catalytic activity of TpRu(=C=CHPh)(Cl)(PPh$_3$) was enhanced by the presence of additives, such as PdCl$_2$(NCMe)$_2$ or BF$_3$·Et$_2$O, and retarded by additives, such as Al(OPr^i)$_3$, CuCl, or AgOTf.[892] Complexes Tp^{x}RuH(H$_2$) and Tp^{x}RuH(COD) (Tp^x = Tp, Tp*) were catalytically active in the reduction of unactivated ketones, using either H$_2$ directly, or by transferring H$_2$ from alcohols.[869]

The polymerization of ethylene was studied using the complex Tp^{x}NbMe$_2$(PhC≡CMe), activated with one equivalent of B(C$_6$F$_5$)$_3$, the reaction being done in toluene at room temperature, and under 1 atm of ethylene. The Tp^x ligands were Tp, Tp*, and Tp*^{Cl}, their activity increasing in this order, as measured by the amount (in kg) of polyethylene produced per mol of catalyst: 20, 100, and 130, respectively. The catalytic activity was discussed in terms of electrochemical data for the dichloro complexes Tp^{x}NbCl$_2$(PhC≡CMe), and also in terms of the rates of alkyl migration in complexes Tp^{x}NbCl(R)(PhC≡CMe) (R = Et, CH$_2$SiMe$_3$). It was found that at a given electron density on the metal, the bulky Tp^x ligand yielded a much more active catalyst than the parent Tp.[407]

Another potentially very promising scorpionate-based system for ethylene polymerization involved yttrium complexes, Tp*YR$_2$(THF), obtained by treating Tp*YCl$_2$(THF) with the appropriate RLi reagents (R = Me, Bu^t, Ph, CH$_2$SiMe$_3$). The corresponding hydride complex was also synthesized. All the alkyl and hydride species polymerized ethylene, yielding linear, high-molecular-weight polyethylene, despite the fact that THF was coordinated to the initial yttrium complexes. Least active were complexes with R = Me or CH$_2$SiMe$_3$ (30 and 15 moles PE/moles of Y, respectively). For R = phenyl, the number was about 1100, and for R = t-butyl, over 1900. It was also possible to obtain active catalysts by replacing Tp* with Tp or with Tp^{Ph} in the above system (although the Tp^{Ph} analog was less active than the complex with Tp*), and to replace Y^{III} with La^{III}, Nd^{III} or Sm^{III}. It is noteworthy that Lewis acid co-catalysts, such as aluminoxane were not required to activate this system.[343]

Still another olefin-polymerization catalyst was Tp*V(=NAr)Cl$_2$, where Ar was 2,6-bis(isopropyl)phenyl, in combination with aluminoxane. This system polymerized ethylene at room temperature in toluene to a high molecular weight polymer. Propylene was polymerized under pressure to yield only viscous oily atactic oligomers.[380] A patent application cited the use of the reaction product of KTp with VOCl$_3$, in the presence of methyl aluminoxane and ethylene, to produce polyethylene.[1498] Another ethylene polymerization catalyst was based on the complex of TiCl$_4$ with the reaction product of KBH$_4$ with Hpz, used in the presence of methyl aluminoxane.[1499] A different catalytic system for olefin polymerization was claimed on the basis of lanthanide or actinide complexes of Tp^x ligands, containing a substituent of at least 3 carbons in the 3-position. A typical catalyst was prepared by the reaction of TiCl$_4$ with KTp^{Ph,Me}.[1500] Complexes of both, Bp^x and Tp^x, with group VIII metals, in addition coordinated to Group VA elements, were claimed as catalysts for the oligomerization and polymerization of olefins, and the copolymerization of olefins with carbon monoxide.[1501]

A nickel-based olefin polymerization catalyst has been reported, involving, among other ligands, also scorpionates, and possessing the general structure [R,R'B(pz^x)$_2$]Ni(L^1)(L^2), where pz^x was a diversely subsituted pyrazolyl group, R, R' were hydrogen, hydrocarbyl or substituted hydrocarbyl, L^1 was a neutral monodentate ligand which could be displaced by the olefin to be polymerized, and L^2 was a monoanionic monodentate ligand, or L^1 and L^2 taken together were a monoanionic bidentate ligand, provided that the monoanionic ligands could add to the olefin. The use of a Lewis acid in this system was optional.[1502] The pzTp ligand was also used as an additive in a system for the dimerization of acrylonitrile.[1503] Stereoregular copolymerization of ethylene with carbon monoxide at good rates was catalyzed above 20 °C by the structurally characterized complex Tp^{Ph}Ni(PPh$_3$)(o-Tol), and to a lesser extent by Tp^{Tol}Ni(PPh$_3$)(o-Tol). This reaction was not poisoned by pressuring the system with CO before the addition of ethylene. Attempts to use related

5.2 CATALYSIS

Tp^{x}ligands, such as Tp, Tp*, Tp^{tBu}, and Tp^{Tol,Me} were unsuccessful for a variety of reasons.[1264]

Polymerization of neat methyl methacrylate at 70 °C to a high polymer (M_w = 931000, M_n =196000, M_w/M_n = 4.74) was effected by the hindered complex [Tp^{tBu,Me}]CoMe, and a control experiment established that the cobalt complex was required for polymerization under these conditions. This rection was thought to be of radical nature, initiated by methyl radicals resulting from homolysis of the cobalt—carbon bond.[1529]

The compound Tp^{Ph}Cd(OAc) was converted to a variety of cyclic ether or thioether derivatives, Tp^{Ph}Cd(OAc)L, where L was THF, dioxane, propylene oxide, cyclohexene oxide, and propylene sulfide, and several of these complexes were structurally characterized by X-ray crystallography. The ligands L were labile, and dissociated in solution, generating a five-coordinate species. It served as a model for the initiation step in the copolymerization of epoxides with carbon dioxide, catalyzed by metal carboxylates.[1271,1272]

5.2.2   Carbene/Nitrene Transfer

The copper(I) complex Tp*Cu(CH$_2$=CH$_2$) was found to catalyze under mild conditions the reaction of ethyl diazoacetate with alkenes to form cyclopropanes in high yields (76-96 %). This reaction was successful with styrene, *cis*-cyclooctene, and 1-hexene, but it failed with *trans*-stilbene. In the analogous reaction with alkynes (1-hexyne, 3-hexyne, 1-phenyl-1-propyne) the yield of cyclopropenes was lower (41-64%). Tp*Cu(CH$_2$=CH$_2$) also catalyzed nitrene transfer from PhI=NTs to alkenes, forming aziridines in 40-90% yield.[1067] Cyclopropanation of styrene, catalyzed by the complexes **166** (p. 123) and **198** (p. 172), was also studied.[1257,1258]

A Bp-based system for the cyclopropanation of olefins, capable of operating under homogeneous and heterogeneous conditions, utilized complexes BpCu, BpCuL (L = bipy, PCy$_3$, Ph$_2$CH$_2$CH$_2$PPh$_2$) and BpCuL$_2$ (L = py, PPh$_3$). These compounds could be used either in a homogeneous system, or supported on silica gel. Thus, BpCu, along with an olefin (styrene, cyclooctene, or 1-hexene) and ethyl diazoacetate, gave rise to the corresponding cyclopropanes in high yield, with no excess of olefin employed. The active species was thought to be the 14-electron BpCu fragment. The heterogeneous system gave somewhat better yields than the homogeneous one, and the catalyst could be recovered and reused several times up to 6-12 cycles.[1416]

A large scope of activity was exhibited by the electrophilic carbene complex [Tp*W(=CH$_2$)(CO)(PhC≡CMe)]$^+$. This species was found to transfer its methylene to olefins, such as styrene, cyclohexene, or 4-methylstyrene, producing cyclopropanes. It also served as a catalyst for aziridine formation from imines and ethyl diazoacetate.[51]

5.2.3 Oxidation

The oxidation of adamantane, cyclohexane and cyclohexene with dioxygen was catalyzed by the dinuclear Fe^{III} complex, $[TpFe]_2(\mu\text{-}OAc)_2(\mu\text{-}O)$. The products from adamantane were adamantan-1-ol, adamantan-2-ol, and adamantan-2-one, while from cyclohexane and cyclohexene, the products were cyclohexanol, and cyclohex-2-en-1-ol, respectively.[846] Oxidation of methyl linoleate was effectively catalyzed by Tp_2Fe, which was much more active than a variety of other iron complexes. The oxidation mimicked the action of lipoxygenase.[1504] 3-R substituted Tp^x ligands were claimed as ligands in manganese complexes, used for bleaching and oxidation purposes.[1505,1506] The complex TpFe(PhCOCOOO), a model of α-ketoglutarate enzyme was active as olefin epoxidation catalyst,[847] while TpCu(py) activated dioxygen for the oxidation of organic substrates under mild conditions.[1567]

5.2.4 Miscellaneous Catalysis

The complexes $Tp^{Ph}Zn(OOCCH_2CN)$,[1268] and $Tp^{Ph}Zn(OAc)$,[1246,1269] were studied as catalysts for the decarboxylation of cyanoacetic acid, and for malonate decarboxylation, respectively. Several ruthenium(II) complexes catalyzed the addition of benzoic acid to phenylacetylene, producing the *trans*-adduct, PhCH=CHOOCPh, along with smaller amounts of the *cis*-adduct and H_2C=C(Ph)OOCPh, the latter being formed only in the case of $TpRu(COD)(MeCN)(CF_3SO_3)$ being the catalyst. The active compounds were TpRu(COD)Cl, $TpRu(py)_2Cl$, TpRu(tmeda)Cl, and $TpRu(COD)(MeCN)(CF_3SO_3)$, while $TpRu(COD)(H_2O)(CF_3SO_3)$ was inactive.[894] Complexes $Tp^xRuCl(NCPh)_2$ (Tp^x = Tp, pzTp) catalyzed olefin hydrogenation.[806]

5.3 Enzyme Modelling

Scorpionate ligands have been used by many researchers in modelling studies for various bioinorganic systems, in particular for enzymes in which the metal is coordinated to three imidazolyl nitrogen atoms from three histidine ligands. For that purpose, homoscorpionates with a variety of 3-substituents have been used, adjusting the size and geometry of the hydrophobic pocket by the choice of the 3-R substituents. Furthermore, heteroscorpionates combining two N and one S donor function, could be used to mimic the coordination by two histidines and one methionine or cysteine ligand to the biologically active metal ion. These studies encompassed a variety of metals, and the use of many types of Tp^x and Bp^x ligands. Some of these complexes succeeded in mimicking quite closely the spectral, and certain chemical properties of the particular enzymes. Since the functions of the enzymes being modeled varied, this section is organized according to the metal present at the active centers.

5.3 ENZYME MODELLING

5.3.1 Vanadium

A series of vanadium(V) phenolate complexes of general structure Tp*VO(OAr)$_2$ has been synthesized, and the structure of Tp*VO(OC$_6$H$_4$-p-Br)$_2$ was determined by X-ray crystallography. These complexes were easily hydrolyzed to the dimeric [Tp*VO(OH)]$_2$O species, and were discussed as possible models for the active site in bromoperoxidase.[372] Other vanadyl compounds, based on Tp and Tp* ligands were synthesized in this context, and they were studied by ^{1}H, ^{13}C, and ^{51}V NMR, electrochemistry, optical spectroscopy, and EXAFS. Two types of complexes were prepared: TpVO(OR)$_2$ and TpVO(OR)Cl. The latter complexes contained the shortest V—OR bonds hitherto reported, and this was of relevance to some of the properties found in bromoperoxidase. The above alkoxides readily underwent hydrolysis, forming various oxo-bridged multinuclear species, and one of these was characterized by X-ray crystallography as the tetrameric [TpVO$_2$]$_4$.[367]

5.3.2 Molybdenum

Molybdenum, usually in its higher oxidation forms, is an essential nutrient for all forms of life. It is present in the molybdopterin enzymes xanthine oxidase/dehydrogenase, sulfite oxidase, nitrate reductase and dimethyl sulfoxide reductase, which possess mononuclear molybdenum cofactors, with the metal bound by one or two 1,2-dithiolate links to one or two pterin systems, as shown in **215**. One frequently encountered reaction is the oxygen transfer reaction (OAT), of the type $M^{VI} + X \longrightarrow M^{IV} + XO$.[68, 1523]

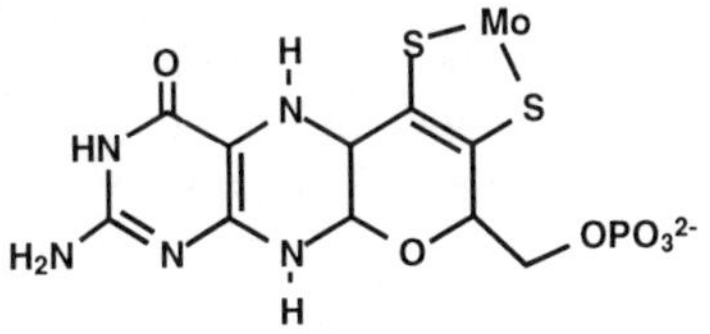

215

An example of OAT involving Tp^x ligands was the oxidation of PPh$_3$ to PPh$_3$O, or the reduction of Me$_2$SO to Me$_2$S, catalyzed by the complex Tp*MoO$_2$[SP(S)R$_2$], which in the former reaction was reduced to Tp*Mo(O)(S$_2$PR$_2$), with the latter complex being oxidized in the second reaction to Tp*MoO$_2$[SP(S)R$_2$].[632] The complex Tp*MoO$_2$(SPh), and related compounds of general structure Tp*MoO$_2$(Y), where Y was an anionic ligand, were synthetic models for the sulfite oxidase family of enzymes, which contain Mo^{VI}O$_2$ sites with one pterin 1,2-dithiolate ligand.[622,624,643,665]

The xanthine oxidase family of enzymes contains $Mo^{VI}OS$ sites with one pterin 1,2-dithiolate ligand. A structurally relevant complex containing a cis-$Mo^{VI}OS$ center, Tp*MoOS($S_2PPr^i_2$), **216**, has been synthesized.[650] The Mo=S distance of 2.227(2) Å was close to that estimated for xanthine oxidase/dehydrogenase, and the stabilizing S···S interaction was an important structural feature, possibly analogous to the one found in aldehyde oxidoreductase. Just as xanthine oxidase/dehydrogenase gets deactivated by cyanide ion, with the formation of a "desulfo" $Mo^{VI}O_2$ species and thiocyanate ion, the complex Tp*MoOS($S_2PPr^i_2$) reacted with cyanide ion, forming quantitatively thiocyanate ion, by way of a sulfur atom transfer. It produced under anaerobic conditions Tp*MoO($S_2PPr^i_2$), and Tp*MoO$_2$($S_2PPr^i_2$) in the presence of water and oxygen.[650]

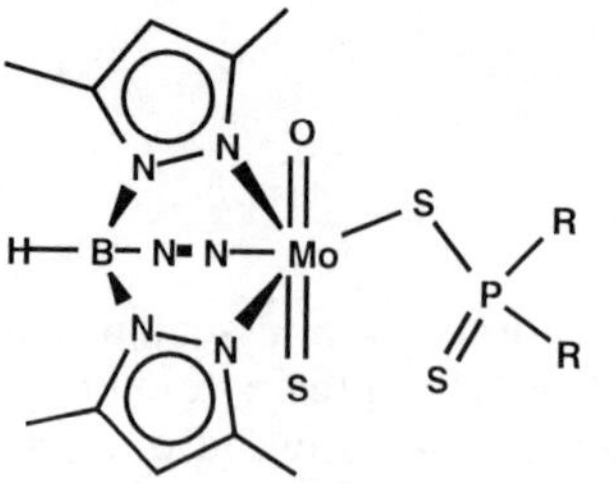

216 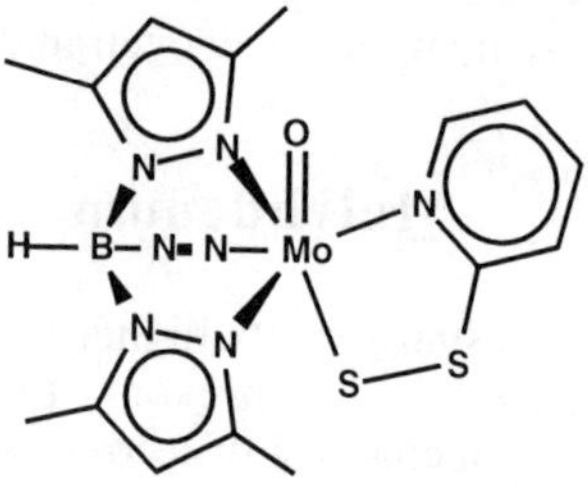   **217**

The structurally characterized complex **217** was prepared by a sulfur atom transfer reaction.[628] On the basis of Mo-S and S-S distances, it was best formulated as an oxodithiomolybdenum(IV) species. Both of these complexes generated EPR-active Mo^V species upon either reduction or oxidation. They also both produced oxothiomolybdenum(V) anions upon reduction, and oxodithiomolybdenum(V) cations upon oxidation.[628]

5.3.3   Tungsten

The tungsten enzymes include aldehyde oxido-reductase, formate dehydrogenase and formylmethanofuran dehydrogenase. Their exact mode of functioning is less known than that of the molybdenum enzymes, although they frequently contain sulfur. In an approach to such structural features, a number of Tp*WOS and Tp*WS$_2$ derivatives has been synthesized.[730,734,737] Such complexes reacted with $W^{IV}O$ and $W^{IV}S$ species via transfer reactions of the sulfur atom. In contrast to the $Mo^{VI}OS$ core of xanthine oxidase/dehydrogenase and Tp*MoOS(S_2PR_2), which react with cyanide ion via formation of thiocyanate, such deactivation was not observed in some of the tungsten enzymes, nor in the complex Tp*WS$_2$Cl.

5.3 ENZYME MODELLING

Complexes of general structure Tp*WS$_2$(X), such as **218**, reacted readily with variously substituted acetylenes, including acetylene itself, under very mild conditions, yielding enedithiol W^{IV} derivatives, exemplified by **219**. The molecular dimensions of these complexes were close to those found in the enzyme formate dehydrogenase.[737]

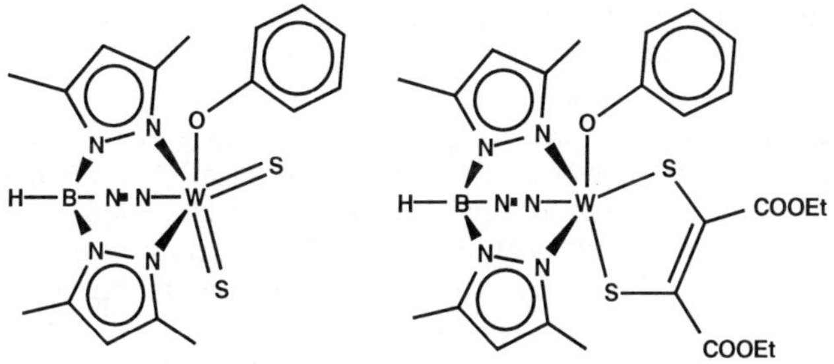

218 **219**

5.3.4 Manganese

Two scorpionate complexes, Tp^{iPr2}Mn(OBz), and the structurally characterized Tp^{iPr2}Mn(OBz)(Hpz^{Pr2}), were investigated as mimics of manganese superoxide dismutase. Activity was observed in the xanthine oxidase-NBT assay, leading to the conclusion that these compounds have the activity of superoxide dismutase, but not that of diformazan superoxidase or that of diformazan peroxidase.[1299] Oxidation of the dinuclear complex [Tp^{iPr2}Mn(OH)]$_2$ with molecular oxygen yielded a complex in which both oxygen atoms were incorporated into the tertiary position of one isopropyl group per ligand, along with the loss of one water molecule, converting the double (OH) bridge, to a single —O— bridge. This was regarded as a dioxygenase type of oxygen insertion.[1303] Dinuclear manganese(III) complexes TpMn(µ-O)(µ-O$_2$CR)$_2$MnTp (R = Me, Et, H) were synthesized as possible models for known binuclear manganese enzymes, which are active in various redox functions in living systems. These complexes were studied spectroscopically, electochemically, by NMR, and the structure of TpMn(µ-O)(µ-OAc)$_2$MnTp was determined by X-ray crystallography. It was concluded that substitution of Fe(III) with Mn(III) in proteins containing the [Fe$_2$O]$^{4+}$ core would permit more effective investigation of these centers in iron-oxo proteins, such as ribonucleotide reductase.[758]

5.3.5 Iron

The earliest example of using scorpionate ligands in modelling enzymes, was the work of Lippard aimed at hemerythrin, an oxo-bridged diiron enzyme, in which each iron was coordinated to three histidine nitrogens and, in addition, there were two carboxylato bridges derived from aspartate and glutamate. Several complexes of the structure [TpFe]$_2$(μ-O)(μ-OOCR)$_2$ (R = H, Me and Ph) were prepared, those with R = H and Me were structurally characterized, and the geometry of the [Fe$_2$O(OOCR)$_2$ core was found to resemble closely that of hemerythrin.[829] The hydroxo-bridged cation, {[TpFe]$_2$(μ-OH)(μ-OAc)$_2$}$^+$, was also prepared and structurally characterized. Its spectral and magnetic properties differed sufficiently from the oxo-bridged complex, to suggest that oxo-, rather than hydroxo bridges are present in the methemerythrins.[830] The complex [TpFe]$_2$(μ-O)(μ-OAc)$_2$ and its ^{18}O analog were studied by Resonance Raman, and their excitation profiles were obtained, permitting the assignment of some bands in azidomethemerythrin.[835] In another study, the XANES and EXAFS spectra of the model complex were compared with those of the purple acid phosphatase from beef spleen.[836] The structure of a more sterically hindered complex, [Tp^{iPr2}Fe]$_2$(μ-OH)(μ-OOCPh), **220**, was compared with that of

220

deoxyhemerythrin. While the Fe-OH distances of 1.96 and 1.99Å were comparable to the average Fe-OH distance in deoxyhemerythrin (2.02 Å), the Fe-(OH)-Fe core in the enzyme is asymmetric (2.15 and 1.88 Å, respectively), and thus not quite similar to the above complex.[1311] Another use of the binuclear complex [TpFe]$_2$(μ-O)(μ-OAc)$_2$ was in the oxidation of alkanes, regarding it as a possible model for methane mono-oxygenase.[846]

A monuclear, structurally characterized complex, Tp^{iPr2}Fe(OOCPh)(MeCN), was regarded as a synthetic model for dioxygen binding sites of non-heme iron proteins, based on its electronic spectral changes under argon and/or dioxygen.[1310] Using more heavily substituted homoscorpionate ligands, Tp^{iPr2}, and Tp^{tBu,iPr} the substituted catecholato complexes **221** and **222** were prepared, by way of oxidizing with O$_2$ their tetrahedral iron(II) precursors, containing just one aryloxy oxygen bonded to iron. Further reaction with oxygen gave, in the case of the less hindered complex, **221**, the extradiol and intradiol cleavage products: the two α-pyrones **223**

5.3 ENZYME MODELLING

and **224**, and the anhydride **225**, while in the case of the more hindered **222**, there was no further reaction with oxygen. The above complexes were viewed as representing structural and functional models of the catechol dehydrogenases.[1309]

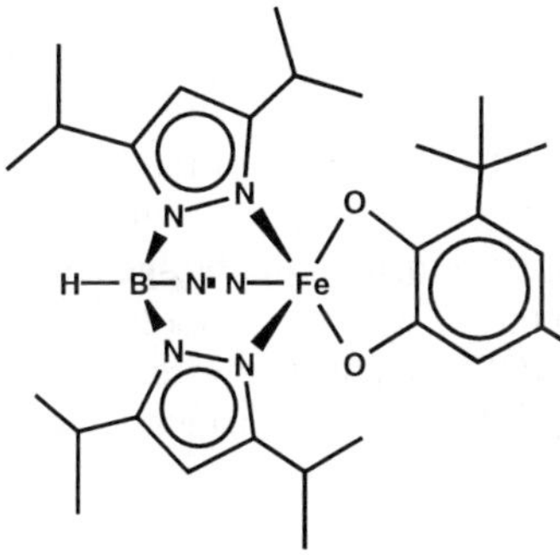

221

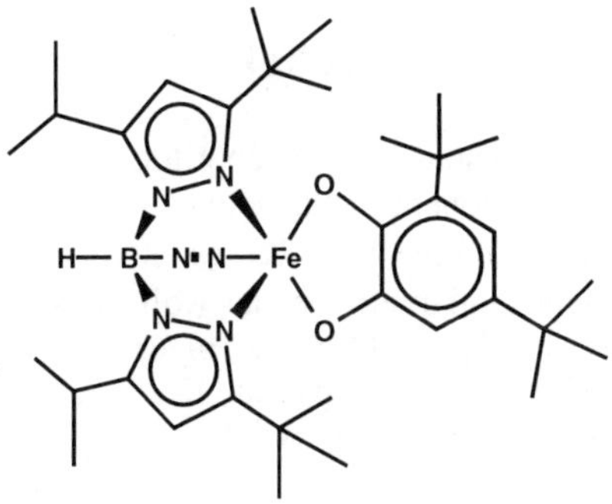

222

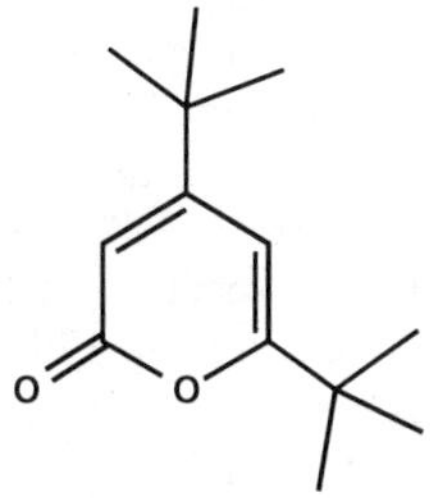

223

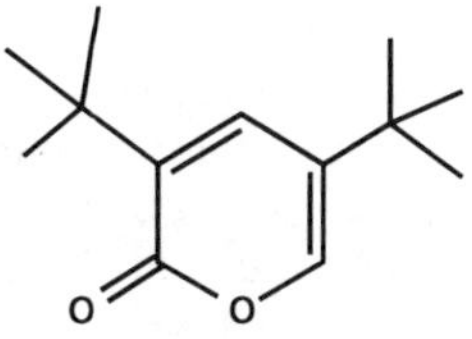

224

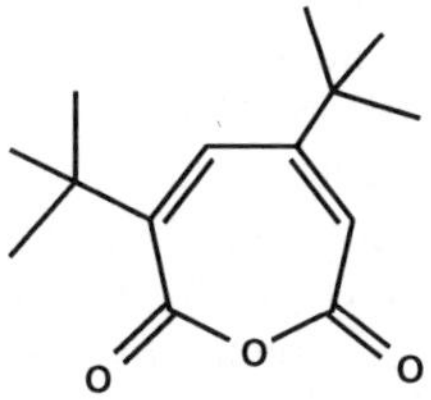

225

A model for the peroxo intermediate in the methane monooxygenase hydroxylase reaction cycle was prepared by way of oxidizing $Tp^{iPr2}Fe(OOCCH_2Ph)$ with oxygen, which yielded the structurally characterized binuclear complex $Tp^{iPr2}Fe(\mu-1,2-O_2)(\mu-OOCCH_2Ph)_2$. Its optical, Mössbauer and Raman spectra resembled closely those of the peroxo intermediate of methanol monooxygenase.[1312] The complex $Tp^{iPr2}Fe(OAr)_2$ reacted with *meta*-chloroperbenzoic acid at -78 °C which resulted in *ortho*-hydroxylation of the 4-substituted phenolate ligand, yielding a catecholato derivative, from which the catechol was isolated through methanolysis of the complex. This was regarded as mimicking the activity of tyrosine hydroxylase.[1308] In order to model the reactivity of α-ketoglutarate-dependent non-heme iron(II)-containing enzymes, the complex $Tp*Fe(OOCCOPh)(MeCN)$ was synthesized, and used in the reaction of cyclohexene with oxygen in the presence and absence of tributyltin hydride (a radical trap). Without Bu_3SnH, the products were the epoxide, 2-cyclohexene-1-one, and 2-cyclohexene-1-ol, but when Bu$_3$SnH was present, only the epoxide was isolated, in yields that were the same as without it (42-43 %). Epoxidation of stilbene was also achieved.[847]

5.3.6   Nickel

As part of a study of five-coordinate nickel-cysteinato complexes, of relevance to the nickel component of the active site in several hydrogenase enzymes, which catalyze reversibly the reaction: H$_2$ ↔ 2H$^+$ + 2 e$^-$, and participate in the bio-generation of hydrogen and methane, as well as nitrogen fixation, the five coordinate nickel complexes Tp*Ni(OEt-cysteinato) and Tp^{Ph,Me}Ni(OEt-cysteinato) were synthesized, studied by spectroscopy, and the structure of Tp*Ni(OEt-cysteinato) was determined by X-ray crystallography.[1534]

5.3.7   Copper

A number of scorpionate copper derivatives, synthesized in order to model various properties of the copper-containing natural products, have been discussed in a review devoted to the structure and function of copper proteins.[1524] The structurally characterized Tp^{Ph}Cu(pterin) complex was prepared in an approach towards a model for the active metal site in phenylalanine hydroxylase.[1263] A monuclear (acylperoxo)copper(II) complex, **226**, was obtained by the reaction of the dimer [Tp^{iPt2}Cu(OH)]$_2$ with *m*-chloroperbenzoic acid. It oxidized PPh$_3$ to PPh$_3$O, being reduced to the structurally characterized, and symmetrically bonded *m*-chlorobenzoate complex, **227**.[44] As part of a study to assess the role of copper in the ethylene effect on plants, a number of copper(I) complexes of the type TpCu(olefin) and Tp*Cu(olefin) was prepared, some of them were structurally characterized, and it was

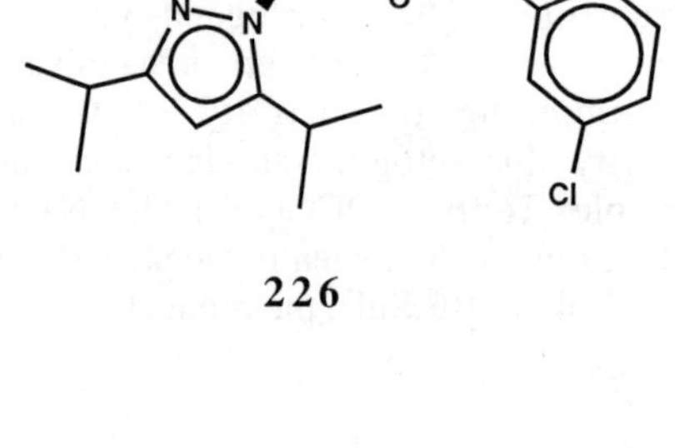

226 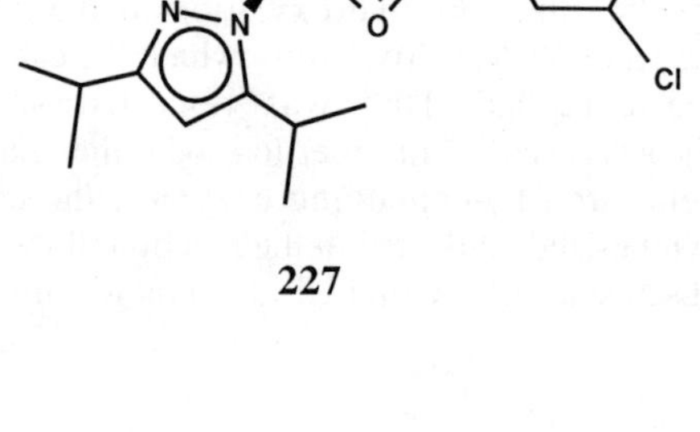   **227**

5.3 ENZYME MODELLING

demonstrated for the first time that the chemistry observed was consistent with the proposed role of copper at the ethylene binding site of plants.[1065] The complex **228**, which contained five-coordinate copper, was prepared as a simple model for the copper center of galactose oxidase, and it was studied by electrochemistry.[1375]

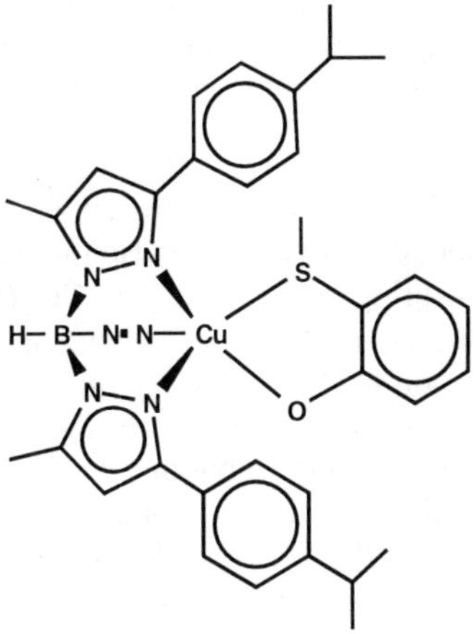

228

Heteroscorpionate complexes (p-TolS)Bp*Cu(L), where L was O-ethylcysteinato, p-nitrophenylthiolato, and pentafluorophenylthiolato, were synthesized, and their spectroscopic properties were compared with those of the N_2S_2 active site in poplar plastocyanin.[139] The compounds K[Tp*CuSAr] and Tp*Cu(SAr) were prepared, aiming for models for the active sites in blue copper proteins, as were complexes Tp*CuL (L = SR, OAr or Me_2NCS_2).[1063,1064] Several other thiolato derivatives, such as $Tp^{iPr2}CuSBu^t$,[1322] and $Tp^{iPr2}CuSC_6F_5$, were synthesized, and their spectra were found to be a better match for those of the blue copper proteins, than those derived from Tp*-based complexes. The structure of $Tp^{iPr2}CuSC_6F_5$ was determined by X-ray crystallography.[1323]

In a more detailed study, four such thiolate complexes of structure $Tp^{iPr2}CuSR$ (R = tert-Bu, sec-Bu, CPh_3 and C_6F_5) were synthesized, and the one with R = CPh_3 was stucturally characterized. These compounds were studied by Resonance Raman spectra, modelling the spectra of the blue copper proteins and providing insight into to coordinative environment of copper in such proteins.[1324] An unusual tricopper (I,II,I) complex of the Bp* ligand has been structurally characterized, in which all three coppers were linked through sulfide bridges as shown in structure **191** (see p. 161). This complex was thought to be of some relationship to the proposed mixed valence dicopper electron transfer sites in the enzymes nitrous to the proposed mixed valence dicopper electron transfer sites in the enzymes nitrous oxide reductase, or in cytochrome *c* oxidase.[1436] Two complexes, regarded as possible intermediates in the nitrite reduction by the copper nitrite reductase, and models for NO coordination to isolated copper sites in proteins and in zeolites, were synthesized. They were the 11-electron Tp^{tBu}CuNO and Tp^{Ph2}CuNO, each prepared by action of NO upon the dimeric copper(I) complex, [Tp^{tBu}Cu]$_2$, and on Tp^{Ph2}Cu(Hpz^{Ph2}), respectively. The former complex was structurally characterized. Their spectroscopic features were discussed, and compared with those of analogous copper protein species.[1239]

A dicopper complex, claimed to be an accurate synthetic model of oxyhaemocyanin, was prepared by hydrogen peroxide oxidation of [Tp*Cu]$_2$(μ-O), which yielded a new peroxo-bridged complex, [Tp*Cu]$_2$(μ-O$_2$). It was characterized by field desorption mass spectroscopy and Resonance Raman, and its O—O stretching frequency was close to that of oxyhaemocyanin (744-752 cm^{-1}).[1069,1070] The reactivity of this complex towards various substrates was tested, and it was found not to oxidize PPh$_3$ or CO, but to form mononuclear copper(I) complexes Tp*CuL. Cyclohexene was oxidized only under aerobic conditions, and the incorporated oxygen came only from the exogenous dioxygen, and not from [Tp*Cu]$_2$(μ-O$_2$). Phenols were oxidatevely coupled under anaerobic conditions, and under a dioxygen atmosphere there was both, coupling, and quinone formation, rationalized in terms of homolytic O—O bond cleavage in [Tp*Cu]$_2$(μ-O$_2$), followed by free radical chain reactions with dioxygen.[1071] A detailed study of dioxygen binding and the core isomerization between the Cu$_2$(μ-η^2:η^2-O$_2$) and Cu$_2$(μ-O)$_2$ was carried out on [TpCu]$_2$O$_2$ – as a mimic of the enzymes oxyhemocyanin (a reversibly binding oxygen carrier) and oxytyrosinase (a monooxygenase oxidizing phenols) – by gradient-corrected density functional methods. The calculated oxygen binding energy was - 184 kJ/mol, implying irreversible binding.[1073]

5.3.8 Zinc

A considerable amount of effort in the area of modelling zinc-based enzymes was centered on carbonic anhydrase, in which zinc is coordinated to three histidine imidazolyl groups. Thie enzyme catalyses the reversible hydration of carbon dioxide by a process which involves deprotonation at neutral pH of a zinc-bonded water molecule, followed by a nucleophilic attack of a zinc-bound hydroxide at the carbon dioxide substrate to produce a bicarbonate intermediate. In the final step of the cycle, water displaces the bicarbonate ion, to regenerate the starting material.

In order to model this enzyme, Tp^x ligands were required with sufficient steric hindrance to support a monomeric Tp^{x}ZnOH complex, without forming the

5.3 ENZYME MODELLING

$Tp^X{}_2Zn$ species. One such ligand was $Tp^{tBu,Me}$. Thus, the complex $Tp^{tBu,Me}ZnOH$ was found to react rapidly and reversibly with carbon dioxide, forming the bicarbonate complex $Tp^{tBu,Me}Zn(OCO_2H)$, which was characterized by IR spectroscopy. This complex was slowly converted to the carbonate complex $[Tp^{tBu,Me}Zn]_2(CO_3)$, in which the carbonate ion bridged the two zinc centers in a symmetric, unidentate mode. That the complex $Tp^{tBu,Me}ZnOH$ was a functional model of carbonic anhydrase was also demonstrated by its catalysis of oxygen atom exchange between carbon dioxide and $H_2{}^{17}O$. Another piece of supporting evidence for the bicarbonate complex was obtained by the synthesis of the well-characterized $Tp^{tBu,Me}ZnOC(O)OMe$ species, which was obtained upon the reaction of $Tp^{tBu,Me}ZnOH$ with $[MeOC(O)]_2O$.[1353]

In contrast to $Tp^{tBu,Me}ZnOH$, $Tp^{iPr2}ZnOH$ yielded readily the carbonate complex, in which the carbonate ion was bonded bidentate to one zinc, and monodentate to the other.[1329] In order to correlate the activity of metal-substituted carbonic anhydrase with the mode of coordination of the bicarbonate ligand, structural studies were carried out, establishing the modes of binding of nitrate and carbonate ligands in complexes of the type $Tp^{tBu}M(NO_3)$ and $[Tp^{iPr2}]_2CO_3$. The binding modes were metal-dependent, and in the case of the nitrate complexes ranged from symmetrical bidentate (Ni, Cu) to asymmetric bidentate (Co), to monodentate (Zn).[61,1296] On the basis of these data, a progressive ground state stabilization of the bicarbonate ligand was anticipated in line with the series Zn , Co < Cu, Ni, and Cd. In the case of the carbonate complexes the bonding was symmetrical bis-bidentate for Ni and Cu, asymmetric bis-bidentate for Fe and Co, and monodentate/bidentate for Zn.

Modelling of the catalytic cycle of liver alcohol dehydrogenase (LAD) was also served by a $[Tp^{tBu,Me}Zn]L$ species. The alkoxide complex, $Tp^{tBu,Me}ZnOR$, which was prepared by the reaction of the hydride $Tp^{tBu,Me}ZnH$ with ROH, was found to equilibrate with the hydroxide, $Tp^{tBu,Me}ZnOH$, in the presence of water, and equilibrium constants for this reaction, and relative Zn-OH *versus* Zn-OR bond energies were obtained. Upon the reaction of various $Tp^{tBu,Me}ZnOR$ species with *p*-nitrobenzaldehyde, the complexes $Tp^{tBu,Me}ZnOCH_2Ar$ were formed, along with an aldehyde or ketone, derived from the original OR ligand. This reaction did provide evidence for zinc-mediated hydride transfer from ROH to ArCHO, of relevance to LAD activity.[1526]

Another zinc enzyme mimicked by a Tp^X ligand was alkaline phosphatase, which catalyses non-specific hydrolysis of phosphate monoesters. Treatment of $Tp^{iPr2}ZnOH$ with mono(*p*-nitrophenyl)phosphate yielded the dinuclear complex $[Tp^{iPr2}Zn]_2[OP(O)(OC_6H_4NO_2)O]$, the structure of which was established by X-ray crystallography. This was the first example of a dinuclear zinc complex bridged solely by a phosphate ligand, and thus resembling the alkaline phosphatase enzyme.[1328]

Modelling of cobalamine independent methionine synthase, which takes part in the biosynthesis of methionine by transferring a methyl group from an appropriate source (for instance, methylcobalamine) to homocysteine, was aimed at employing the complexes Tp*ZnSR and Tp^{Ph,Me}ZnSR (R = Et, CH$_2$Ph). All four complexes reacted with methyl iodide under mild conditions yielding the appropriate methyl thioethers, but the rates of methylation were much higher with Tp^{Ph,Me}ZnSR than with Tp*ZnSR, which was ascribed to higher hydrophobicity of the Tp^{Ph,Me} ligand. The Tp^{Ph,Me}ZnSEt complex also formed the methyl ether upon reacting with trimethylsulfonium iodide, modelling the reaction of S-adenosyl methionine. These methylations were regarded as the closest analogs of their biological models.[1089]

A thorough and detailed kinetic study of the hydrolytic cleavage reaction of various triorganophosphates OP(OR)$_2$(OR') by several Tp^{x}ZnOH complexes was carried out, leading to Tp^{x}ZnOP(O)(OR)$_2$ and HOR' as products. The scorpionate ligands used were Tp^{Ph,Me}, Tp^{Cum,Me} and Tp^{Pic,Me}, while the OR groups were OPh, OC$_6$H$_4$-p-NO$_2$, OEt, and also those in which the R group was 2,3-isopropylidene-5-methylribosyl. Similar results obtained during ester cleavage by the same Tp^{x}ZnOH complexes led to the conclusion that a four-center arrangement (ZnOPO or ZnOCO, respectively) in the activated complex is in operation, akin to that operating in zinc enzymes. A detailed mechanism was proposed for the reaction trajectory in such hydrolytic reactions, activated by [L$_3$ZnOH] species.[1431]

5.4　C—H Bond Activation

Scorpionate complexes, primarily of rhodium and iridium, were studied in conjunction with the activation of aliphatic and aromatic C—H bonds. Such activation was achieved either photolytically or thermally, resulting in oxidative addition of H and R or Ar entities from the solvent to the metal, or the formation of M—C bonds to specific substituents on the homoscorpionate ligand. The most commonly used complexes were those of rhodium(I) and iridium(I).

The first example of such activation was reported in 1987 by Graham, who found that irradiation of Tp*Rh(CO)$_2$ in benzene proceeded with loss of CO, and formation of the benzene oxidative addition product, Tp*Rh(CO)H(Ph), which underwent exchange with C$_6$D$_6$, producing Tp*Rh(CO)D(C$_6$D$_5$). An analogous reaction took place in cyclohexane, yielding Tp*Rh(CO)H(Cy).[945] The same activation of benzene or cyclohexane occurred, with concomitant addition of H and of Ph or R to rhodium, when the corresponding solutions of Tp*Rh(CO)$_2$ were purged in the dark with N$_2$O.[946] The complex Tp*Rh(CO)$_2$ was converted by benzene to Tp*Rh(CO)H(Ph), and it equilibrated with methane, forming Tp*Rh(CO)H(Me). Thermolysis of Tp*Rh(CO)$_2$ in benzene at 140 °C also yielded Tp*Rh(CO)H(Ph),

5.4 C—H BOND ACTIVATION

but this was accompanied by decomposition.[945] On the other hand, benzene solutions of Tp*Rh(CO)(olefin) complexes (olefin = ethylene, propene, cyclooctene), when thermolyzed in the dark at 70-100 °C, gave rise to Tp*Rh(CO)H(Ph) in high yield.[947] Photolysis at room temperature of Tp*Rh(CO)(CH$_2$=CH$_2$) yielded Tp*Rh(CO)H(Ph) and Tp*Rh(CO)(Et)(Ph) in about 1:1 ratio. These products were not interconvertible.[948]

A detailed infrared study of Tp*Rh(CO)$_2$ photolysis in pentane, which yielded Tp*Rh(CO)H(pentyl), permitted the determination of the absolute quantum efficiencies for intermolecular C—H bond activation at several excitation wavelengths, and showed that the precess is highly wavelength-dependent, and that high quantum efficiencies can be attained.[949,950] The best conversion efficiency (ϕ_{CH} = 0.31-0.34) was achieved upon near-UV excitation at 313 or 366 nm. In trimethylsilane, the product was Tp*Rh(CO)H(SiMe$_3$), in toluene it was Tp*Rh(CO)H(Tol), but in pyridine, only decomposition took place. Photolysis of BpRh(CO)$_2$ did not lead to C—H bond activation. These studies also indicated that the thermal, or longer-wavelength chemistry, is based on a species with bidentate Tp* ligand.[951] Further investigation of this reaction in aromatic (benzene, toluene, mesitylene, xylene) and aliphatic (pentane, hexane, heptane and isooctane) solvents permitted refinement of the previous data, and also showed that the process is not retarded by the presence of excess carbon monoxide. This was interpreted to mean that the intermediates are very short-lived and become solvated before CO is able to coordinate.[952]

When instead of Tp*, the Tp^{Ph} ligand was used, different results were obtained. Irradiation of Tp^{Ph}Rh(CO)$_2$ in benzene led to internal C—H bond activation, and oxidative addition of the 3-phenyl group, yielding **231**. It reacted with

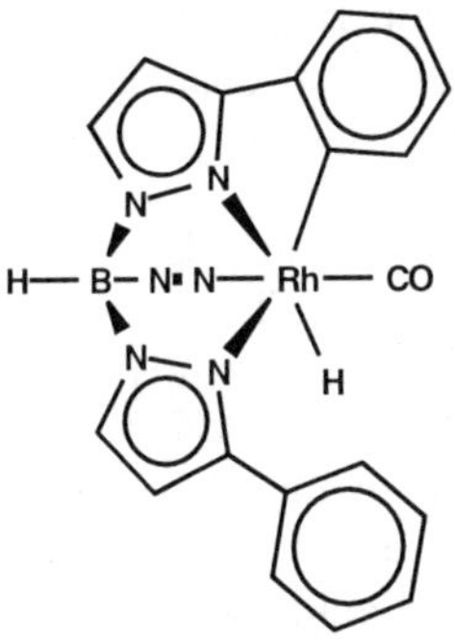

231 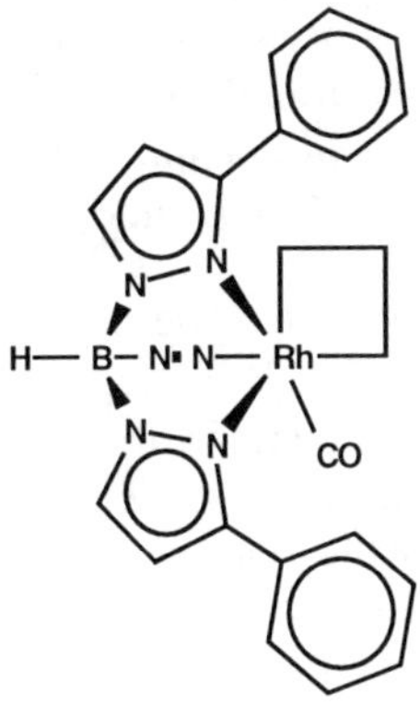   **232** 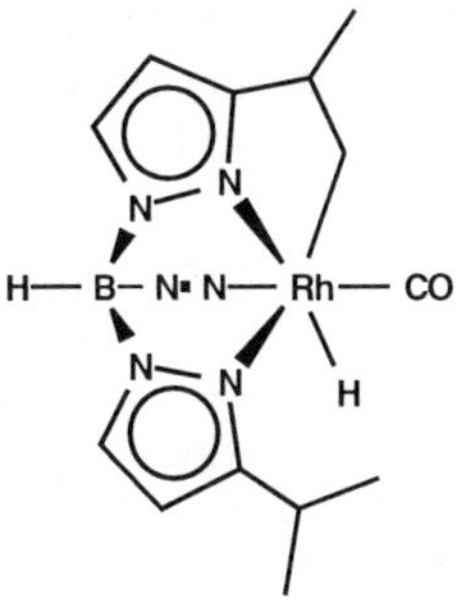   **233**

CO, regenerating the starting material, which led to the conclusion that **231** is in equilibrium with $Tp^{Ph}Rh(CO)$. When the substrate was cyclopropane, the product was **232**, presumably arising from a rearrangement of the initial cyclopropyl hydride species, $Tp^{Ph}Rh(CO)H(Cpr)$. Using the Tp^{iPr} ligand, different results were obtained, depending on the solvent used. In benzene, $Tp^{iPr}Rh(CO)_2$ was converted to $Tp^{iPr}Rh(CO)H(Ph)$, but in cyclohexane, the reaction yielded the intramolecular oxidative addition product **233**. Complex **233** reacted with cyclopropane at room temperature, being converted to the Tp^{iPr} analog of **232**.[1199] Photolysis of $Tp^{tBu}Rh(CO)_2$ yielded only decomposition products. On the other hand, both complexes, $Tp^{CF_3,Me}Rh(CO)_2$ and $Tp^{CF_3,Me}Rh(CO)(ethylene)$, when irradiated in benzene, gave the same product, $Tp^{CF_3,Me}Rh(CO)H(Ph)$, although at a slower rate than in the case of the Tp* analog. Formation of $Tp^{CF_3,Me}Rh(CO)(Et)(Ph)$ was not observed.[1199]

Tp*Rh(CO)$_2$ photolysis was also studied in argon, perfluorokerosene, Nujol and methane matrices at about 12 K. There was indication that the reaction started by CO loss, followed by dechelation of one pz* arm of the Tp* ligand. In the nitrogen matrix, a N_2 complex was formed, but no activation of C—H bonds was noted. Only at higher temperatures was C—H bond activation observed.[953] Bergman studied this reaction by means of ultra-fast IR spectroscopy, which permitted to establish the femtosecond dynamics of the C—H activation process. It proceeds through the loss of CO, producing a short-lived tetrahedral monocarbonyl complex, which gets rapidly solvated by the RH solvent. This is followed by dechelation of one pz* arm of the Tp* ligand, followed by R-H activation and addition to Rh, with final rechelation of the detached pz* arm. The energy barriers for each specific reaction step were calculated. For comparison, the photolysis of Bp*Rh(CO)$_2$ was also studied, but in contrast to the Tp* complex, it did not result in C—H bond activation.[954] Details of the photochemical reaction of Tp*Rh(CO)$_2$ which activates C—H bonds were analyzed, and compared with those of the Cp and Cp* analogs.[955] Photolysis of the complex Tp*Rh(PMe$_3$)(C$_2$H$_4$) in thiophene at lower temperatures yielded two products, **99** and **100** (see p. 76), corresponding to C—H and S—C bond activation, respectively. At higher temperatures, **100** became the almost exclusive product.[956]

Photolytic reactions, similar to those of Tp*Rh(CO)$_2$, were also observed with isonitrile analogs, Tp*Rh(CNR)$_2$ (R = neopentyl, 2,6-xylyl). Thus, irradiation of Tp*Rh(CNR)$_2$ in benzene, resulted in the loss of one isonitrile molecule, followed by oxidative addition of benzene, to yield Tp*Rh(CNR)H(Ph).[958,959] Instead of Tp*Rh(CNR)$_2$, it was also possible to use the carbodiimide complex, Tp*Rh(CN-CH$_2$Bu^t)(PhN=C=N-CH$_2$Bu^t), which upon irradiation in a variety of alkanes or arenes yielded products arising from the elimination of the carbodiimide ligand, and the formation of a C—H oxidative addition product.[957] The thermolysis of one of such products, Tp*Rh(CN-CH$_2$Bu^t)(Ph)(H), in benzene in the presence of excess isocyanide produced Tp*Rh(CN-CH$_2$Bu^t)$_2$, and the mechanistic aspects of this

5.4 C—H BOND ACTIVATION

reaction were studied in detail.[960] 1,3-dipolar cycloaddition of PhN$_3$ to Tp*Rh(CNR)$_2$ gave rise to Tp*Rh(CNR)(η^2-PhN=C=NR), the photolysis of which in benzene resulted in C—H bond activation, and produced Tp*Rh(CNR)H(Ph). These reactions, and related ones, were investigated.[962] Along the same lines, photolysis of the isonitrile/carbodiimide complex, Tp*Rh(NCR)(RN=C=NR), generated the transient species Tp*Rh(NCR). Cyclopropane added to this intermediate, yielding the hydrido cyclopropyl derivative **104**, which rearranged to the structurally characterized rhodiacyclobutane **105**, which added stepwise isonitrile, CNR, being converted to **106** (see p. 78). The propylene complex, Tp*Rh(CNCH$_2$CMe$_3$)(H$_2$C=CHMe), was produced upon pyrolysis of the rhodacyclobutane species. In contrast to the cyclopropane reaction, cyclobutane produced only the cyclobutyl hydride complex, Tp*Rh(CNR)H(cyclobutyl), which did not rearrange to the rhodiacyclopentane, but instead reductively eliminated cyclobutane.[971] The energetics of homogeneous intermolecular vinyl and allyl carbon–hydrogen bond activation by the 16-electron species Tp*Rh(CNCH$_2$Bu^t) were assessed by studying the rearrangements of hydrido vinyl complexes, such as Tp*Rh(CN-CH$_2$Bu^t)(CH=CH$_2$)(H) to the η^2-ethylene complex, or of the allylic complex Tp*Rh(CN-CH$_2$Bu^t)(CH$_2$CH=CH$_2$)(H) to the η^2-propylene derivative, and similar reactions. On the basis of these experiments, the relative Rh—C bond strengths were calculated, and were found to parallel the strengths of C—H bonds: Rh—Ph > Rh—vinyl > Rh—Me > Rh—(primary alkyl) > Rh—(cycloalkyl) > Rh—benzyl > Rh—allyl, although the differences in Rh—C bond strengths usually exceeded the corresponding differences in C—H bond strengths.[1528]

The question whether the inter- and intramolelcular C—H bond activation by Tp*Rh complexes proceeds through Rh^I or Rh^{III} complexes was addressed in a study of the reactions of Tp*Rh(CH$_2$=CH$_2$)$_2$. While ligands such as CO or phosphines yielded Tp*Rh(CH$_2$=CH$_2$)(L) complexes, MeCN or pyridine gave rise to products of the type Tp*Rh(Et)(CH=CH$_2$)(L). The chemistry of such species suggested the presence of an equilibrium between Tp*Rh(CH$_2$=CH$_2$)$_2$ and an undetectable amount of the hydrido vinyl species, Tp*Rh(H)(CH=CH$_2$)(CH$_2$=CH$_2$). Thermolysis of Tp*Rh(Et)(CH=CH$_2$)(NCMe) in benzene at 60 °C produced Tp*Rh(Et)(NCMe)(Ph). It was concluded that Rh^I species were responsible for the activation of benzene.[963]

Similar studies of C—H bond activation were also carried out with the analogous iridium complexes. In the first such study, the mere reaction of TpK with [IrCl(COE)]$_2$ (COE = cyclooctene), yielded a C—H bond activation product, the cyclooctenyl hydride, TpIrH(η^3-C$_8$H$_{13}$), while the photolysis of TpIr(CH$_2$=CH$_2$)$_2$ gave rise to to the hydrido vinyl derivative, TpIr(H)(CH=CH$_2$)(CH$_2$=CH$_2$). The latter rearrangement could not be achieved thermally.[987]

Thermolysis of Tp*Ir(C_2H_4)$_2$ produced at first the hydrido vinyl intermediate, Tp*Ir(H)(CH=CH_2)(CH_2=CH_2), which underwent intramolecular coupling, to form the hydrido allyl complex Tp*IrH(η^3-CH_2CHCHCMe).[988,989] The same intermediate was also capable of regioselectively activating the oxygen-adjacent methylene hydrogens in cyclic ethers (THF, 2-methyltetrahydrofuran, 1,3-dioxapentane, tetrahydropyran and dioxane) forming in each instance a Fischer-type carbene derivative, also containing a hydrogen and butyl group on iridium, as exemplified by the structurally characterized THF-derived product, **234**.[990] Another

234   **235**

reaction of the iridium(III) complex Tp*Ir(H)(CH=CH_2)(CH_2=CH_2) was the activation of two molecules of benzene, during its thermal reaction under nitrogen, forming the structurally characterized nitrogen complex **235** which was shown to undergo a variety of other transformations. These included formation of a Fischer carbene, conversion to the C_6D_5 analog, and to the dinuclear [Tp*Ir(Ph)$_2$]$_2$(μ-N_2) complex, as well as replacement of N_2 with other ligands, such as CO or PR_3.[991]

Thermal activation of the diethylene complex Tp*Ir(C_2H_4)$_2$ in thiophene also led to double activation of thiophene, producing Tp*Ir(2-thienyl)$_2$(thiophene), from which the thiophene molecule could be displaced by various other ligands. Its hydrogenation resulted in loss of the C-bonded thiophene molecules, producing a dihydride, which on thermolysis was converted to **109** (see p. 80).[992] This C—H activation reaction was expanded to include 2-methylthiophene and 3-methylthiophene, which were activated just like thiophene. The S-bonded thiophenes could be readily replaced with CO or with PMe$_3$, and the structure of Tp*Ir(2-SC$_4$H$_3$)CO was established by X-ray crystallography. Thermal activation of several substituted thiophenes by Tp*Ir(η^4-H_2C=C(Me)C(Me)=CH_2) was also studied.[993] A 3 + 2 cycloaddition reaction, in which coordinated acetonitrile added to the vinyl group on iridium in the complex Tp*Ir(C_2H_3)(Et)(MeCN), with formation of the iridapyrrole **110** (see p. 80), has been demonstrated. This reaction was catalyzed by traces of water, and was applicable to a variety of nitriles and olefins coordinated to iridium.[994]

5.4 C—H BOND ACTIVATION

An uncommon example of C—H bond activation in the Tp ligand itself was provided by the reaction of Tp*Ir(C_2H_4)(PPh_3), **236**, with excess PPh_3. This yielded a product in which one pyrazolyl ring added oxidatively its 5-CH to iridium to produce the species **237**, which remained in equilibrium with the starting material. The cyclometallated product was not structurally characterized, but it was identified unambiguously by NMR.[997] Protonation of **237** with HBF_4, which took place at the free nitrogen atom of the cyclometallated pyrazolyl ring, produced an isolable salt, the structure of which was determined by X-ray crystallography.[1537]

236   **237**

In a detailed study of photochemical C—H bond activation in a series of $Tp^xIr(\eta^4$-1,3-diene) complexes (Tp^x = Tp or Tp*; diene = butadiene, 2-methylbutadiene, 2,3-dimethylbutadiene, cyclopentadiene and 1,3-cyclohexadiene), it was demonstrated that the products are 1,2,3-η^3-butadienyl derivatives. For instance, photolysis of Tp*Ir(H_2C=C(Me)C(Me)=CH_2) produced in good yield the η^3-allyl complex [Tp*Ir(H)(η^3-CH_2C(C(Me)=CH_2)CH_2], while thermal activation of benzene by Tp*Ir(H_2C=(Me)C(Me)=CH_2) yielded the N_2-bridged [Tp*IrH(Ph)]$_2$(μ-N_2).[998] The major findings of the preceding investigations, dealing with the formation of hydrido-η^3-allyl complexes of Ir^{III} by sequential olefinic C—H bond activation and C—C coupling of alkenyl and olefin ligands, have been summarized.[999] Various aldehydes were also activated by the Tp*Ir(H_2C=(Me)C(Me)=CH_2) complex. For instance, its reaction with p-anisaldehyde yielded initially the Fischer hydroxycarbene, **112**, which was converted by heating to the complex **111** (see p. 81).[1000] Another type of activation by a homoscorpionate iridium complex was that of carbon monoxide. Thus, the reaction of TpIr(CO)$_2$ with water or with alcohols produced complexes TpIrH(CO)(COOH), and TpIrH(CO)(COOR), respectively,[995] while a similar reaction with amines gave rise to species such as TpIrH(CO)(CONHR).[996]

Apart from the rhodium and iridium scorpionate complexes, there were few examples of C—H bond activation with other metals. One such example involved a rhenium complex, TpReO(Cl)I, which could be photochemically activated in a variety of aromatic solvents, with resultant replacement of iodine by an aryl group. The yields for this reaction were improved by the presence of pyridine.[779,780] There was also one report using ruthenium. In that example, regio- and stereoselective C—C bond formation with terminal acetylenes was achieved through facile γ-C—H bond activation in phosphinoamine ligands of TpRuCl(L) complexes. The novel products were exemplified by TpRuCl[κ^3(P,C,C)-Ph$_2$PCH$_2$CH(NEt$_2$)CH=CHR)], and by related structures, such as **238** (Fc = ferrocenyl) and **239**. These couplings took place, depending on the steric requirements of the R substituent, either at the internal

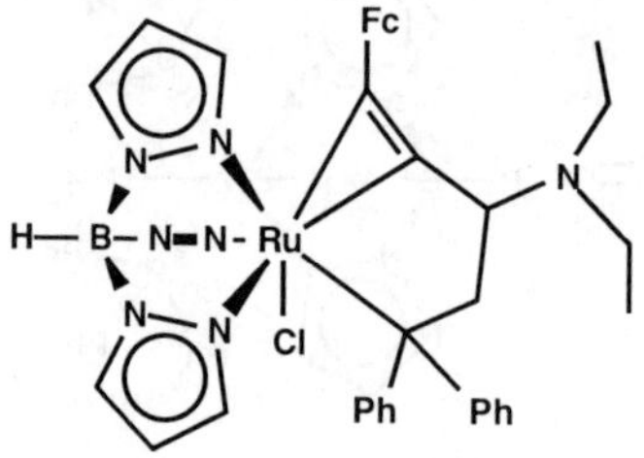

238 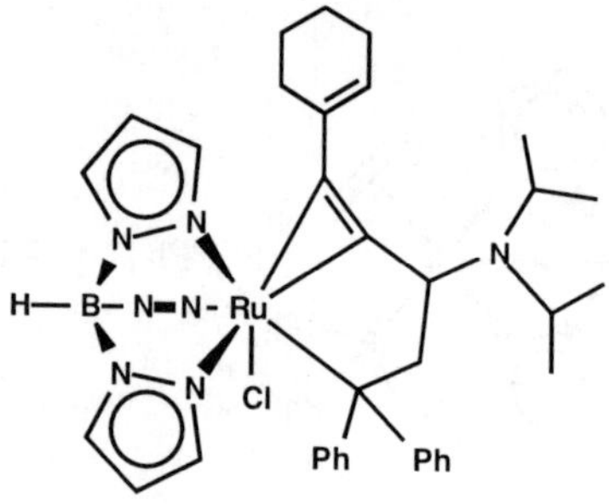   **239**

or at the terminal carbon atom of the acetylene molecule.[900] In the other example, a coupling reaction of terminal acetylenes with a ruthenium-coordinated ligand, Ph$_2$PCH$_2$CH$_2$NMe$_2$, was achieved, and this involved C—H activation of the -CH$_2$CH$_2$- chain. In this reaction the cation [(TpRu(Ph$_2$PCH$_2$CH$_2$NMe$_2$)(solvent)]$^+$ reacted with monosubstituted acetylenes HC≡CR (R = COOEt, Ph, CH$_2$Ph), losing dimethylamine, and forming the novel coupling products of general structure TpRu(Cl)[κ^3-(P,C,C)-Ph$_2$PCH=CHC(R)=CH$_2$)].[899]

Finally, an example of C—H bond activation by a platinum complex involved the activation of benzene. In this reaction the anion [Tp*PtMe$_2$]$^-$ was monodemethylated with B(C$_6$F$_5$)$_3$, and this was followed by oxidative addition of benzene to the resulting [TpPtMe] species, producing the structurally characterized octahedral complex TpPtMe(H)Ph.[1046]

5.5   Metal Deposition

Copper(I) scorpionates were used in preparing thin copper films by the MOCVD (Metal Organic Chemical Vapor Deposition) method. It was important that the complexes be oxygen-free, and that the decomposition products in a hydrogen atmosphere be volatile, and thus not be incorporated in the film. The complexes used were TpCuPMe$_3$ and TpCuPEt$_3$, and in selectivity studies it was found that selective deposition of copper occurs on metal coated films, such as Pt and W.[1507] In a more detailed investigation, a wider range of copper complexes Tp^{x}CuL was tested, including those with Tp^x = Tp, Tp^{Me} and Tp*, and L = PMe$_3$, PEt$_3$, PMe$_2$Ph, Ph$_2$PCH$_2$PPh$_2$, Ph$_2$PCH$_2$CH$_2$PPh$_2$, PPh$_3$, CNBu^t, CNPh-3,5-Me$_2$ and PCy$_3$, some of which were novel compounds. Polycrystalline copper phases were obtained in the 150-350 °C range. Selective deposition on metal-seeded surfaces was noted on Pt, Au, Al and W using SiO$_2$ as standard.[1508] It was also possible to deposit thin TiO$_2$ films by the MOCVD technique, using TpTi(OPr^i)$_3$ as the source of titanium. Amorphous TiO$_2$ films were prepared successfully in the 300-800 °C range on quartz, sapphire and Si[100].[1509]

5.6 Metal Extraction

The use of Tp and pzTp ligand in the extraction of divalent first row transition metals, and alkaline earth metals was studied as a function of pH, and it was found that at pH > 1.0, divalent nickel, cobalt and copper were quantitatively extracted into chloroform from an aqueous phase, while magnesium and calcium were extracted in the neutral pH range.[1510] These studied were continued, including also the Bp ligand, which did not extract any IIa metals, but did so in the presence of diphenylphosphinylmethane synergist.[1511] Studying the extractability of alkaline earth cations with scorpionate ligands, it was found that Tp is an effective extractant for Be^{2+}, Mg^{2+}, and Ca^{2+}, while pzTp was effective only for Be^{2+} and Mg^{2+}.[320]

5.7 Miscellaneous Studies

A number of scorpionate complexes of the *d*- and *f*-block metals has been included in tables listing bond lengths, which were determined by X-ray crystallography and by neutron diffraction.[1512] In a negative ion electrospray mass spectrometric study it was found that solutions of KTp show ions [Tp]$^-$ and K[Tp$_2$]$^-$, and that alkali metal halides do not form observable adducts with the Tp anion.[1513] Negative-ion fast-atom bombardment mass spectra of scorpionates with various matrices (glycerol, *m*-nitrobenzyl alcohol, magic bullet, tetraethyleneglycol dimethyl ether) have been explored, and information on ion structure was obtained from the collisionally-activated dissociation spectra.[1514] Seven different paramagnetic octahedral homoscorpionate complexes of Co^{II} were investigated by FAB mass spectrometry, and by ^{1}H and ^{13}C NMR spectroscopy.[78] The nature of metal-ligand bonding was studied by gas-phase ultraviolet photoelectron spectroscopy on a series of octahedral Tp$_2$M complexes, where M was Zn, Fe, Co, Ni and Cu.[1515] The structures of Bp, Tp and pzTp salts (Na and K), also containing water of hydration, were studied in the solid state by X-ray crystallography, and in solution by ^{1}H, ^{11}B, ^{13}C and ^{15}N NMR, and the σ_p values for [BH$_3$]$^-$, [pzBH$_2$]$^-$, [pz$_2$BH]$^-$ and [Bpz$_3$]$^-$ were estimated.[318] The structure of the hydrated free acid, pzTpH, was determined by X-ray crystallography, and was found to be similar to that of the alkali metal salts.[312] A comparison of a series of Tp$_2$M complexes with related triazolylborates was carried out on the basis of electrochemical data, and of their orbital energy levels.[158]

It was shown that the apparent Tl–H and Tl–C coupling constants in Tp^{x}Tl complexes can be modulated, as a result of nuclear shift relaxation due to chemical shift anisotropy. This could be seen as a reduction of coupling constants at higher magnetic field strengths, or at lower temperatures. Such behavior was particularly noticeable in complexes of structure Tp^{tBu,R}Tl.[1516] The thermodynamics of heterolytic and homolytic M—H bond cleavage reactions of 18-electron and 17-electron complexes Tp^{x}M(CO)$_3$H, (Tp^x = Tp, Tp*; M = Cr, Mo, W) were studied, and compared with available data for their Cp analogs. It was found by means of proton transfer equilibrium measurments that the pK$_a$ values for the metal hydrides decrease in the order CpM(CO)$_3$H > TpM(CO)$_3$H > Tp*M(CO)$_3$H. The homolytic bond dissociation energies, obtained on the basis of the known thermochemical cycle based on the pK$_a$ and anion oxidation potential data, were found to decrease in the same order, CpM(CO)$_3$H > TpM(CO)$_3$H > Tp*M(CO)$_3$H.[1517] Related to his, was a study of the rates of degenerate transfer of electrons, protons and hydrogen atoms between Tp*Mo(CO)$_3$H and of the derived 17-electron radical, Tp*Mo(CO)$_3$.[423]

A series of tungsten homoscorpionate hydrides of structure Tp^{x}W(CO)$_3$H, where Tp^x was Tp, Tp*, Tp^{Ph}, Tp^{iPr} and Tp^{tBu}, was synthesized, and the structure of TpW(CO)$_3$H was determined by X-ray crystallography. It was assumed on the basis of IR data, that in all these complexes tungsten exhibits a 3:4 coordination.[680] A series of complexes having the structure Tp^{x}RhCl$_2$L (Tp^x = Tp^{Me}, Tp*, Tp^{Me3}, Tp*^{Cl}, Tp^{iPr}, Tp^{iPr,4Br}, Tp^{CF3,Me},, L = MeOH, MeCN, or Hpz^x) has been prepared starting with NaTp^x and RhCl$_3$ in MeOH or in MeCN. The Tp^{Ph,Me} ligand was degraded under these conditions, while no reaction occurred with Tp$^{(CF3)2}$, presumably because of non-bonding interaction of the 5-CF$_3$ groups.[977] Another multi-ligand study involved the synthesis of Tp^{x}Ir(COD) complexes, for Tp^x = Tp, Tp^{Me}, Tp^{iPr}, Tp*, Tp^{CF3,Me}, Tp^{Ph,Me}, Tp^{iPr2}, Tp^{Me3}, Tp*^{Cl} and Tp*^{Br}, and the structure of the κ^3-bonded Tp*^{Cl}Ir(COD) was determined by X-ray crystallography. The bulkiness of these complexes resulted in the isolation of some products containing rearranged Tp^x ligands. In the case of Tp^{iPr}, the product was Tp^{iPr*}Ir(COD), while in the case of Tp^{Me}, the final product was [HB(pz^{3Me})(pz^{5Me})$_2$]Ir(COD), thus far a unique example of two pz^R moieties undergoing rearrangment within the ligand.[979]

The reduction of twenty different aldehydes and ketones to the corresponding alcohols by KBp, Bp$_2$Cu, and by pyrazabole, H$_2$B(μ-pz)$_2$BH$_2$, has been studied in ethanol solution. In general, yields ranging up to 93 % were obtained with KBp or

5.8 MISCELLANEOUS STUDIES

with the pyrazabole, while Bp_2Cu was only half as active. In the work-up of the reaction residue from the reduction with pyrazabole, using HCl, some 4,4,8,8-tetrachloropyrazabole was isolated.[1518] The salt KpzTp was investigated as a mobile phase additive for reverse phase High Performance Liquid Chromatography of metal chelating compounds, and KpzTp was found to be definitely superior to previously employed eluent modifiers, including NaTp.[1519] Some scorpionate derivatives of ruthenium were used for the inhibition of the immune response, thus significantly reducing graft rejection, and for treating hyperproliferative vascular diseases.[1520]

In probing the geometric and electronic structure of cobalt(II)-substituted proteins by means of magnetic circular dichroism spectroscopy, in which ground-state zero-field splitting was a coordination number indicator, several cobalt(II) scorpionates were used as model compounds. They included $Tp^{tBu}CoNCO$, $Tp^{tBu}CoNO_3$, $Tp^{iPr,4Br}CoCl$, $Tp^{iPr,4Br}CoNCS$, $Tp^{iPr,4Br}Co[(BBN)Bp]$, $Tp^{Ph}CoN_3$, $Tp^{Ph}CoNCS$, $Tp^{Ph}CoNCO$, $Tp^{Ph}CoNCS(THF)$, and $[pz^oTp^{iPr}]_2Co$.[1521] The M—N stretching bands of $Tp*_2Fe$, and of other related scorpionates have been observed and assigned in the region below 400 cm^{-1}. In a study of electron transfer in transition metal-pteridine systems, complexes of the type $Tp^xCu(H_4pterin)$ ($Tp^x = Tp^{Ph}$, $Tp^{iPr,4Br}$, Tp^{tBu}) were synthesized and their electronic spectral data and lifetimes were determined.[1522] A detailed study of the low-frequency (650-150 cm^{-1}) infrared spectra of octahedral Tp^x_2M complexes (Tp^x = Tp, pzTp, Tp* and Tp^{Me3}; M = Fe, Co, Ni, Cu and Zn) was carried out, and the M—N stretching bands were assigned, based on the metal-isotope substitution method.[1525]

An abiguity often exists in assigning the hapticity to homoscorpionate ligands of rhodium, iridium and platinum complexes in solution. It was found in a study of a large number of $Tp*ML_n$ species that good correlation exists between the hapticity of the Tp* ligand and the ^{11}B NMR chemical shifts of these complexes. Thus, κ^3-bonded Tp* ligands had ^{11}B NMR chemical shifts in the -8.4 to -9.8 ppm range, while for the κ^2-bonded Tp* ligands, the range was between -5.9 and -7.0 ppm.[1536] This method complements the one based on the infrared B—H stretch frequency as a criterion for hapticity assignment: complexes with the B—H stretch frequency above 2500 cm^{-1} contained the Tp^R ligand bonded in κ^3 fashion, while frequencies below 2500 cm^{-1} were typical of κ^2-bonding.[1318] Results obtained by both of these methods were in agreement, and led to reassignment of the solution structure of $Tp*Rh(CO)(CH_2=CH_2)$ from κ^2 to κ^3 bonding of the Tp* ligand.[947]

It was discovered that solid Tp_2Ru absorbed chlorine gas in redox fashion, and released it on heating in vacuo, these events being accompanied by a well-defined color change. This gave rise to the possibility that such property might lead to the development of a solid-state sensor for chlorine gas.[886] The carbonyl complex TpCuCO was used as a sensitizer for the valence isomerization of norbornadiene to quadricyclene.[1062] A considerable number of $TpMo(CO)_2(\eta^3\text{-allyl})$ complexes, where the η^3-allyl ligand contained a large variety of substituents, or was part of a cyclic system, was synthesized,[444,445] and these complexes were investigated as useful reagents or catalysts for organic synthesis.[446-450] The complex [pz°Tp]$_2$Fe was evaluated in a study of size-exclusion chromatography.[814] Some Sn^{IV} compounds of general structure TpSnR$_3$ were investigated for antimutagenic activity,[1133] while Tp*SnBu$_2$Cl was found to be a useful transfer agent for the Tp* ligand to Zr, Nb, and Ta compounds.[356] Tp and Tp* complexes of Rh, Ir, and Ru organometallic derivatives catalyzed regioselective homogeneous hydrogenation of quinoline, the best catalyst being TpRh(COD).[1566]

5.8 Concluding Remarks

At the current strong rate of growth in scorpionate chemistry, possibly aided by the commercial availability (Aldrich, Acros) of the first generation, and of some second generation ligands, further development of novel homoscorpionates is to be expected. In addition to introducing novel Tp substituents, one can anticipate a bonanza in heteroscorpionates, especially in devising ligands with new types of donor arms, in addition to the two pz^x rings. Some of the possibilities may be illustrated by R(R'NH)Bp^x, where deprotonation of the R'NH substituent would yield a dianion [R(R'N)Bp^x]$^{2-}$, adding a novel twist to the traditional Tp^x chemistry. Other types of complexes will be accesible through the already demonstrated method of adding the B—H bond in Bp^x complexes to multiple bonds, as in RR'C=E species, which then become modifiers of the original ligand.

The neutral cousins of polypyrazolylborates, polypyrazolylmethanes, the coordination ability of which has been known for almost as long as that of the former,[187] are likely to continue their recently more expanding development.

Another area which might see some activity is the "organic" chemistry of octahedral Tp^x$_2$M complexes, akin to that of ferrocene. Triple functionalization of the 4-position, could lead to interesting dendrimers, crosslinked structures, and the like. The Tp^x$_2$M moiety could be attached to amino acids, and inserted into proteins. Other possibilities will become obvious, as researchers from neighboring sub-disciplines become more familiar with scorpionate ligands. In the final analysis, the scorpionate field is wide open, and can be extended in almost any direction, being restricted only by the creativity of the scientist.

References

1 S. Trofimenko, *J. Am. Chem. Soc.* **88** (1966) 1842.
2 J. C. Calabrese, S. Trofimenko and J. S. Thompson, *J. Chem. Soc., Chem. Commun.* (1986) 1122.
3 S. Trofimenko, J. C. Calabrese and J. S. Thompson, *Inorg. Chem.* **26** (1987) 1507.
4 S. Trofimenko, *J. Am. Chem. Soc.* **91** (1969) 588.
5 S. Trofimenko, J. C. Calabrese, P. J. Domaille and J. S. Thompson, *Inorg. Chem.* **28** (1989) 1091.
6 M. D. Curtis, K.-B. Shiu and W. M. Butler, *Organometallics* **2** (1983) 1475.
7 M. D. Curtis, K.-B. Shiu and W. M. Butler, *J. Am. Chem. Soc.* **108** (1986) 1550.
8 S. Trofimenko, *Chem. Revs.* **93** (1993) 943.
9 S. Trofimenko, *J. Am. Chem. Soc.* **89** (1967) 3170.
10 S. Trofimenko, *J. Am. Chem. Soc.* **89** (1967) 3165.
11 S. Trofimenko, *J. Am. Chem. Soc.* **89** (1967) 6288.
12 S. Trofimenko, *J. Am. Chem. Soc.* **89** (1967) 4948.
13 J. P. Jesson, S. Trofimenko and D. R. Eaton, *J. Am. Chem. Soc.* **89** (1967) 3148.
14 J. P. Jesson, J. F. Weiher and S. Trofimenko, *J. Chem. Phys.* **48** (1968) 2058.
15 M. R. Churchill, K. Gold and C. E. Maw, Jr., *Inorg. Chem.* **9** (1970) 1597.
16 S. Trofimenko, *J. Am. Chem. Soc.* **89** (1967) 3904.
17 S. Trofimenko, *J. Am. Chem. Soc.* **91** (1969) 3183.
18 S. Trofimenko, *Inorg. Chem.* **8** (1969) 2675.
19 S. Trofimenko, *Inorg. Chem.* **10** (1971) 504.
20 T. Desmond, F. J. Lalor, G. Ferguson, B. Ruhl and M. Parvez, *J. Chem. Soc., Chem. Commun.* (1983) 55.
21 S. Trofimenko, *J. Am. Chem. Soc.* **90** (1968) 4754.
22 S. Trofimenko, *Inorg. Chem.* **9** (1970) 2493.
23 C. A. Kosky, P. Ganis and G. Avitabile, *Acta Cryst.* **B27** (1971) 1859.
24 F. A. Cotton, T. LaCour and A. G. Stanislowski, *J. Am. Chem. Soc.* **96** (1974) 754.

25	F. A. Cotton and A. G. Stanislowski, *J. Am. Chem. Soc.* **96** (1974) 5074.
26	F. A. Cotton, B. A. Frenz and C. A. Murillo, *J. Am. Chem. Soc.* **97** (1980) 2118.
27	J. A. McCleverty, *Inorg. Chim. Acta* **62** (1982) 67.
28	W. W. Greaves and R. J. Angelici, *J. Organomet. Chem.* **191** (1980) 49.
29	W. W. Greaves and R. J. Angelici, *Inorg. Chem.* **20** (1981) 2983.
30	T. Desmond, F. J. Lalor, G. Ferguson, and M. Parvez, *J. Chem. Soc., Chem. Commun.* (1983) 457.
31	T. Desmond, F. J. Lalor, G. Ferguson, and M. Parvez, *J. Chem. Soc., Chem. Commun.* (1984) 75.
32	R. Han and G. Parkin, *Inorg. Chem.* **32** (1993) 4968.
33	R. Han and G. Parkin, *Organometallics* **10** (1991) 1010.
34	I. B. Gorrell, A. Looney and G. Parkin, *J. Chem. Soc., Chem. Commun.* (1990) 220.
35	J. W. Egan, Jr., B. S. Haggerty, A. L. Rheingold, S. C. Sendlinger and K. H. Theopold, *J. Am. Chem. Soc.* **112** (1990) 2445.
36	O. M. Reinaud, G. P. A. Yap, A. L. Rheingold and K. H. Theopold, *Angew. Chem. Int. Ed. Engl.* **34** (1995) 2051.
37	S. M. Carrier, C. E. Ruggiero and W. B. Tolman, *J. Am. Chem. Soc.* **114** (1995) 4407.
38	W. B. Tolman, *Inorg. Chem.* **30** (1991) 4877.
39	U. Hartmann and H. Vahrenkamp, *Chem. Ber.* **127** (1994) 2381.
40	M. Ruf, R. Burth, K. Weis and H. Vahrenkamp, *Chem. Ber.* **129** (1996) 1251.
41	L. Hasinoff, J. Takats, X. W. Zhang, A. H. Bond and R. D. Rogers, *J. Am. Chem. Soc.* **116** (1994) 8833.
42	X. Zhang, R. McDonald and J. Takats, *New J. Chem.* **19** (1995) 573.
43	N. Kitajima, K. Fujisawa and Y. Moro-oka, *J. Am. Chem. Soc.* **111** (1989) 8975.
44	N. Kitajima, K. Fujisawa and Y. Moro-oka, *Inorg. Chem.* **29** (1990) 357.
45	N. Kitajima, K. Fujisawa, T. Koda, S. Hikichi and Y. Moro-oka, *J. Chem. Soc., Chem. Commun.* (1990) 1357.
46	D. L. Reger, S. S. Mason and A. L. Rheingold, *J. Am. Chem. Soc.* **115** (1993) 10406.
47	M. Etienne, *Coord. Chem. Revs.* **156** (1996) 201.
48	S. G. Feng and J. L. Templeton, *Organometallics* **11** (1992) 1295.
49	S. A. O'Reilly, P. S. White, and J. L. Templeton, *J. Am. Chem. Soc.* **118** (1996) 5684.
50	B. E. Woodworth, P. S. White, and J. L. Templeton, *J. Am. Chem. Soc.* **119** (1997) 828.
51	T. Brent Gunnoe, P. S. White, J. L. Templeton, and L. Casarrubios, *J. Am. Chem. Soc.* **119** (1997) 3171.
52	S. Trofimenko, *Acc. Chem. Research* **4** (1971) 17.
53	S. Trofimenko, *Chem. Rev.* **72** (1972) 497.

REFERENCES

54 K. Niedenzu and S. Trofimenko, *Gmelin Handbuch der Anorganischen Chemie* **23/5** (1975) 1.
55 K. Niedenzu and S. Trofimenko, *Topics in Current Chemistry* **131** (1986) 1.
56 A. Shaver, *Organomet. Chem. Libr.* **3** (1977) 157.
57 J. A. McCleverty, *Chem. Soc. Rev.* **12** (1983) 331.
58 S. Trofimenko, *Progress Inorg. Chem.* **34** (1986) 115.
59 K. Niedenzu, Advances in Boron and the Boranes, **5** (1988) 357. *Molecular Structure and Energetics,* J. F. Liebman, A. Greenberg and R. E. Williams, editors, VXH Publishers.
60 P. K. Byers, A. J. Canty and R. T. Honeyman, *Adv. Organomet. Chem.* **34** (1992) 1.
61 R. Han, A. Looney, K. McNeill, G. Parkin, A. L. Rheingold and B. S. Haggerty, *J. Inorg. Biochem.* **49** (1993) 105.
62 N. Kitajima and Y. Moro-oka, *Chem. Rev.* **94** (1994) 737.
63 S. Santos and N. Marques, *New J. Chem.* **19** (1995) 551.
64 G. Parkin, *Adv. Inorg. Chem.* **42** (1995) 291.
65 N. Kitajima and W. B. Tolman, *Progress Inorg. Chem.* **43** (1995) 419.
66 D. L. Reger, *Coord. Chem. Rev.* **147** (1996) 571.
67 G. Parkin, *Handbook of Grignard Reagents,* G. S. Silverman, P. E. Rakita, Editors, Marcel Dekker, Inc. (1996) 291.
68 C. G. Young and A. G. Wedd, *Chem. Commun.* (1997) 1251.
69 K. H. Theopold, O. M. Reinaud, D. Doren and R. Konecny, *Stud. Surf. Sci. Catal.* **110** (3rd World Congress on Oxidation Catalysis) (1997)1081.
70 C. Janiak, *Coord. Chem. Rev.* **163** (1997) 107-216.
71 D. Janiak, *Main Group Met. Chem.* **21** (1998) 2.
72 J. A. McCleverty and M. D. Ward, *Acc. Chem. Res.* **31** (1998) 842.
73 C. Slugovc, R. Schmid and K. Kirchner, *Coord. Chem. Rev.* in print.
74 S. Trofimenko, *Inorg. Synth.* **12** (1970) 99.
75 F. Jäkle, K. Polborn and M. Wagner, *Chem. Ber.* **129** (1996) 603.
76 D. L. Reger and M. E. Tarquini, *Inorg. Chem.* **21** (1982) 840.
77 U. E. Bucher, T. F. Fässler, M. Hunzicker, R. Nesper, H. Rüegger and L. M. Venanzi, *Gazz. Chim. Ital.* **125** (1995) 181.
78 J.-L. Aubagnac, R. M. Claramunt, J. Elguero, I. Gilles, D. Sanz, S. Trofimenko and A. Virgili, *Bull. Chem. Soc. Belg.* **104** (1995) 473.
79 K. Niedenzu, P. M. Niedenzu and K. R. Warner, *Inorg. Chem.* **24** (1985) 1604.
80 J. C. Calabrese and S. Trofimenko, *Inorg. Chem.* **31** (1992) 4810.
81 A. L. Rheingold, L. M. Liable-Sands and S. Trofimenko, *Chem. Commun.* (1997) 1691.
82 A. L. Rheingold, G. P. A. Yap, L. M. Liable-Sands, I. A. Guzei and S. Trofimenko, *Inorg. Chem.* **36** (1997) 6261.
83 S. Trofimenko, unpublished results.

84 A. L. Rheingold, B. S. Haggerty and S. Trofimenko, *Angew. Chem. Int. Ed. Engl.* **33** (1994) 1983.
85 D. J. Darensbourg, E. L. Maynard, M. W. Holtcamp, K. K. Klausmeyer and J. H. Reibenspies, *Inorg. Chem.* **35** (1996) 2682.
86 S. Trofimenko, J. C. Calabrese, J. K. Kochi, S. Wolowiec, F. B. Hulsbergen and J. Reedijk, *Inorg. Chem.* **31** (1992) 3943.
87 P. L. Jones, K. L. V. Mann, J. C. Jeffery, J. A. McCleverty and M. D. Ward, *Polyhedron* **16** (1997) 2435.
88 E. R. Humphrey, N. C. Harden, L. H. Rees, J. C. Jeffery, J. A. McCleverty and M. D. Ward, *J. Chem. Soc., Dalton Trans.* (1998) 3353.
89 R. Han, G. Parkin and S. Trofimenko, *Polyhedron* **14** (1995) 387.
90 M. A. Halcrow, E. J. L. McInnes, F. E. Mabbs, I. J. Scowen, M. McPartlin, H. R. Powell and J. E. Davies, *J. Chem. Soc., Dalton Trans.* (1997) 4025.
91 J. C. Calabrese, P. J. Domaille, S. Trofimenko and G. J. Long, *Inorg. Chem.* **30** (1991) 2795.
92 C. Lopez, D. Sanz, R. M. Claramunt, S. Trofimenko and J. Elguero, *J. Organomet. Chem.* **503** (1995) 265.
93 A. L. Rheingold, C. B. White and S. Trofimenko, *Inorg. Chem.* **32** (1993) 3471.
94 H. V. R. Dias, W. Jin, H.-J. Kim and H.-L. Lu, *Inorg. Chem.* **35** (1996) 2317.
95 H. V. R. Dias, H.-J. Kim, H.-L. Lu, K. Rajeshwar, N. R. de Tacconi, A. Derecskei-Kovacs and D. S. Marynick, *Organometallics* **15** (1996) 2994.
96 H.V. R. Dias and H.-J. Kim, *Organometallics* **15** (1996) 5374.
97 T. Fillebeen, T. Hascall and G. Parkin, *Inorg. Chem.* **36** (1997) 3787.
98 D. D. LeCloux, M. C. Keyes, M. Osawa, V. Reynolds and W. B. Tolman, *Inorg. Chem.* **33** (1994) 6361.
99 G. G. Lobbia, G. Valle, S. Calogero, P. Cecchi, C. Santini and F. Marchetti, *J. Chem. Soc., Dalton Trans.* (1996) 2475.
100 G. G. Lobbia, B. Bovio, C. Santini, P. Cecchi, C. Pettinari and F. Marchetti, *Polyhedron* **17** (1998) 17.
101 A. L. Rheingold, B. S. Haggerty, G. P. A. Yap and S. Trofimenko, *Inorg. Chem.* **36** (1997) 5097.
102 A. L. Rheingold, L. M. Liable-Sands, G. P. A. Yap and S. Trofimenko, *Chem. Commun.* (1996) 1233.
103 A. L. Rheingold, B. S. Haggerty and S. Trofimenko, *J. Chem. Soc., Chem. Commun.* (1994) 1973.
104 A. L. Rheingold, R. L. Ostrander, B. S. Haggerty and S. Trofimenko, *Inorg. Chem.* **33** (1994) 3666.
105 D. C. Bradley, M. B. Hursthouse, J. Newton and N. P. C. Walker, *J. Chem. Soc., Chem. Commun.* (1984) 188.
106 M. Cano, J. V. Heras, S. Trofimenko, A. Monge, E. Gutierrez, C. J. Jones and J. A. McCleverty, *J. Chem. Soc., Dalton Trans.* (1990) 3577.

REFERENCES

107 N. Kitajima, K. Fujisawa, C. Fujimoto, Y. Moro-oka, S. Hashimoto, T. Kitagawa, K. Toriumi, K. Tatsumi and A. Nakamura, *J. Am. Chem. Soc.* **114** (1992) 1277.
108 C. M. Dowling, D. Leslie, M. H. Chisholm and G. Parkin, *Main Group Chem.* **1** (1995) 29.
109 W. Kläui, W. Schilde and M. Schmidt, *Inorg. Chem.* **36** (1997) 1598.
110 M. Ruf, K. Weis and H. Vahrenkamp, *J. Chem. Soc., Chem. Commun* (1994) 135.,
111 P. Ghosh and G. Parkin, *Inorg. Chem.* **35** (1996) 1429.
112 K. Weis and H. Vahrenkamp, *Inorg. Chem.* **36** (1997) 5589.
113 E. Libertini, K. Yoon and G. Parkin, *Polyhedron* **12** (1993) 2539.
114 U. E. Bucher, A. Currao, R. Nesper, H. Rüegger, L. M. Venanzi and E. Younger, *Inorg. Chem.* **34** (1995) 66.
115 R. Han, P. Ghosh, P. J. Desrosiers, S. Trofimenko and G. Parkin, *J. Chem. Soc., Dalton Trans.* (1997) 3713.
116 H. V. R. Dias, H.-L. Lu, R. E. Ratcliff and S. G. Bott, *Inorg. Chem.* **34** (1995) 1975.
117 O. Renn, L. M. Venanzi, A. Martelletti and V. Gramlich, *Helv. Chim. Acta* **78** (1995) 993.
118 A. L. Rheingold, G. Yap and S. Trofimenko, *Inorg. Chem.* **34** (1995) 759.
119 F. J. Lalor, S. M. Miller and N. Garvey, *Polyhedron* **9** (1990) 63.
120 J. A. McCleverty, D. Seddon, N. A. Bailey and N. W. J. Walker, *J. Chem. Soc., Dalton Trans.* (1976) 898.
121 J. C. Jeffery, S. S. Kurek, J. A. McCleverty, E. Psillakis, R. M. Richardson, M. D. Ward and A. Wlodarczyk, *J. Chem. Soc., Dalton Trans.* (1994) 2559.
122 Y. Takahashi, M. Akita, S. Hikichi and Y. Moro-oka, *Organometallics* **17** (1998) 4884.
123 J. Huang, L. Lee, B. S. Haggerty, A. L. Rheingold and M. A. Walters, *Inorg. Chem.* **34** (1995) 4268.
124 S. Trofimenko, *J. Am. Chem. Soc.* **91** (1969) 5410.
125 P. J. Domaille, *J. Am. Chem. Soc.* **102** (1980) 5392.
126 A. A. Bothner-By, P. J. Domaille and C. Gayathri, *J. Am. Chem. Soc.* **103** (1981) 5602.
127 D. L. White and J. W. Faller, *J. Am. Chem. Soc.* **104** (1982) 1548.
128 F. F. de Biani, F. Jäkle, M. Spiegler, M. Wagner and P. Zanello, *Inorg. Chem.* **36** (1997) 2103.
129 K. Niedenzu, S. S. Seelig and W. Weber, *Z. Anorg. Allg. Chem.* **483** (1981) 51.
130 K. Niedenzu and S. Trofimenko, *Inorg. Chem.* **24** (1985) 4222.
131 C. P. Brock, M. K. Das, R. P. Minton and K. Niedenzu, *J. Am. Chem. Soc.* **110** (1988) 817.
132 W. H. McCurdy, Jr., *Inorg. Chem.* **14** (1975) 2292.

133 K. Niedenzu, J. Serwatowski and S. Trofimenko, *Inorg. Chem.* **30** (1991) 524.
134 H. V. R. Dias and J. D. Gorden, *Inorg. Chem.* **35** (1996) 318.
135 K. R. Breakell, D. J. Patmore and A. Storr, *J. Chem. Soc., Dalton Trans.* (1975) 749.
136 W. J. Layton, K. Niedenzu, P. M. Niedenzu and S. Trofimenko, *Inorg. Chem.* **24** (1985) 1454.
137 S. Trofimenko, J. C. Calabrese and J. S. Thompson, *Inorg. Chem.* **31** (1992) 974.
138 M. Bortolin, U. E. Bucher, H. Rüegger, L. M. Venanzi, A. Albinati, F. Lianza and S. Trofimenko, *Organometallics* **11** (1992) 2514.
139 J. S. Thompson, J. L. Zitzman, T. J. Marks and J. A. Ibers, *Inorg. Chim. Acta* **46** (1980) L101.
140 A. F. Hill and J. M. Malget, *J. Chem. Soc. Dalton Trans.* (1997) 2003.
141 S. A. A. Zaidi and M. A. Neyazi, *Trans. Met. Chem.* **4** (1979) 164.
142 Z. A. Siddiqi, S. Khan and S. A. A. Zaidi, *Synth. React. Inorg. Met.-Org. Chem.* **12** (1982) 433.
143 G. Agrifoglio, *Inorg. Chim. Acta* **197** (1992) 159.
144 P. Ghosh, T. Hascall, C. Dowling and G. Parkin, *J. Chem. Soc. , Dalton Trans.* (1998) 3355.
145 I. B. Gorrell, A. Looney and G. Parkin, *J. Am. Chem. Soc.* **112** (1990) 4068.
146 C. Dowling and G. Parkin, *Polyhedron* **15** (1996) 2463.
147 D. A. Bardwell, J. C. Jeffery, P. L. Jones, J. A. McCleverty, E. Psillakis, Z. Reeves and M. D. Ward, *J. Chem. Soc. Dalton* (1997) 2079.
148 G. G. Lobbia, F. Bonati and P. Cecchi, *Synth. React. Inorg. Met.-Org. Chem.* **21** (1991) 1141.
149 C. Janiak and L. Esser, *Z. Naturforsch.* **48b** (1993) 394.
150 C. Janiak and H. Hemling, *J. Chem. Soc., Dalton Trans.* (1994) 2497.
151 C. Janiak, *J. Chem. Soc., Chem. Commun.* (1994) 545.
152 C. Janiak, *Chem. Ber.* **127** (1994) 1379.
153 C. Janiak and H. Hemling, *J. Chem. Soc., Dalton Trans.* (1994) 2947.
154 C. Janiak, T. G. Scharmann, H. Hemling, D. Lentz and J. Pickardt, *Chem. Ber.* **128** (1995) 235.
155 C. Janiak, T. G. Scharmann, K.-W. Brzezinka and P. Reich, *Chem. Ber.* **128** (1995) 323.
156 C. Janiak, T. G. Scharmann, W. Günther, F. Girgsdies, H. Hemling, W. Hinrichs and D. Lentz, *Chem. Eur. J.* **1** (1995) 637.
157 C. Janiak, T. G. Scharmann, P. Albrecht, F. Marlow and R. Macdonald, *J. Am. Chem. Soc.* **118** (1996) 6307.
158 C. Janiak, T. G. Scharmann, J. C. Green, R. P. G. Parkin, M. J. Kolm, E. Riedel, W. Mickler, J. Elguero, R. M. Claramunt and D. Sanz, *Chem. Eur. J.* **2** (1996) 992.

REFERENCES

159 K.-B. Shiu, W.-N. Guo, S.-M. Peng and M.-C. Cheng, *Inorg. Chem.* **33** (1994) 3010.
160 F. J. Lalor, S. M. Miller and N. Garvey, *J. Organomet. Chem.* **356** (1998) C57.
161 J. Cartwricht and A. F. Hill, *J. Organomet. Chem.* **429** (1992) 229.
162 S. Anderson, A. Harman and A. F. Hill, *J. Organomet. Chem.* **498** (1995) 251.
163 P. Cecchi, G. G. Lobbia, D. Leonesi, C. Pettinari, C. Sepe and V. Vinciguerra, *Gazz. Chim. Ital.* **123** (1993) 569.
164 I. T. Macleod, E. R. T. Tiekink and C. G. Young, *J. Organomet. Chem.* **506** (1996) 301.
165 Z. Xiao, R. W. Gable, A. G. Wedd and C. G. Young, *J. Chem. Soc., Chem. Commun.* (1994) 1295.
166 T. G. Hodgkins and D. R. Powell, *Inorg. Chem.* **35** (1996) 2140.
167 M. Garner, J. Reglinski, I. Cassidy, M. D. Spicer and A. R. Kennedy, *Chem. Commun.* (1996) 1975.
168 P. Ge, B. S. Haggerty, A. L. Rheingold and C. G. Riordan, *J. Am. Chem. Soc.* **116** (1994) 8406.
169 C. Ohrenberg, P. Ge, P. Schebler, C. G. Riordan, G. P.A. Yap and A. L. Rheingold, *Inorg. Chem.* **35** (1996) 749.
170 K. R. Breakell, D. J. Patmore and A. Storr, *J. Chem. Soc., Dalton Trans.* (1979) 139.
171 D. F. Rendle, A. Storr and J. Trotter, *J. Chem. Soc., Chem. Commun.* (1974) 406.
172 D. F. Rendle, A. Storr and J. Trotter, *J. Chem. Soc., Dalton Trans.* (1975) 176.
173 D. J. Patmore, D. F. Rendle, A. Storr and J. Trotter, *J. Chem. Soc., Dalton Trans.* (1975) 718.
174 B. M. Louie and A. Storr, *Can. J. Chem.* **62** (1984) 633.
175 K. S. Chong, S. J. Rettig, A. Storr and J. Trotter, *Can. J. Chem.* **59** (1981) 996.
176 S. E. Anslow, K. S. Chong, S. J. Rettig, A. Storr and J. Trotter, *Can. J. Chem.* **59** (1981) 3123.
177 B. M. Louie, S. J. Rettig, A. Storr and J. Trotter, *Can. J. Chem.* **62** (1984) 1057.
178 B. M. Louie, S. J. Rettig, A. Storr and J. Trotter, *Can. J. Chem.* **62** (1984) 503.
179 A. Nussbaum, S. J. Rettig, A. Storr and J. Trotter, *Can. J. Chem.* **63** (1985) 692.
180 K. R. Breakell, S. J. Rettig, D. L. Singbeil, A. Storr and J. Trotter, *Can. J. Chem.* **56** (1978) 2099.
181 S. J. Rettig, A. Storr and J. Trotter, *Can. J. Chem.* **57** (1979) 182.
182 B. M. Louie, S. J. Rettig, A. Storr and J. Trotter, *Can. J. Chem.* **62** (1984) 633.

183	K. S. Chong, S. J. Rettig, A. Storr and J. Trotter, *Can. J. Chem.* **57** (1979) 3107.
184	K. R. Breakell, S. J. Rettig, A. Storr and J. Trotter, *Can. J. Chem.* **61** (1983) 1659.
185	E. C. Onyiriuka, S. J. Rettig and A. Storr, *Can. J. Chem.* **64** (1986) 321.
186	E. C. Onyiriuka and A. Storr, *Can. J. Chem.* **65** (1987) 2464.
187	S. Trofimenko, *J. Am. Chem. Soc.* **92** (1970) 5118.
188	S. Trofimenko, *U. S. Pat.* 3,808,228 (1974).
189	K. I. The and L. K. Peterson, *J. Chem. Soc., Chem. Commun.* (1972) 841.
190	K. I. The, L. K. Peterson and E. Kiehlmann, *Can. J. Chem.* **51** (1973) 422.
191	D. J. O'Sullivan and F. J. Lalor, *J. Organomet. Chem.* **57** (1973) C58.
192	I. N. Kremenskaya, N. K. Evseeva, V. M. Dziomko, O. V. Ivanov and G. A. Kumaneva, *Radiokhimia* **16** (1974) 455.
193	V. N. Dziomko, N. K. Evseeva, O. V. Ivanov, I. N. Kremenskaya and G. A. Kumaneva, *Radiokhimia* **16** (1974) 588.
194	N. K. Evseeva, I. N. Kremenskaya, N. S. Smirnova, Yu. V. Garnovskii, V. M. Dziomko, L. N. Komisarova, O. V. Ivanov and G. A. Kumaneva, *Zh. Prikl. Khim.* **48** (1975) 1130.
195	J. Reedijk and J. Verbiest, *Trans. Met. Chem.* **3** (1978) 51.
196	A. I. Sukhanovskaya, I. N. Kremenskaya, N. J. Evseeva and V. N. Stepanova, *Zh. Anal. Khim.* **33** (1978) 1539.
197	F. Mani, *Inorg. Nucl. Chem. Lett.* **45** (1979) 297.
198	F. Mani and R. Morassi, *Inorg. Chim. Acta* **36** (1979) 63.
199	J. Reedijk and J. Verbiest, *Trans. Met. Chem.* **4** (1979) 239.
200	J. C. Jansen, H. van Koningsveld, J. A. C. van Ooijen and J. Reedijk, *Inorg. Chem.* **19** (1980) 170.
201	J. Verbiest, J. A. C. van Ooijen and J. Reedijk, *J. Inorg. Nucl. Chem.* **42** (1980) 971.
202	R. W. M. ten Hoedt and J. Reedijk, *J. Chem. Soc., Chem. Commun.* (1980) 844.
203	L. K. Peterson and H. E. W. Rhodes, *Inorg. Chim. Acta* **45** (1980) L95.
204	F. Mani, *Inorg. Chim. Acta* **38** (1981) 97.
205	H. C. Clark and M. A. Mesubi, *J. Organomet. Chem.* **215** (1981) 131.
206	A. J. Canty, N. J. Minchin, J. M. Patrick and A. H. White, *J. Chem. Soc., Dalton. Trans.* (1982) 1795.
207	M. A. Mesubi and R. E. Enemo, *Spectrochim. Acta* **38A** (1982) 599.
208	H. C. Clark, G. Ferguson, V. K. Jain and M. Parvez, *Organometallics* **2** (1983) 806.
209	M. A. Mesubi, *Trans. Met. Chem.* **9** (1984) 181.
210	M. A. Mesubi, *J. Coord. Chem.* **13** (1984) 179.
211	L. A. Oro, M. Esteban, R. M. Claramunt, J. Elguero, C. Foces-Foces and F. H. Cano, *J. Organomet. Chem.* **276** (1984) 79.

REFERENCES

212 M. A. Mesubi and P. I. Ekemenzie, *Trans. Met. Chem.* **9** (1984) 91.
213 H. C. Clark, G. Ferguson, V. K. Jain and M. Parvez, *J. Organomet. Chem.* **270** (1984) 365.
214 A. Lorenzotti, A. Cingolani, D. Leonesi and F. Bonati, *Gazz. Chim. Ital.* **115** (1985) 619.
215 G. Minghetti, M. A. Cinellu, A. L. Bandini, G. Banditelli, F. Demartin and M. Manassero, *J. Organomet. Chem.* **315** (1986) 387.
216 A. J. Canty, R. T. Honeyman, B. W. Skelton and A. H. White, *Inorg. Chim. Acta* **114** (1986) L39.
217 H. C. Clark, G. Ferguson, V. K. Jain and M. Parvez, *Inorg. Chem.* **25** (1986) 3808.
218 G. G. Lobbia, A. Cingolani, D. Leonesi, A. Lorenzotti and F. Bonati, *Inorg. Chim. Acta* **130** (1987) 203.
219 A. Lorenzotti, A. Cingolani;, G. G. Lobbia, D. Leonesi and F. Bonati, *Gazz. Chim. Ital.* **117** (1987) 191.
220 D. Leonesi, A. Cingolani, G. G. Lobbia and A. Lorenzotti, *Gazz. Chim. Ital.* **117** (1987) 491.
221 A. Cingolani, A. Lorenzotti, G. G. Lobbia, D. Leonesi, F. Bonati and B. Bovio, *Inorg. Chim. Acta* **132** (1987) 167.
222 K.-B. Shiu and K.-S. Liou, *J. Chin. Chem. Soc.* **35** (1988) 187.
223 K.-B. Shiu, C. J. Chang, Y. Wang and M.-C. Chen *J. Chin. Chem. Soc.* **36** (1989) 26.
224 K.-B. Shiu, K.-S. Liou, S.-L. Wang, C. P. Cheng and F.-J. Wu, *J. Organomet. Chem.* **359** (1989) C1.
225 O. Juanes, J. de Mendoza and J. C. Rodriguez-Ubis, *J. Organomet. Chem.* **363** (1989) 393.
226 G. G. Lobbia, F. Bonati, A. Ciongolani, D. Leonesi and A. Lorenzotti, *J. Organomet. Chem.* **359** (1989) 21.
227 G. G. Lobbia and F. Bonati, *J. Organomet. Chem.* **366** (1989) 127.
228 P. K. Byers, A. J. Canty and R. T. Honeyman, *J. Organomet. Chem.* **385** (1990) 417.
229 P. K. Byers and A. J. Canty, *Organometallics* **9** (1990) 210.
230 K.-B. Shiu, C.-C. Chou, S.-L. Wang and S.-C. Wei, *Organometallics* **9** (1990) 286.
231 A. Lorenzotti, F. Bonati, A. Cingolani, G. G. Lobbia, D. Leonesi and B. Bovio, *Inorg. Chim Acta* **170** (1990) 199.
232 A. J. Canty and R. T. Honeyman, *J. Organomet. Chem.* **387** (1990) 247.
233 K.-B. Shiu, K.-S. Liou, S.-L. Wang and S.-C. Wei, *Organometallics* **9** (1990) 669.
234 F. Bonati, A. Cingolani, G. G. Lobbia, D. Leonesi, A. Lorenzotti and C. Pettinari, *Gazz. Chim. Ital.* **120** (1990) 341.
235 A. J. Canty, R. T. Honeyman, B. W. Skelton and A. H. White, *J. Organomet. Chem.* **389** (1990) 277.

236 A. J. Canty, R. T. Honeyman, B. W. Skelton and A. H. White, *J. Organomet. Chem.* **396** (1990) 105.
237 M. A. Mesubi and B. A. Omotowa, *Transition Met. Chem.* **16** (1991) 348.
238 T. Astley, A. J. Canty, M. A. Hitchman, G. L. Rowbotton, B. W. Skelton and A. H. White, *J. Chem. Soc., Dalton Trans.* (1991) 1981.
239 K.-B. Shiu, C. J. Chang, S.-L. Wang and F.-L. Liao, *J. Organomet. Chem.* **407** (1991) 225.
240 V. S. Joshi, A. Sarkar and P. R. Rajamohanan, *J. Organomet. Chem.* **409** (1991) 341.
241 G. G. Lobbia, A. Cingolani, P. Cecchi, S. Calogero and F. E. Wagner, *J. Organomet. Chem.* **436** (1992) 35.
242 A. J. Canty, R. T. Honeyman, B. W. Skelton and A. H. White, *J. Chem. Soc., Dalton Trans.* (1992) 2663.
243 P. K. Byers, A. J. Canty, R. T. Honeyman, B. W. Skelton and A. H. White, *J. Organomet. Chem.* **433** (1992) 223.
244 A. J. Canty, R. T. Honeyman, B. W. Skelton and A. H. White, *J. Chem. Soc., Dalton Trans.* (1992) 2663.
245 M. C. López, R. M. Claramunt and P. Ballesteros, *J. Org. Chem.* **57** (1992) 5240
246 B. Bovio, A. Cingolani and F. Bonati, *Z. Anorg. Allg. Chem.* **610** (1992) 151.
247 C. Pettinari, A. Lorenzotti, A. Cingolani, D. Leonesi, M. Marra and F. Marchetti, *Gazz. Chim. Ital.*, **123** (1993) 481.
248 V. S. Joshi, M. Nandi, H. Zhang, B. S. Haggerty and A. Sarkar, *Inorg. Chem.* **32** (1993) 1301.
249 M. C. L. Gallego-Preciado, P. Ballesteros, R. M. Claramunt, M. Cano, J. V. Heras, E. Pinilla and A. Monge, *J. Organomet. Chem.* **450** (1993) 237.
250 K.-B. Shiu, L.-Y. Ueh, S.-M. Peng and M.-C. Cheng, *J. Organomet. Chem.* **460** (1993) 203.
251 R. Alsfasser and H. Vahrenkamp, *Inorg. Chim. Acta* **209** (1993) 19.
252 K.-B. Shiu, K.-S. Liou, Y. Wang, M.-C. Cheng and G.-H. Lee, *J. Organomet. Chem.* **453** (1993) 201.
253 M. Fajardo, A. de la Hoz, E. Diéz-Barra, F. A. Jalón, A. Otero, A. Rodriguez, J. Tejeda, D. Belletti, M. Lanfranchi and M. A. Pellinghelli, *J. Chem. Soc., Dalton Trans.* (1993) 1935.
254 C. Pettinari, C. Santini, D. Leonesi and P. Cecchi, *Polyhedron* **13** (1994) 1553.
255 J. Fernandez-Baeza, F. A. Jalón, A. Otero and M. E. Rodrigo-Blanco, *J. Chem. Soc., Dalton Trans.* (1995) 1015.
256 C. Pettinari, G. G. Lobbia, A. Lorenzotti and A. Cingolani, *Polyhedron* **14** (1995) 793.
257 C. Pettinari, G. G. Lobbia, G. Sclavi, D. Leonesi, M. Colapietro and G. Portalone, *Polyhedron* **14** (1995) 1709.

258	C. Pettinari, A. Lorenzotti, G. Sclavi, A. Cingolani, E. Rivarola, M. Colapietro and A. Cassetta, *J. Organomet. Chem.* **496** (1995) 69.
259	K.-B. Shiu, S.-T Lin, D.-W. Fung, T.-J. Chan, S.-M. Peng, M.-C. Cheng and J.-L. Chou, *Inorg. Chem.* **34** (1995) 854.
260	C. Pettinari, A. Cingolani and B. Bovio, *Polyhedrton* **15** (1996) 115.
261	L. Tang, Z. Wang, Y. Su, J. Wang, H. Wang and X. Yao, *Polyhedron* **17** (1998) 3765.
262	C. Pettinari, F. Marchetti, A. Cingolani, D. Leonesi, M. Colapietro and S. Margadonna, *Polyhedron* **17** (1998) 4145.
263	M. Onishi, K. Sugimura and K. Hiraki, *Bull. Chem. Soc. Jpn.* **51** (1978) 3209.
264	A. J. Canty and N. J. Minchin, *J. Organomet. Chem.* **226** (1982) C14.
265	A. J. Canty, N. J. Minchin, J. M. Patrick and A. H. White, *J. Chem. Soc., Dalton. Trans.* (1983) 1253.
266	J. M. Calvert, R. H. Schmehl, B. P. Sullivan, J. S. Facci, T. J. Meyer and R. W. Murray, *Inorg. Chem.* **22** (1983) 2151.
267	J.-P. Declercq and M. Van Meerssche, *Acta Cryst.* **C40** (1984) 1098.
268	P. K. Byers and A. J. Canty, *Inorg. Chim. Acta* **104** (1985) L13.
269	M. A. Mesubi and F. O. Anumba, *Transition Met. Chem.* **10** (1985) 5.
270	R. Visalakshi V. K. Jain, S. K. Kulshreshtha and G. S. Rao, *Inorg. Chim. Acta* **118** (1986) 119.
271	G. G. Lobbia, D. Leonesi, A. Cingolani, A. Lorenzotti and F. Bonati, *Synth. React. Inorg. Met.-Org. Chem.* **17** (1987) 909.
272	P. K. Byers, A. J. Canty, B. W. Skelton and A. H. White, *J. Chem. Soc., Chem.Commun.* (1987) 1093.
273	K.-B. Shiu and C.-J. Chang, *J. Chin. Chem. Soc.* **34** (1987) 297.
274	A. Llobet, P. Doppelt and T. J. Meyer, *Inorg. Chem.* **27** (1988) 514.
275	M. A. Esteruelas, L. A. Oro, M. C. Apreda, C. Foces-Foces, F. H. Cano, R. M. Claramunt, C. López, J. Elguero and M. Begtrup, *J. Organomet. Chem.* **344** (1988) 93.
276	K. R. Barqawi, A. Llobet and T. J. Meyer, *J. Am. Chem. Soc.* **110** (1988) 7751.
277	M. A. Esteruelas, L. A. Oro, R. M. Claramunt, C. López, J. L. Lavandera and J. Elguero, *J. Organomet.Chem.* **366** (1989) 245.
278	G. G. Lobbia, F. Bonati, A. Cingolani and D. Leonesi, *Synth. React. Inorg. Met.-Org. Chem.* **19** (1989) 827.
279	R. A. Doyle and R. J. Angelici, *Organometallics* **8** (1989) 2214.
280	P. K. Byers and F. G. A. Stone, *J. Chem. Soc., Dalton Trans.* (1990) 3499.
281	A. Llobet, D. J. Hodgson and T. J. Meyer, *Inorg. Chem.* **29** (1990) 3760.
282	D. G. Brown, P. K. Byers, B. W. Skelton and A. J. Canty, *Organometallics* **9** (1990) 826.
283	D. G. Brown, P. K. Byers and A. J. Canty, *Organometallics* **9** (1990) 1231.

284 P. K.Byers, A. J. Canty, R. T. Honeyman, R. M. Claramunt, C. López, J. L. Lavandera and J. Elguero, *Gazz. Chim. Ital.* **122** (1992) 341.
285 A. J. Canty, R. T. Honeyman, B. W. Skelton and A. H. White, *J. Organomet. Chem.* **430** (1992) 245.
286 A. J. Canty, R. T. Honeyman, B. W. Skelton and A. H. White, *J. Organomet. Chem.* **424** (1992) 381.
287 M. Fajardo, A. de la Hoz, E. Diez-Barra, F. A. Jalon, A. Otero, A. Rodriguez, J. Tajeda, D. Baletti, M. Lanfranchi and M. A. Pellinghelli, *J. Chem.Soc., Dalton Trans.* (1993) 1935.
288 T. Astley, J. M. Gulbis, M. A. Hitchman and E. R. T. Tiekink, *J. Chem. Soc., Dalton Trans.* (1993) 509.
289 A. J. Canty, P. R. Traill, R. Colton and I. M. Thomas, *Inorg. Chim. Acta* **210** (1993) 91.
290 J. J. McGarvey, H. Toftlund, A. H. R. Al-Obaidi, K. P. Taylor and S. E. J. Bell, *Inorg. Chem.* **32** (1993) 2469.
291 W. E. Jones, C. A. Bignozzi, P. Chen and T. J. Meyer, *Inorg. Chem.* **32** (1993) 1167.
292 J. J. McGarvey, H. Toftlund, A. H. R. Al-Obaidi, K. P. Taylor and S. E. J. Bell, *Inorg. Chem.* **32** (1993) 2469.
293 P. Ballesteros, C. López, C. López, R. M. Claramunt, J. A. Jimenez, M. Cano, J. V. Heras, E. Pinilla and A. Monge, *Organometallics* **13** (1994) 289.
294 E. S. Zvargulis, I. E. Buys and T. W. Hambley **14** (1995) 2267
295 F. A. Jalón, B. R. Manzano, A. Otero and M. C. Rodríguez-Pérez, *J. Organomet. Chem.* **494** (1995) 179.
296 I. K. Dhavan, M. A. Bruck, B. Schilling, C. Grittini and J. H. Enemark, *Inorg. Chem.* **34** (1995) 3801.
297 D. L. Reger, J. E. Collins, A. L. Rheingold and L. M. Liable-Sands *Organometallics* **15** (1996) 2029.
298 C. Titze, J. Hermann and H. Vahrenkamp, *Chem. Ber.* **128** (1995) 1095.
299 D. L. Reger, J. E. Collins, S. M. Myers, A. L. Rheingold and L. M. Liable-Sands, *Inorg. Chem.* **35** (1996) 4904.
300 P. K. Byers and F. G. A. Stone, *J. Chem. Soc., Dalton Trans.* (1991) 93.
301 D. L. Reger, J. E. Collins, A. L. Rheingold, L. M. Liable-Sands, and G. P. A. Yap, *Organometallics* **16** (1997) 349.
302 S. Vepachedu, R. T. Stibrany, S. Knapp, J. A. Potenza and H. J. Schugar, *Acta Cryst.* **51C** (1995) 423.
303 A. Steiner and D. Stalke, *J. Chem. Soc., Chem. Commun.* (1993) 1702.
304 A. Steiner and D. Stalke, *Inorg. Chem.* **34** (1995) 4846.
305 D. Carmona, F. J. Lahoz, R. Atencio, A. J. Edwards, L. A. Oro, M. P. Lamata, M. Esteban and S. Trofimenko, *Inorg. Chem.* **35** (1996) 2549.
306 M. Scotti, M. Valderrama, R. Moreno, R. López and D. Boys, *Inorg. Chim. Acta* **219** (1994) 67.

REFERENCES

307 V. S. Joshi, V. K. Kale, M. Sathe, A. Sarkar, S. S. Tavale and C. G. Suresh, *Organometallics* **10** (1991) 2898.
308 S. G. N. Roundhill, D. M. Roundhill, D. R. Bloomquist, C. Landee, R. D. Willett, D. M. Dooley and H. B. Gray, *Inorg. Chem.* **18** (1979) 831.
309 A. J. Canty, H. Jin, A. S. Roberts, P. R. Traill, B. W. Skelton and A. H. White, *J. Organomet. Chem.* **489** (1995) 153.
310 A. J. Canty, N. J. Minchin, L. M. Engelhardt, B. W. Skelton and A. H. White, *J. Chem. Soc., Dalton Trans.* (1986) 645.
311 S. Trofimenko, *J. Coord. Chem.* **2** (1972) 75.
312 C. López, R. M. Claramunt, C. Foces-Foces, F. H. Cano and J. Elguero, *Revue Roum. Chim.* **9** (1994) 795.
313 R. W. Thomas, G. W. Estes, R. C. Elder and E. Deutsch, *J. Am. Chem. Soc.* **101** (1979) 4581.
314 J. A. Thomas and A. Davison, *Inorg. Chim. Acta* **190** (1991) 231.
315 A. Looney, G. Parkin and A. L. Rheingold, *Inorg. Chem.* **30** (1991) 3099.
316 P. K. Byers, A. J. Canty, N. J. Minchin, J. M. Patrick, B. W. Skelton and A. H. White, *J. Chem. Soc., Dalton Trans.* (1985) 1183.
317 A. J. Canty, N. J. Minchin, J. M. Patrick, and A. H. White, *Aust. J. Chem.* **36** (1983) 1107.
318 C. López, R. M. Claramunt, D. Sanz, C. F. Foces, F. H. Cano, R. Faure, E. Cayon and J. Elguero, *Inorg. Chim. Acta* **176** (1990) 195.
319 J. Reglinski, M. D. Spicer, M. Garner and A. R. Kennedy, *J. Am. Chem. Soc.* **121** (1999) 2317.
320 Y. Sohrin, H. Kokusen, S. Kihara, M. Matsui, Y. Kushi and M. Shiro, *J. Am. Chem. Soc.* **115** (1993) 4128.
321 Y. Sohrin, H. Kokusen, S. Kihara, M. Matsui, Y. Kushi and M. Shiro, *Chem. Letters* (1992) 1461.
322 Y. Sohrin, M. Matsui, Y. Hata, H. Hasegawa and H. Kokusen, *Inorg. Chem.* **33** (1994) 4376.
323 R. Han and G. Parkin, *J. Organomet. Chem.* **393** (1990) C43.
324 R. Han and G. Parkin, *Organometallics* **10** (1991) 1010.
325 R. Han and G. Parkin, *J. Am. Chem. Soc.* **114** (1992) 748.
326 S. G. Dutremez, D. B. Leslie, W. E. Streib, M. H. Chisholm and K. G. Caulton, *J. Organomet. Chem.* **462** (1993) C1.
327 T. R. Belderrain, L. Contreras, M. Paneque, E. Carmona, A. Monge and C. Ruiz, *Polyhedron* **15** (1996) 3453.
328 T. R. Belderrain, L. Contreras, M. Paneque, E. Carmona, A. Monge and C. Ruiz, *J. Organomet. Chem.* **474** (1994) C5.
329 C. Apostolidis, J. Rebizant, B. Kanellakopulos, R. von Ammon, E. Dornberger, J. Müller, B. Powietzka and B. Nuber, *Polyhedron* **16** (1997) 1057.
330 S.-Y. Liu, G. H. Maunder, A. Sella, M. Stevenson and D. A. Tocher, *Inorg. Chem.* **35** (1996) 76.
331 M. A. J. Moss and C. J. Jones, *Polyhedron* **8** (1989) 555.

332   M. A. J. Moss and C. J. Jones, *Polyhedron* **9** (1990) 1119.
333   M. A. J. Moss and C. J. Jones, *Polyhedron* **8** (1989) 117.
334   M. A. J. Moss and C. J. Jones, *J. Chem. Soc., Dalton Trans.* (1990) 581.
335   M. A. J. Moss and C. J. Jones, *Polyhedron* **8** (1989) 2367.
336   M. A. J. Moss, C. J. Jones and A. J. Edwards, *J. Chem. Soc., Dalton Trans.* (1989) 1393.
337   M. A. J. Moss, C. J. Jones and A. J. Edwards, *Polyhedron* **7** (1988) 79.
338   M. A. J. Moss, C. J. Jones and A. J. Edwards, *Polyhedron* **9** (1990) 697.
339   R. G. Lawrence, C. J. Jones and R. A. Kresinski, *J. Chem. Soc., Dalton Trans.* (1996) 501.
340   D. L. Reger, J. A. Lindeman and L. Lebioda, *Inorg. Chim. Acta* **139** (1987) 71.
341   D. L. Reger, J. A. Lindeman and L. Lebioda, *Inorg. Chem.* **27** (1988) 3923.
342   R. G. Lawrence, C. J. Jones and R. A. Kresinski, *Polyhedron* **15** (1996) 2011.
343   D. P. Long and P. A. Bianconi, *J. Am. Chem. Soc.* **118** (1996) 12453.
344   L. E. Manzer, *J. Organomet. Chem.* **102** (1975) 167.
345   P. Burchill and M. G. H. Wallbridge, *Inorg. Nucl. Chem. Lett.* **12** (1976) 93.
346   J. K. Kouba and S. S. Wreford, *Inorg. Chem.* **15** (1976) 2313.
347   J. Ipaktschi and W. Sulzbach, *J. Organomet. Chem.* **426** (1992) 59.
348   D. L. Hughes, G. J. Leigh and D. G. Walker, *J. Chem. Soc., Dalton Trans.* (1989) 1413.
349   D. L. Hughes, G. J. Leigh and D. G. Walker, *J. Chem. Soc., Dalton Trans.* (1988) 1153.
350   S. C. Dunn, P. Mountford and O. V. Shishkin, *Inorg. Chem.* **35** (1996) 1006.
351   S. C. Dunn, A. S. Batsanov and P. Mountford, *J. Chem. Soc., Chem. Commun.* (1994) 2007.
352   U. Kilimann, M. Noltemeyer, M. Schäfer, R. Herbst-Irmer, H.-G. Schmidt and F. T. Edelmann, *J. Organomet. Chem.* **469** (1994) C27.
353   K. M. Chi, S. R. Frerichs, B. K. Stein, D. W. Blackburn and J. E. Ellis, *J. Am. Chem. Soc.* **110** (1988) 163.
354   P. J. Fischer, K. A. Ahrendt, V. G. Young, Jr. and J. E. Ellis, *Organometallics* **17** (1998) 13.
355   D. L. Reger and M. E. Tarquini, *Inorg. Chem.* **21** (1982) 840.
356   K. Mashima, T. Oshiki and K. Tani, *Organometallics* **16** (1997) 2760.
357   R. Kresinski, T. A. Hamor, L. Isam, C. J. Jones and J. A. McCleverty, *Polyhedron* **8** (1989) 845.
358   R. Kresinski, L. Isam, T. A. Hamor, C. J. Jones and J. A. McCleverty, *J. Chem. Soc. Dalton Trans.* (1991) 1835.
359   D. L. Reger, M. E. Tarquini and L. Lebioda, *Organometallics* **2** (1983) 1763.

REFERENCES

360   D. L. Reger and M. E. Tarquini, *Inorg. Chem.* **22** (1983) 1064.
361   D. L. Reger, R. Mahtab, J. C. Baxter and L. Liebioda, *Inorg. Chem.* **25** (1986) 2046.
362   R. Kresinski, T. A. Hamor, C. J. Jones and J. A. McCleverty, *J. Chem. Soc. Dalton Trans.* (1991) 603.
363   R. Kresinski, C. J. Jones and J. A. McCleverty, *Polyhedron* **9** (1990) 2185.
364   P. Dapporto, F. Mani and C. Mealli, *Inorg. Chem.* **17** (1978) 1323.
365   D. Rehder, H. Gailus and H. Schmidt, *Acta Crystallogr.* **C54** (1998) 1590.
366   E. Kime-Hunt, K. Spartalian, M. DeRusha, C. M. Nunn and C. J. Carrano, *Inorg. Chem.* **28** (1989) 4392.
367   C. J. Carrano, M. Mohan, S. M. Holmes, R. de la Rosa, A. Butler, J. M. Charnock and C. D. Garner, *Inorg. Chem.* **33** (1994) 646.
368   R. L. Beddoes, D. Collison, F. E. Mabbs and M. A. Passand, *Polyhedron* **9** (1990) 2483.
369   R. L. Beddoes, D. R. Eardley, F. E. Mabbs, D. Moorcroft and M. A. Passand, *Acta Crystallogr.* **C49** (1993) 1923.
370   R. Marsh, *Acta Crystallogr.* **C50** (1994) 1596.
371   M. Mohan, S. M. Holmes, R. J. Butcher, J. P. Jasinski and C. J. Carrano, *Inorg. Chem.* **31** (1992) 2029.
372   S. Holmes and C. J. Carrano, *Inorg. Chem.* **30** (1991) 1231.
373   N. E. Heimer and W. E. Cleland, Jr., *Acta Crystallogr.* **C46** (1990) 2049.
374   D. Collison, F. E. Mabbs and S. S. Turner, *J. Chem. Soc. Faraday Trans.* **89** (1993) 3705.
375   D. Collison, F. E. Mabbs, K. Rigby and W. E. Cleland, Jr., *J. Chem. Soc. Faraday Trans.* **89** (1993) 3695.
376   D. Collison, F. E. Mabbs, M. A. Passand, K. Rigby and W. E. Cleland, Jr., *Polyhedron* **8** (1989) 1827.
377   D. Collison, F. E. Mabbs and K. Rigby, *Polyhedron* **8** (1989) 1830.
378   J. Sundermeyer, J. Putterlik, M. Foth, J. S. Field and N. Ramesar, *Chem. Ber.* **127** (1994) 1201.
379   M. Herberhold, G. Frohmader, T. Hofmann, W. Milius and J. Darkwa, *Inorg. Chim. Acta* **267** (1998) 19.
380   S. Scheuer, J. Fischer and J. Kress, *Organometallics* **14** (1995) 2627.
381   M. Köppen, G. Fresen, K. Wieghardt, R. M. Llusar, B. Nuber and J. Weiss, *Inorg. Chem.* **27** (1988) 721.
382   C. J. Carrano, R. Verastegue and M. R. Bond, *Inorg. Chem.* **32** (1993) 3589.
383   D. Collison, D. R. Eardley, F. E. Mabbs, A. K. Powell and S. S. Turner, *Inorg. Chem.* **32** (1993) 664.
384   D. Collison, F. E. Mabbs, S. S. Turner, A. K. Powell, E. J. L. McInnes and L. J. Yellowlees, *J. Chem. Soc., Dalton Trans.* (1997) 1201.
385   M. R. Bond, L. M. Mokry, T. Otieno, J. Thompson and C. J. Carrano, *Inorg. Chem.* **34** (1995) 1894.

386 N. S. Dean, M. R. Bond, C. J. O'Connor and C. J. Carrano, *Inorg. Chem.* **35** (1996) 7643.
387 T. Otieno, L. M. Mokry, M. R. Bond, C. J. Carrano and N. S. Dean, *Inorg. Chem.* **35** (1996) 850.
388 N. S. Dean, L. M. Mokry, M. R. Bond, C. J. O'Connor and C. J. Carrano, *Inorg. Chem.* **35** (1996) 2818.
389 N. S. Dean, L. M. Mokry, M. R. Bond, C. J. O'Connor and C. J. Carrano, *Inorg. Chem.* **35** (1996) 3541.
390 M. R. Bond, R. S. Czernuszewicz, B. C. Dave, Q. Yan, M. Mohan, R. Verastegue and C. J. Carrano, *Inorg. Chem.* **34** (1995) 5857.
391 L. G. Hubert-Pfalzgraf and J. G. Riess, *Inorg. Chim. Acta* **47** (1980) 7.
392 L. G. Hubert-Pfalzgraf and M. Tsunoda, *Polyhedron* **47** (1983) 203.
393 S. Minhas and D. T. Richens, *J. Chem. Soc., Dalton Trans.* (1996) 703.
394 S. Minhas, A. Devlin, D. T. Richens, A. C. Benyei and P. Lightfoot, *J. Chem. Soc., Dalton Trans.* (1998) 953.
395 A. Antiñolo, F. Carillo-Hermosilla, J. Fernández-Baeza, M. Lanfranchi, A. Lara-Sánchez, A. Otero, E. Palomares, M. A. Pellinghelli and A. M. Rodriguez, *Organometallics* **17** (1998) 3015.
396 M. Etienne, P. S. White and J. L. Templeton, *Organometallics* **10** (1991) 3801.
397 M. Etienne, B. Donnadieu, R. Mathieu, J. F. Baeza, F. Jalón, A. Otero and M. E. Rodrigo-Blanco, *Organometallics* **15** (1996) 4597.
398 M. Etienne, P. Zéline, J. L. Templeton and P. S. White, *New J. Chem.* **17** (1993) 515.
399 M. Etienne, *Organometallics* **13** (1994) 410.
400 M. Etienne, F. Biasotto and R. Mathieu, *J. Chem. Soc., Chem. Commun.* (1994) 1661.
401 M. Etienne, F. Biasotto, R. Mathieu and J. L. Templeton, *Organometallics* **15** (1996) 1106.
402 M. Etienne, P. S. White and J. L. Templeton, *Organometallics* **12** (1993) 4010.
403 F. Biasotto, M. Etienne and F. Dahan, *Organometallics* **14** (1995) 1870.
404 P. Lorente, C. Carfagna, M. Etienne and B. Donnadieu, *Organometallics* **15** (1996) 1090.
405 P. Lorente, M. Etienne and B. Donnadieu, *Anales de Quimica Int. Ed.* **92** (1996) 88.
406 M. Etienne, R. Mathieu and B. Donnadieu, *J. Am. Chem.Soc.* **119** (1997) 3218.
407 J. Jaffart, C. Nayral, R. Choukroun, R. Mathieu and M. Etienne, *Eur. J. Inorg. Chem.* (1998) 425.
408 J. Jaffart, R. Mathieu, M. Etienne, J. E. McGrady, O. Eisenstein and F. Maseras, *Chem. Commun.* (1998) 2011.
409 D. L. Reger, C. A. Swift and L. Lebioda, *Inorg. Chem.* **23** (1984) 349.

410 J. M. Boncella, M. L. Cajigal and K. A. Abboud, *Organometallics* **15** (1996) 1905.
411 T. Beissel, B. S. P. C. Della Vedova, K. Wieghardt and R. Boese, *Inorg. Chem.* **29** (1990) 1736.
412 J. H. MacNeil, W. C. Watkins, M. C. Baird and K. F. Preston, *Organometallics* **11** (1992) 2761.
413 J. H. MacNeil, A. W. Roszak, M. C. Baird, K. F. Preston and A. L. Rheingold, *Organometallics* **12** (1993) 4402.
414 M. J. Abrams, R. Faggiani and C. J. L. Lock, *Inorg. Chim. Acta* **106** (1985) 69.
415 T. Fujihara, T. Schönherr and S. Kaizaki, *Inorg. Chim. Acta* **249** (1996) 135.
416 C.-H. Li, J.-D. Chen, L.-S. Liou and J.-C. Wang, *Inorg. Chim. Acta* **269** (1998) 302.
417 A. C. Filippou, K. Wanninger and C. Mehnert, *J. Organomet. Chem.* **461** (1993) 99.
418 K. Mashima, T. Oshiki, K. Tani, T. Aoshima and H. Urata, *J. Organomet. Chem.* **569** (1998) 15.
419 K.-B. Shiu, J. Y. Lee, Y. Wang, M.-C. Cheng, S.-L. Wang and F.-L. Liao, *J. Organomet. Chem.* **453** (1993) 211.
420 P. Marabella and J. H. Enemark, *J. Organomet. Chem.* **226** (1982) 57.
421 S. P. Nolan, R. Lopez de la Vega and C. D. Hoff, *Organometallics* **5** (1986) 2529.
422 J. Wang, Y. Zhang, Y. Xu and R. Zhou, *Gaodeng Xuexiao Huaxue Xuebao* **13** (1992) 1401; *Chem. Abstr.* **119** 49543u.
423 J. D. Protasiewicz and K. H. Theopold, *J. Am. Chem. Soc.* **115** (1993) 5559.
424 K.-B. Shiu, M. D. Curtis and J. C. Huffman, *Organometallics* **2** (1983) 936.
425 M. D. Curtis, K.-B. Shiu, W. M. Butler and J. C. Huffman, *J. Am. Chem. Soc.* **108** (1986) 3335.
426 D. M. Collins, F. A. Cotton and C. A. Murillo, *Inorg. Chem.* **15** (1976) 1861.
427 C.-T. Lee, J.-D. Chen, L.-S. Liou and J.-C. Wang, *Inorg. Chim. Acta* **249** (1996) 115.
428 G. A. Banta, B. M. Louie, E. Onyiriuka, S. J. Rettig and A. Storr, *Can. J. Chem.* **64** (1986) 373.
429 Y.-Y. Liu, A. Mar, S. J. Rettig, A. Storr and J. Trotter, *Can. J. Chem.* **64** (1988) 1997.
430 S. J. Rettig, A. Storr and J. Trotter, *Can. J. Chem.* **66** (1988) 2194.
431 K. B. Shiu and L.-Y. Lee, *J. Organomet. Chem.* **348** (1988) 357.
432 D. L. Lichtenberger and J. L Hubbard, *Inorg. Chem.* **23** (1984) 2718.
433 C. G. Young, S. A. Roberts and J. H. Enemark, *Inorg. Chim. Acta* **114**, (1986) L7.

434 C. G. Young, S. A. Roberts and J. H. Enemark, *Inorg. Chem.* **25** (1985) 3667.
435 M. Cano, J. A. Campo, J. V. Heras, E. Pinilla and A. Monge, *Polyhedron* **15** (1996) 1705.
436 A. S. Gamble, P. S. White and J. L Templeton, *Organometallics* **10** (1991) 693.
437 S. Lincoln, S.-L. Soong, S. A. Koch, M. Sato and J. H. Enemark, *Inorg. Chem.* **24** (1985) 1355.
438 T. Begley, D. Condon, G. Ferguson, F. J. Lalor and M. A. Khan, *Inorg. Chem.* **20** (1981) 3420.
439 E. M. Holt, S. L. Holt and K. J. Watson, *J. Chem. Soc., Dalton Trans.* (1973) 2444.
440 V. S. Joshi, K. M. Sathe, M. Nandi, P. Chakrabarti and A. Sarkar, *J. Organomet. Chem.* **485** (1995) C1.
441 Y. D. Ward, L. A. Villanueva, G. D. Allred, S. C. Payne, M. A. Semones and L. S. Liebeskind, *Organometallics* **14** (1995) 4132.
442 K. Mauthner, C. Slugovc, K. Mereiter, R. Schmid and K. Kirchner, *Organometallics* **15** (1996) 181.
443 E. M. Holt and S. L. Holt, *J. Chem. Soc., Dalton Trans.* (1973) 1893.
444 J. Ipaktschi and A. Hartmann, *J. Organomet. Chem.* **431** (1992) 303.
445 J. Ipaktschi, A. Hartmann and R. Boese, *J. Organomet. Chem.* **434** (1992) 303.
446 Y. D. Ward, L. A. Villanueva, G. D. Allred and L. S. Liebeskind, *J. Am. Chem. Soc.* **118** (1996) 897.
447 Y. D. Ward, L. A. Villanueva, G. D. Allred and L. S. Liebeskind, *Organometallics* **15** (1996) 4201.
448 A. J. Pearson, I. B. Neagu, A. A. Pinkerton, K. Kirschbaum and M. J. Hardie, *Organometallics* **16** (1997) 4346.
449 A. J. Pearson and E. Schoffers, *Organometallics* **16** (1997) 5365.
450 A. J. Pearson and A. R. Douglas, *Organometallics* **17** (1998) 1446.
451 V. N. Sapunov, C. Slugovc, K. Mereiter, R. Schmid and K. Kirchner, *J. Chem. Soc., Dalton Trans.* (1997) 3599.
452 P. Meakin, S. Trofimenko and J. P. Jesson, *J. Am. Chem. Soc.* **94** (1972) 5677.
453 S. K. Chowdhury, M. Nandi, V. S. Joshi and A. Sarkar, *Organometallics* **16** (1997) 1806.
454 R. A. Clement, U. Klabunde and G. W. Parshall, *J. Mol. Catal.* **4** (1978) 87.
455 K. B. Shiu and L.-Y. Lee, *J. Chin. Chem. Soc.* **36** (1989) 31.
456 L. Contreras, A. Pizzano, L. Sanchez and E. Carmona, *J. Organomet. Chem.* **500** (1995) 61.
457 G. Ferguson, B. Kaitner, T. Desmond, F. J. Lalor and B. O'Sullivan, *Acta Crystallogr.* **C47** (1991) 2651.

REFERENCES

458   T. Desmond, F. J. Lalor, B. O'Sullivan and G. Ferguson, *J. Organomet. Chem.* **381** (1990) C33.
459   M. D. Curtis and K.-B. Shiu, *Inorg. Chem.* **24** (1985) 1213.
460   A. A. Saleh, B. Pleune, J. C. Fettinger and R. Poli, *Polyhedron*, **16** (1997) 1391.
461   S. Wolowiec and J. K. Kochi, *Inorg. Chem.* **30** (1991) 1215.
462   K.-B. Shiu, J. Y. Lee, Y. Wang and M.-C. Cheng, *Inorg. Chem.* **32** (1993) 3565.
463   C. A. Rusik, T. L. Tonker and J. L. Templeton, *J. Am. Chem. Soc.* **108** (1986) 4652.
464   C. A. Rusik, M. A. Collins, A. S. Gamble, T. L. Tonker and J. L. Templeton, *J. Am. Chem. Soc.* **111** (1989) 2550.
465   D. M. Schuster and J. L Templeton, *Organometallics* **17** (1998) 2707.
466   D. C. Brower, M. Stoll and J. L. Templeton, *Organometallics* **8** (1989) 2786.
467   G. M. Jamison, A. E. Bruce, P. S. White and J. L. Templeton, *J. Am. Chem. Soc.* **113** (1991) 5057.
468   A. F. Hill, J. M. Malget, A. J. P. White and D. J. Williams, *Inorg. Chem.* **37** (1998) 598.
469   H. Wadepohl, U. Arnold and H. Pritzkow, *Angew. Chem. Int. Ed. Engl.* **36** (1997) 974.
470   F. J. Lalor, T. J. Desmond, G. M. Cotter, C. A. Shanahan, G. Ferguson, M. Parvez and B. Ruhl, *J. Chem. Soc., Dalton Trans.* (1995) 1709.
471   M. Etienne, P. S. White and J. L. Templeton, *J. Am. Chem. Soc.* **113** (1991) 2324.
472   S. Anderson and A. F. Hill, *J. Chem. Soc., Dalton Trans.* (1993) 587.
473   L. Weber, G. Dembeck, R. Boese and D. Bläser, *Chem. Ber./Recueil* **130** (1997) 1305.
474   L. Weber, G. Dembeck, H.-G. Stammler, B. Neumann, M. Schmidtmann and A. Müller, *Organometallics* **17** (1998) 5254.
475   L. Weber, G. Dembeck, H.-G. Stammler and B. Neumann, *Eur. J. Inorg. Chem.* (1998) 579.
476   G. M. Jamison, P. S. White and J. L Templeton, *Organometallics* **10** (1991) 1954.
477   S. Chaona, F. J. Lalor, G. Ferguson and M. M. Hunt, *J. Chem. Soc., Chem. Commun.* (1988) 1606.
478   A. F. Hill and J. M. Malget, *Chem. Commun.* (1996) 1177.
479   I. Baxter, A. F. Hill, J. M. Malget, A. J. P. White and D. J. Williams, *Chem. Commun.* (1997) 2049.
480   B. E. Woodworth, D. S. Frohnapfel, P. S. White and J. L. Templeton, *Organometallics* **17** (1998) 1655.
481   B. E. Woodworth and J. L Templeton, *J. Am. Chem. Soc.* **118** (1996) 7418.

482	B. E. Woodworth, P. S. White and J. L Templeton, *J. Am. Chem. Soc.* **120** (1998) 9028.
483	D. M. Schuster and J. L Templeton, *Organometallics* **17** (1998) 2707.
484	F. J. Lalor, D. M. Condon, G. Ferguson, M. Parvez and P. Y. Siew, *J. Chem. Soc., Dalton Trans.* (1986) 103.
485	D. Sutton, *Can. J. Chem.* **52** (1974) 2634.
486	W. E. Carroll, M. E. Deane and F. J. Lalor, *J. Chem. Soc., Dalton Trans.* (1974) 1837.
487	M. E. Deane and F. J. Lalor, *J. Organomet. Chem.* **67** (1974) C19.
488	D. Condon, G. Ferguson, F. J. Lalor, M. Parvez and T. Spalding, *Inorg. Chem.* **21** (1982) 188.
489	M. E. Deane and F. J. Lalor, *J. Organomet. Chem.* **57** (1973) C61.
490	M. E. Deane, F. J. Lalor, G. Ferguson, B. L. Ruhl and M. Parvez, *J. Organomet. Chem.* **381** (1990) 213.
491	G. Ferguson, B. L. Ruhl, M. Parvez, F. J. Lalor and M. E. Deane, *J. Organomet. Chem.* **381** (1990) 357.
492	F. Ferguson, B. L. Ruhl, F. J. Lalor and M. E. Deane, *J. Organomet. Chem.* **282** (1985) 75.
493	S. A. Roberts and J. H. Enemark, *Acta Crystallogr.* **C45** (1989) 1292.
494	E. M. Holt, S. L. Holt, F. Cavalito and K. J. Watson, *Acta Chem. Scand.* **A30** (1976) 225.
495	E. R. de Gil, A. V. Rivera and H. Noguera, *Acta Crystallogr.* **B33** (1977) 2653.
496	E. Frauendorfer and H. Brunner, *J. Organomet. Chem.* **240** (1982) 371.
497	E. Frauendorfer and J. Puga, *J. Organomet. Chem.* **265** (1984) 257.
498	M. Minelli, J. L. Hubbard, K. A. Christensen and J. H. Enemark, *Inorg. Chem.* **22** (1983) 2652.
499	H. Brunner, P. Beier, E. Frauendorfer, M. Muschiol, D. K. Rastogi, J. Wachter, M. Minelli and J. H. Enemark, *Inorg. Chim. Acta* **96** (1985) L5.
500	M. Minelli, J. L. Hubbard, D. L. Lichtenberger and J. H. Enemark, *Inorg. Chem.* **23** (1984) 2721.
501	T. B. Gunnoe, P. S. White and J. L. Templeton, *Organometallics* **16** (1997) 370.
502	M. Herberhold, G.-X. Jin and A. L. Rheingold, *Z. Naturforsch.* **51b** (1996) 681.
503	J. Huang, R. L. Ostrander, A. L. Rheingold, Y. Leung and M. A. Walters, *J. Am. Chem. Soc.* **116** (1994) 6769.
504	J. Huang, R. L. Ostrander, A. L. Rheingold and M. A. Walters, *Inorg. Chem.* **34** (1995) 1090.
505	J. A. McCleverty and N. E. Murr, *J. Chem. Soc., Chem. Commun.* (1981) 960.
506	A. Wlodarczyk, J. P. Maher, S. Coles, D. E. Hibbs, M. H. B. Hursthouse and K. M. Abdul Malik, *J. Chem. Soc., Dalton Trans.* (1997) 2597.

507 S. L. W. McWhinnie, S. M. Charsley, C. J. Jones, J. A. McCleverty and L. J. Yellowlees, *J. Chem. Soc., Dalton Trans.* (1993) 413.
508 S. L. W. McWhinnie, C. J. Jones, J. A. McCleverty, T. A. Hamor and J.-D. Foulon, *Polyhedron* **12** (1993) 37.
509 N. Alobaidi, C. J. Jones and J. A. McCleverty, *Polyhedron* **8** (1989) 371.
510 J. A. McCleverty, D. Seddon, N. A. Bailey and N. W. Walker, *J. Chem. Soc., Dalton Trans.* (1976) 898.
511 A. S. Drane and J. A. McCleverty, *Polyhedron*, **2** (1983) 53.
512 J. A. McCleverty, A. S. Drane, N. A. Bailey and J. M. A. Smith, *J. Chem. Soc., Dalton Trans.* (1983) 91.
513 S. J. Reynolds, C. F. Smith, C. J. Jones and J. A. McCleverty, *Inorg. Synth.* **23** (1985) 4.
514 J. A. McCleverty, A. E. Rae, I. Wolochowicz, N. A. Bailey and J. M. A. Smith, *J. Chem. Soc., Dalton Trans.* (1982) 951.
515 J. A. McCleverty and A. Wlodarczyk, *Polyhedron* **7**, (1988) 449.
516 E. M. Coe, C. J. Jones and J. A. McCleverty, *Polyhedron* **11** (1992) 655.
517 A. Wlodarczyk, S. S. Kurek, J.-D. Foulon, T. A. Hamor and J. A. McCleverty, *J. Chem. Soc., Dalton Trans.* (1992) 981.
518 B. J. Coe, C. J. Jones, J. A. McCleverty and D. W. Bruce, *Polyhedron* **11** (1992) 3007.
519 B. J. Coe, C. J. Jones and J. A. McCleverty, *Polyhedron* **11** (1992) 3129.
520 A. Wlodarczyk, S. S. Kurek, M. A. J. Moss, M. S. Tolley, A. S. Batsanov, J. A. K. Howard and J. A. McCleverty, *J. Chem. Soc., Dalton Trans.* (1993) 2027.
521 G. Denti, J. A. McCleverty and A. Wlodarczyk, *J. Chem. Soc., Dalton Trans.* (1981) 2021.
522 H. Adams, N. A. Bailey, A. S. Drane and J. A. McCleverty, *Polyhedron* **2** (1983) 465.
523 H. Adams, N. A. Bailey, G. Denti, J. A. McCleverty, J. M. A. Smith and A. Wlodarczyk, *J. Chem. Soc., Chem. Commun.* (1981) 348.
524 H. Adams, N. A. Bailey, G. Denti, J. A. McCleverty, J. M. A. Smith and A. Wlodarczyk, *J. Chem. Soc., Dalton Trans.* (1983) 2287.
525 P. D. Beer, C. J. Jones, J. A. McCleverty and R. P. Sidebotham, *J. Organomet. Chem.* **325** (1987) C19.
526 S. L. W. McWhinnie, C. J. Jones, J. A. McCleverty and L. J. Yellowlees, *J. Chem. Soc., Dalton Trans.* (1996) 4401.
527 N. Al Obaidi, C. J. Jones and J. A. McCleverty, *J. Chem. Soc., Dalton Trans.* (1990) 3329.
528 E. M. Coe, C. J. Jones and J. A. McCleverty, *J. Chem. Soc., Dalton Trans.* (1990) 1429.
529 J. A. McCleverty, G. Denti, S. J. Reynolds, A. S. Drane, N. E. Murr, A. E. Rae, N. A. Bailey and J. M. A. Smith, *J. Chem. Soc. Dalton Trans.* (1983) 81.

530 B. J. Coe, C. J. Jones, J. A. McCleverty and D. W. Bruce, *Polyhedron* **9** (1990) 687.
531 A. Abdul-Rahman, A. A. Amoroso, T. N. Branston, A. Das, J. P. Maher, J. A. McCleverty, M. D. Ward and A. Wlodarczyk, *Polyhedron* **16** (1997) 4353.
532 S. L. W. McWhinnie, S. M. Charsley, C. J. Jones, J. A. McCleverty and L. J. Yellowlees, *J. Chem. Soc., Dalton Trans.* (1993) 413.
533 J. A. Thomas, C. J. Jones, J. A. McCleverty and M. G. Hutchings, *Polyhedron* **15** (1996) 1409.
534 F. McQuillan and C. J. Jones, *Polyhedron* **15** (1996) 1553.
535 F. S. McQuillan, H. Chen, T. A. Hamor and C. J. Jones, *Polyhedron* **15** (1996) 3909.
536 F. S. McQuillan, C. J. Jones and J. A. McCleverty, *Polyhedron* **14** (1995) 3157.
537 F. S. McQuillan, H. Chen, T. A. Hamor, C. J. Jones and K. Paxton, *Inorg. Chem.* **36** (1997) 4458.
538 C. J. Jones, J. A. McCleverty, B. D. Neaves, S. J. Reynolds, H. Adams, N. A. Bailey and G. Denti, *J. Chem. Soc., Dalton Trans.* (1986) 733.
539 A. Wlodarczyk, A. J. Edwards and J. A. McCleverty, *Polyhedron* **7** (1988) 103.
540 J. A. McCleverty, A. E. Rae, I. Wolochowicz, N. A. Bailey and J. M. A. Smith, *J. Chem. Soc., Dalton Trans.* (1982) 429.
541 N. Al Obaidi, T. A. Hamor, C. J. Jones, J. A. McCleverty and K. Paxton, *J. Chem. Soc., Dalton Trans.* (1987) 1063.
542 N. El Murr, A. Sellami and J. A. McCleverty, *New J. Chem.* **12** (1988) 209.
543 B. J. Coe, C. J. Jones and J. A. McCleverty, *Polyhedron* **11** (1992) 547.
544 B. J. Coe, C. J. Jones, J. A. McCleverty and D. W. Bruce, *Polyhedron* **12** (1993) 45.
545 A. Wlodarczyk, S. S. Kurek, J.-D. Foulon, T. A. Hamor and J. A. McCleverty, *Polyhedron* **11** (1992) 217.
546 J. A. McCleverty, A. E. Rae, I. Wolochowicz, N. A. Bailey and J. M. A. Smith, *J. Organomet. Chem.* **168** (1979) C1.
547 J. A. McCleverty, A. E. Rae, I. Wolochowicz, N. A. Bailey and J. M. A. Smith, *J. Chem. Soc., Dalton Trans.* (1983) 71.
548 N. Alobaidi, C. J. Jones, J. A. McCleverty, A. J. Howes and M. B. Hursthouse, *Polyhedron* **7** (1988) 235.
549 N. Alobaidi, T. A. Hamor, C. J. Jones, J. A. McCleverty, K. Paxton, A. J. Howes and M. B. Hursthouse, *Polyhedron* **7** (1988) 1931.
550 N. Alobaidi, T. A. Hamor, C. J. Jones, J. A. McCleverty and K. Paxton, *J. Chem. Soc., Dalton Trans.* (1986) 1525.
551 N. Alobaidi, C. J. Jones and J. A. McCleverty, *Polyhedron* **8** (1989) 1033.
552 N. J. Al Obaidi, S. L. W. McWhinnie, T. A. Hamor, C. J. Jones and J. A. McCleverty, *J. Chem. Soc., Dalton Trans.* (1992) 3299.

553	G. Denti, C. J. Jones, J. A. McCleverty, B. D. Neaves and S. J. Reynolds, *J. Chem. Soc., Chem. Commun.* (1983) 474.
554	R. Cook, J. P. Maher, J. A. McCleverty, M. D. Ward and A. Wlodarczyk, *Polyhedron* **12** (1993) 2111.
555	A. Wlodarczyk, J. P. Maher, J. A. McCleverty and M. D. Ward, *J. Chem. Soc., Chem. Commun.* (1995) 2397.
556	S. Chiappetta, G. Denti and J. A. McCleverty, *Transition Met. Chem.* **14**, (1989) 449.
557	P. D. Beer, C. J. Jones, J. A. McCleverty and S. S. Salam, *J. of Inclusion Phenomena* **5** (1987) 521.
558	N. Al Obaidi, P. D. Beer, J. P. Bright, C. J. Jones, J. A. McCleverty and S. S. Salam, *J. Chem. Soc., Chem. Commun.* (1986) 239.
559	N. Al Obaidi, S. S. Salam, P. D. Beer, C. D. Bush, T. A. Hamor, F. S. McQuillan, C. J. Jones and J. A. McCleverty, *Inorg. Chem.* **31** (1992) 263.
560	R. P. Sidebotham, P. D. Beer, T. A. Hamor, C. J. Jones and J. A. McCleverty, *J. Organomet. Chem.* **371** (1989) C31.
561	N. A. Obaidi, K. P. Brown, A. J. Edwards, S. A. Hollins, C. J. Jones, J. A. McCleverty and B. D. Neaves, *J. Chem. Soc., Chem. Commun.* (1984) 690.
562	G. Denti, M. Ghedini, J. A. McCleverty, H. Adams and N. A. Bailey, *Transition Met. Chem.* **7** (1982) 222.
563	T. N. Briggs, C. J. Jones, J. A. McCleverty, B. D. Neaves, N. E. Murr and H. A. Colquhoun, *J. Chem. Soc., Dalton Trans.* (1985) 1249.
564	S. E. M. Flynn, J. A. McCleverty, M. D. Ward and A. Wlodarczyk, *Polyhedron* **15** (1996) 2247.
565	S. L. W. McWhinnie, C. J. Jones, J. A. McCleverty, D. Collison and F. E. Mabbs, *Polyhedron* **11** (1992) 2639.
566	N. Al Obaidi, A. J. Edwards, C. J. Jones, J. A. McCleverty, B. D. Neaves, F. E. Mabbs and D. Collison, *J. Chem. Soc., Dalton Trans.* (1989) 127.
567	A. Wlodarczyk, S. S. Kurek, J. P. Maher, A. S. Batsanov, J. A. K. Howard and J. A. McCleverty, *Polyhedron,* **12** (1993) 715.
568	J. A. Thomas, C. J. Jones, J. A. McCleverty, D. W. Bruce and M. G. Hutchings, *Polyhedron* **14** (1995) 2499.
569	S. S. Salam, C. J. Lovely, A. G. R. Poole, C. J. Jones and J. A. McCleverty, *Polyhedron* **9** (1990) 527.
570	S. M. Charsley, C. J. Jones and J. A. McCleverty, *Transition Met. Chem.* **11** (1986) 329.
571	N. Al Obaidi, M. Chaudhury, D. Clague, C. J. Jones, J. C. Pearson, J. A. McCleverty and S. S. Salam, *J. Chem. Soc., Dalton Trans.* (1987) 1733.
572	N. Al Obaidi, S. M. Charsley, W. Hussain, C. J. Jones, J. A. McCleverty, B. D. Neaves and S. J. Reynolds, *Transition Metal Chem.* **12** (1987) 143.
573	N. Al Obaidi, T. A. Hamor, C. J. Jones, J. A. McCleverty and K. Paxton, *J. Chem. Soc., Dalton Trans.* (1987) 2653.

574 S. M. Charsley, C. J. Jones and J. A. McCleverty, B. D. Neaves, S. J. Reynolds and G. Denti, *J. Chem. Soc., Dalton Trans.* (1988) 293.
575 S. M. Charsley, C. J. Jones and J. A. McCleverty, B. D. Neaves, S. J. Reynolds and G. Denti, *J. Chem. Soc., Dalton Trans.* (1988) 301.
576 F. S. McQuillan, T. L. Green, T. A Hamor, C. J. Jones, J. P. Maher and J. A. McCleverty, *Chem. Soc., Dalton Trans.* (1995) 3243.
577 A. Wlodarczyk, G. A. Doyle, J. P. Maher, J. A. McCleverty and M. D. Ward, *Chem. Commun.* (1997) 769.
578 J. A. McCleverty, *Proc. Indian Natn. Sci. Acad.* **52A** (1986) 796.
579 M. Minelli, J. H. Hubbard and J. H. Enemark, *Inorg. Chem.* **23** (1984) 970.
580 C. G. Young and J. H. Enemark, *Inorg. Chem.* **24** (1985) 4416.
581 C. G. Young, M. Minelli, J. H. Enemark, W. Hussain, C. J. Jones and J. A. McCleverty, *J. Chem. Soc., Dalton Trans.* (1987) 619.
582 N. M. Atherton, G. Denti, M. Ghedini and C. Oliva, *J. Magn. Reson.* **43** (1981) 167.
583 A. J. Amoroso, J. P. Maher, J. A. McCleverty and M. D. Ward, *J. Chem. Soc., Chem. Commun.* (1994) 1273.
584 S. L. W. McWhinnie, C. J. Jones, J. A. McCleverty, D. Collison and F. E. Mabbs, *J. Chem. Soc., Chem. Commun.* (1990) 940.
585 A. M. W. Cargill Thompson, D. Gateschi, J. A. McCleverty, J. A. Navis, E. Rentschler and M. D. Ward, *Inorg. Chem.* **35** (1996) 2701.
586 J. A. Thomas, C. J. Jones, J. A. McCleverty, D. Collison, F. E. Mabbs, C. J. Harding and M. G. Hutchings, *J. Chem. Soc., Chem. Commun.* (1992) 1796.
587 J. A. Thomas, M. G. Hutchings, C. J. Jones and J. A. McCleverty, *Inorg. Chem.* **35** (1996) 289.
588 S. L. W. McWhinnie, J. A. Thomas, T. A. Hamor, C. J. Jones, J. A. McCleverty, D. Collison, F. E. Mabbs, C. J. Harding, L. J. Yellowlees and M. G. Hutchings, *Inorg. Chem.* **35** (1996) 760.
589 J. Hock, A. M. W. Cargill Thompson, J. A. McCleverty and M. D. Ward, *J. Chem. Soc., Dalton Trans.* (1996) 4257.
590 A. J. Amoroso, A. M. W. Cargill Thompson, J. P. Maher, J. A. McCleverty and M. D. Ward, *Inorg. Chem.* **34** (1995) 4828.
591 A. M. W. Cargill Thompson, J. A. McCleverty and M. D. Ward, *Inorg. Chim. Acta* **250** (1996) 29.
592 A. Wlodarczyk, J. P. Maher, J. A. McCleverty and M. D. Ward, *J. Chem. Soc., Dalton Trans.* (1997) 3287.
593 C. J. Jones, J. A. McCleverty and S. J. Reynolds, *Transition Met. Chem.* **11**, (1986) 138.
594 B. J. Coe, C. J. Jones, J. A. McCleverty, D. Bloor, P. V. Kolinsky and R. J. Jones, *J. Chem. Soc., Chem. Commun.* (1989) 1485.

595	B. J. Coe, S. S. Kurek, N. M. Rowley, J.-D. Foulon, T. A. Hamor, M. E. Harman, M. B. Hursthouse, C. J. Jones, J. A. McCleverty and D. Bloor, *Chemtronics* **5** (1991) 23.
596	N. M. Rowley, S. S. Kurek, M. W. George, S. M. Hubig, P. D. Beer, C. J. Jones, J. M. Kelly and J. A. McCleverty, *J. Chem. Soc., Chem. Commun.* (1992) 497.
597	B. J. Coe, C. J. Jones, J. A. McCleverty, D. Bloor, P. V. Kolinsky and R. J. Jones, *Polyhedron* **14** (1994) 2107.
598	B. J. Coe, J.-D. Foulon, T. A. Hamor, C. J. Jones, J. A. McCleverty, D. Bloor, G. H. Cross and T. L. Axon, *J. Chem. Soc., Dalton Trans.* (1994) 3427.
599	N. M. Rowley, S. S. Kurek, J.-D. Foulon, T. A. Hamor, C. J. Jones, J. A. Yellowlees, *Inorg. Chem.* **34** (1995) 4414.
600	N. M. Rowley, S. S. Kurek, P. R. Ashton, T. A. Hamor, C. J. Jones, N. Spencer, J. A. McCleverty, G. S. Beddard, T. M. Feehan, N. T. H. White, E. J. L. McInnes, N. N. Payne and L. J. Yellowlees, *Inorg. Chem.* **35** (1996) 7526.
601	J. A. McCleverty, J. A. Navas Badiola and M. D. Ward, *J. Chem. Soc., Dalton Trans.* (1994) 2415.
602	M. M. Bhadbhade, A. Das, J. C. Jeffery, J. A. McCleverty, J. A. Navas Badiola and M. D. Ward, *J. Chem. Soc., Dalton Trans.* (1995) 2769.
603	B. J. Coe, C. J. Jones and J. A. McCleverty, *Polyhedron* **13** (1994) 2117.
604	B. J. Coe, T. A. Hamor, C. J. Jones, J. A. McCleverty, D. Bloor, G. H. Cross and T. L. Axon, *J. Chem. Soc., Dalton Trans.* (1995) 673.
605	V. A. Ung, A. M. W. Cargill Thompson, D. A. Bardwell, D. Gatteschi, J. C. Jeffery, J. A. McCleverty, F. Totti and M. D. Ward, *Inorg. Chem.* **36** (1997) 3447.
606	T. E. Berridge, H. Chen, T. A. Hamor and C. J. Jones, *Polyhedron* **16** (1997) 2329.
607	H. A. Hinton, H. Chen, T. A. Hamor, C. J. Jones, F. S. McQuillan and M. S. Tolley, *Inorg. Chem.* **37**, (1998) 2933.
608	S. M. Kagwanja, C. J. Jones, J. P. Maher and J. A. McCleverty, *Polyhedron* **13** (1994) 2615.
609	S. M. Kagwanja, C. J. Jones and J. A. McCleverty, *Polyhedron* **16** (1997) 1439.
610	S. M. Kagwanja, J. C. Jeffery, C. J. Jones and J. A. McCleverty, *Polyhedron* **15** (1996) 2959.
611	M. Millar, S. Lincoln and S. A. Koch, *J. Am. Chem. Soc.* **104** (1982) 288.
612	D. Collison, D. R. Eardley, F. E. Mabbs, K. Rigby and J. H. Enemark, *Polyhedron* **8** (1989) 1833.
613	N. S. Nipales and T. D. Westmoreland, *Inorg. Chem.* **34** (1995) 3374.
614	N. S. Nipales and T. D. Westmoreland, *Inorg. Chem.* **36** (1997) 756.
615	P. Basu and J. H. Enemark, *Inorg. Chim. Acta* **263** (1997) 81.

616 S. A. Koch and S. Lincoln, *Inorg. Chem.* **21** (1982) 2904.
617 S. Lincoln and S. A. Koch, *Inorg. Chem.* **25** (1986) 1594.
618 S. E. Lincoln and T. M. Loehr, *Inorg. Chem.* **29** (1990) 1907.
619 A. A. Eagle, M. F. Mackay and C. G. Young, *Inorg. Chem.* **30** (1991) 1425.
620 S. A. Roberts, C. G. Young, W. E. Cleland, Jr., K. Yamanouchi, R. B. Ortega and J. H. Enemark, *Inorg. Chem.* **27** (1988) 2647.
621 U. Küsthardt and J. H. Enemark, *J. Am. Chem. Soc.* **109** (1987) 7926.
622 Z. Xiao, C. G. Young, J. H. Enemark and A. G. Wedd, *J. Am. Chem. Soc.* **114** (1992) 9194.
623 W. E. Cleland, Jr., K. M. Barnhart, K. Yamanouchi, D. Collison, F. E. Mabbs, R. B. Ortega and J. H. Enemark, *Inorg. Chem.* **26** (1987) 1017.
624 Z. Xiao, M. A. Bruck, C. Doyle, J. H. Enemark, C. Grittini, R. W. Gable, A. G. Wedd and C. G. Young, *Inorg. Chem.* **34** (1995) 5950.
625 C. A. Kipke, W. E. Cleland, Jr., S. A. Roberts and J. H. Enemark, *Acta Crystallogr.* **C45** (1989) 870.
626 D. Collison, D. R. Eardley, F. E. Mabbs, K. Rigby, M. A. Bruck, J. H. Enemark and P. A. Wexler, *J. Chem. Soc., Dalton Trans.* (1994) 1003.
627 C.-S. J. Chang, T. J. Pecci, M. D. Carducci and J. H. Enemark, *Acta Crystallogr.* **C48** (1992) 1096.
628 J. P. Hill, L. J. Laughlin, R. W. Gable and C. G. Young, *Inorg. Chem.* **35**, (1996) 3447.
629 P. Basu, M. A. Bruck, Z. Li, I. K. Dhawan and J. H. Enemark, *Inorg. Chem.* **34** (1995) 405.
630 I. K. Dhawan. A. Pacheco and J. H. Enemark, *J. Am. Chem. Soc.* **116** (1994) 7911.
631 S. A. Roberts, R. B. Ortega, L. M. Zolg, W. E. Cleland, Jr. and J. H. Enemark, *Acta Crystallogr.* **C43** (1987) 51.
632 S. A. Roberts, C. G. Young, W. E. Cleland, Jr., R. B. Ortega and J. H. Enemark, *Inorg. Chem.* **27** (1988) 3044.
633 N. E. Heimer and W. E. Cleland, Jr., *Acta Crystallogr.* **C47** (1991) 56.
634 C. G. Young, S. A. Roberts, R. B. Ortega and J. H. Enemark, *J. Am. Chem. Soc.* **109** (1987) 2938.
635 D. Collison, F. E. Mabbs, J. H. Enemark and W. E. Cleland, Jr., *Polyhedron* **5** (1986) 423.
636 C. G. Young, J. H. Enemark, D. Collison and F. E. Mabbs, *Inorg. Chem.* **26** (1987) 2925.
637 C.-S. J. Chang, T. J. Pecci, M. D. Carducci and J. H. Enemark, *Inorg. Chem.* **32** (1993) 4106.
638 C. S. J. Chang, D. Collison, F. E. Mabbs and J. H. Enemark, *Inorg. Chem.* **29** (1990) 2261.
639 C. S. J. Chang, A. Rai-Chaudhuri, D. L. Lichtenberger and J. H. Enemark, *Polyhedron* **9** (1990) 1965.
640 G. M. Olson and F. A. Schultz, *Inorg. Chim. Acta* **225** (1994) 1.

REFERENCES

641 B. L. Westcott, N. E. Gruhn and J. H. Enemark, *J. Am. Chem. Soc.* **120** (1998) 3382.
642 I. K. Dhavan and J. H. Enemark, *Inorg. Chem.* **35** (1996) 4873.
643 Z. Xiao, M. A. Bruck, J. H. Enemark, C. G. Young and A. G. Wedd, *Inorg. Chem.* **35** (1996) 7508.
644 B. L. Westcott and J. H. Enemark, *Inorg. Chem.* **36** (1997) 5404.
645 G. N. George, W. E. Cleland, Jr., J. H. Enemark, B. E. Smith, C. A. Kipke, S. A. Roberts and S. P. Cramer, *J. Am. Chem. Soc.* **112** (1990) 2541.
646 M. Carducci, C. Brown, E. I. Solomon and J. H. Enemark, *J. Am. Chem. Soc.* **116** (1994) 11856.
647 U. Küsthardt, M. J. La Barre and J. H. Enemark, *Inorg. Chem.* **29** (1990) 3182.
648 W. E. Cleland, Jr., K. M. Barnhart, K. Yamanouchi, D. Collison, F. E. Mabbs, R. B. Ortega and J. H. Enemark, *Inorg. Chem.* **26** (1987) 1017.
649 S. A. Roberts, G. P. Darsey, W. E. Clelland, Jr. and J. H. Enemark, *Inorg. Chim. Acta* **154** (1988) 95.
650 A. A. Eagle, L. J. Laughlin, C. G. Young and E. R. T. Tiekink, *J. Am. Chem. Soc.* **114** (1992) 9195.
651 T. E. Berridge and C. J. Jones, *Polyhedron* **16** (1997) 3695.
652 V. A. Ung, S. M. Couchman, J. C. Jeffery, J. A. McCleverty, M. D. Ward, F. Tutti and D. Gatteschi, *Inorg. Chem.* **38** (1999) 365.
653 J. Beck and J. Strähle, *Z. Naturforsch.* **42b** (1987) 255.
654 C. G. Young, F. Janos, M. A. Bruck, P. A. Wexler and J. H. Enemark, *Aust. J. Chem.* **43** (1990) 1347.
655 C. Manzur, D. Carillo and D. Boys, *Acta Crystallogr.* **C53** (1997) 1401.
656 A. A. Eagle, S. M. Harben, E. R. T. Tiekink and C. G. Young, *J. Am. Chem. Soc.* **116** (1994) 9749.
657 G. Ferguson, B. Kaitner, F. J. Lalor and G. Roberts, *J. Chem. Res. (S)* (1982) 6; *J. Chem. Res. (M)* (1982) 0143C.
658 S. A. Roberts, C. G. Young, C. A. Kipke, W. E. Cleland, Jr., K. Yamanouchi, M. D. Carducci and J. H. Enemark, *Inorg. Chem.* **29** (1990) 3650.
659 M. Onishi, K. Ikemoto, K. Hiraki and R. Koga, *Bull. Chem. Soc. Jpn.* **66** (1993) 1849.
660 J. Sundermeyer, J. Putterlik and H. Pritzkow, *Chem. Ber.* **126** (1993) 289.
661 M. Minelli, K. Yamanouchi, J. H. Enemark, P. Subramanian, B. B. Kaul and J. T. Spence, *Inorg. Chem.* **23** (1984) 2554.
662 K. M. Barnhart and J. H. Enemark, *Acta Crystallogr.* **C40** (1985) 1362.
663 Z. Xiao, J. H. Enemark, A. G. Wedd and C. G. Young, *Inorg. Chem.* **33** (1994) 3438.
664 J. H. Enemark, K. Yamanouchi, K. Barnhart, D. Collison and F. E. Mabbs, *Inorg. Chim. Acta* **79** (1983) 210.

665 Z. Xiao, R. W. Gable, A. G. Wedd and C. G. Young, *J. Am. Chem. Soc.* **118** (1996) 2912.
666 C. G. Young, I. P. McInerney, M. A. Bruck and J. H. Enemark, *Inorg. Chem.* **29** (1990) 412.
667 A. A. Eagle, C. G. Young and E. R. T. Tiekink, *Polyhedron* **9** (1990) 2965.
668 S. A. Roberts, C. G. Young, W. E. Cleland, Jr., R. B. Ortega and J. H. Enemark, *Inorg. Chem.* **27** (1988) 3044.
669 W. M. Vaughan, J. M. Boncella and K. A. Abboud, *Acta Crystallogr.* **C51** (1995) 1075.
670 W. M. Vaughan, K. A. Abboud and J. M. Boncella, *J. Organomet. Chem.* **485** (1995) 37.
671 W. M. Vaughan, K. A. Abboud and J. M. Boncella, *Organometallics* **14** (1995) 1567.
672 C. Manzur, D. Carillo, F. Robert, P. Gouzerh, P. Hamon and J.-R. Hamon, *Inorg. Chim. Acta* **268** (1998) 199.
673 D. D. Ellis, J. M. Farmer, J. M. Malget, D. F. Mullica and F. G. A. Stone, *Organometallics* **17** (1998) 5540.
674 F. A. Cotton, R. Llusar and W. Schwotzer, *Inorg. Chim. Acta* **155** (1989) 231.
675 F. A. Cotton, Z. Dori, R. Llusar and W. Schwotzer, *Inorg. Chem.* **25** (1986) 3529.
676 S. J. Davies, A. F. Hill, M. U. Pilotti and F. G. A. Stone, *Polyhedron* **18** (1989) 2265.
677 I. J. Hart, A. F. Hill and F. G. A. Stone, *J. Chem. Soc., Dalton Trans.* (1989) 2261.
678 D. M. Schuster, P. S. White and J. L. Templeton, *Organometallics* **15** (1996) 5467.
679 L. L. Blosch, A. S. Gamble and J. M. Boncella, *J. Mol. Catal.* **76** (1992) 229.
680 A. J. M. Caffyn, S. G. Feng, A. Dierdorf, A. S. Gamble, P. A. Eldredge, M. R. Vossen, P. S. White and J. L Templeton, *Organometallics* **10** (1991) 2842.
681 S. G. Feng, C. C. Philipp, A. S. Gamble, P. S. White and J. L. Templeton, *Organometallics* **10** (1991) 3504.
682 C. C. Philipp, P. S. White and J. L. Templeton, *Inorg. Chem.* **31** (1992) 3825.
683 C. G. Young, S. Thomas and R. W. Gable, *Inorg. Chem.* **37** (1998) 1299.
684 P. S. Pregosin, A. Macchioni, J. L. Templeton, P. S. White and S. Feng, *Magn. Reson. Chem.* **32** (1994) 415.
685 D. S. Frohnapfel, P. S. White, J. L. Templeton, H. Rüegger and P. S. Pregosin, *Organometallics* **16** (1997) 3737.
686 C. C. Philipp, C. G. Young, P. S. White and J. L. Templeton, *Inorg. Chem.* **32** (1993) 5437.

REFERENCES

687 S. G. Feng, L. Luan, P. S. White, M. S. Brookhart, J. L. Templeton and C. G. Young, *Inorg. Chem.* **30** (1991) 2582.
688 C. G. Young, M. A. Bruck, P. A. Wexler, M. D. Carducci and J. H. Enemark, *Inorg. Chem.* **31** (1992) 587.
689 S. Thomas, E. R. T. Tiekink and C. G. Young, *Inorg. Chem.* **33** (1994) 1416.
690 S. Thomas, E. R. T. Tiekink and C. G. Young, *J. Organomet.Chem.* **560** (1998) 1.
691 H. P. Kim, S. Kim, R. A. Jacobson and R. J. Angelici, *J. Am. Chem. Soc.* **108** (1986) 5154.
692 R. A. Doyle and R. J. Angelici, *J. Am. Chem. Soc.* **112** (1990) 194.
693 J. C. Jeffery, F. G. A. Stone and G. K. Williams, *Polyhedron* **10** (1991) 215.
694 H. P. Kim and R. J. Angelici, *Organometallics* **5** (1986) 2489.
695 R. A. Doyle and R. J. Angelici, *J. Organomet. Chem.* **375** (1989) 73.
696 R. A. Doyle, L. M. Daniels and R. J. Angelici, *J. Am. Chem. Soc.* **111** (1989) 4995.
697 H. P. Kim, S. Kim, R. J. Jacobson and R. J. Angelici, *Organometallics* **3** (1984) 1124.
698 H. P. Kim, S. Kim, R. A. Jacobson and R. J. Angelici, *Organometallics* **5** (1986) 2481.
699 M. Green, J. A. K. Howard, A. P. James, A. N. de M. Jelfs, C. M. Nunn and F. G. A. Stone, *J. Chem. Soc., Chem. Commun.* (1984) 1623.
700 J. C. Jeffery, J. A. McCleverty, M. D. Mortimer and M. D. Ward, *Polyhedron* **13** (1994) 353.
701 A. C. Filippou, C. Wagner, E. O. Fischer and C. Völkl, *J. Organomet. Chem.* **438** 1992) C15.
702 A. E. Bruce, A. S. Gamble, T. L. Tonker and J. L Templeton, *Organometallics* **6** (1987) 1350.
703 D. S. Frohnapfel, S. Reinartz, P. S. White and J. L. Templeton, *Organometallics* **17** (1998) 3759.
704 S. G. Feng, P. S. White and J. L. Templeton, *J. Am. Chem. Soc.* **112** (1990) 8192.
705 S. G. Feng and J. L Templeton, *Organometallics* **11** (1992) 2168.
706 J. L. Templeton, J. L. Caldarelli, S. Feng, C. C. Philipp, M. B. Wells, B. E. Woodworth and P. S. White, *J. Organomet. Chem.* **478** (1994) 103.
707 M. A. Collins, S. G. Feng, P. S. White and J. L. Templeton, *J. Am. Chem. Soc.* **114** (1992) 3771.
708 S. A. O'Reilly, P. S. White and J. L Templeton, *Chem. of Materials* **8** (1996) 93.
709 S. G. Feng and J. L. Templeton, *J. Am. Chem. Soc.* **111** (1989) 6477.
710 S. G. Feng, P. S. White and J. L. Templeton, *J. Am. Chem. Soc.* **114** (1992) 2951.

711 S. G. Feng, P. S. White and J. L. Templeton, *Organometallics* **12** (1993) 2131.
712 J. L. Caldarelli, L. E. Wagner, P. S. White and J. L. Templeton, *J. Am. Chem. Soc.* **116** (1994) 2878.
713 T. B. Gunnoe, J. L. Caldarelli, P. S. White and J. L. Templeton, *Angew. Chem. Int. Ed.* **37** (1998) 2093.
714 L. W. Francisco, P. S. White and J. L. Templeton, *Organometallics* **16** (1997) 2547.
715 T. B. Gunnoe, M. Surgan, P. S. White, J. L. Templeton and L. Casarrubios, *Organometallics* **16** (1997) 4865.
716 M. B. Wells, P. S. White and J. L. Templeton, *Organometallics* **16** (1997) 1857.
717 C. C. Philip, P. S. White and J. L. Templeton, *Inorg. Chem.* **31** (1992) 3825.
718 A. F. Hill, J. M. Malget, A. J. F. White and D. J. Williams, *J. Chem. Soc., Chem. Commun.* (1996) 721.
719 L. Luan, M. Brookhart and J. L. Templeton, *Organometallics* **11** (1992) 1433.
720 P. J. Pérez, P. S. White, M. Brookhart and J. L Templeton, *Inorg. Chem.* **33** (1994) 6050
721 L. W. Francisco, P. S. White and J. L Templeton, *Organometallics* **15** (1996) 5127.
722 K. R. Powell, P. J. Pérez, L. Luan, S. G. Feng, P. S. White, M. Brookhart and J. L. Templeton, *Organometallics* **13** (1994) 1851.
723 W.-H. Leung, M.-C. Wu, J. L. C. Chim and W.-T. Wong, *Polyhedron* **17** (1998) 457.
724 C. G. Young, C. C. Philipp, P. S. White and J. L Templeton, *Inorg. Chem.* **34** (1995) 6412.
725 L. Luan, P. S. White, M. Brookhart and J. L. Templeton, *J. Am. Chem.Soc.* **112** (1990) 8190.
726 S. Thomas, P. J. Lim, R. W. Gable and C. G. Young, *Inorg. Chem.* **37** (1998) 590.
727 S. G. Feng, P. S. White and J. L Templeton, *Organometallics* **14** (1995) 5184.
728 L. L. Blosch, A. S. Gamble, K. Abboud and J. M. Boncella, *Organometallics* **11** (1992) 2342.
729 A. S. Gamble and J. M. Boncella, *Organometallics* **12** (1993) 2814.
730 S. Thomas, E. R. T. Tiekink and C. G.Young, *Organometallics* **15** (1996) 2428.
731 S. Thomas, C. G. Young and E. R. T. Tiekink, *Organometallics* **17** (1998) 182.
732 S. G. Feng, P. S. White and J. L. Templeton, *Organometallics* **12** (1993) 1765.

REFERENCES

733 C. G. Young, M. A. Bruck and J. H. Enemark, *Inorg. Chem.* **31** (1992) 593.
734 A. A. Eagle, E. R. T. Tieking and C. G. Young, *J. Chem. Soc., Chem. Commun.* (1991) 1746.
735 A. A. Eagle, C. G. Young and E. R. T. Tiekink, *Organometallics* **11** (1992) 2934.
736 A. A. Eagle, S. Thomas and C. G. Young, *Transition Metal Sulfur Chemistry, ACS Symposium Series* **653** (1996) 324.
737 A. A. Eagle, S. M. Harben, E. R. T. Tiekink and C. G. Young, *J. Am. Chem. Soc.* **116** (1994) 9749.
738 C. G. Young, R. W. Gable and M. F. Mackay, *Inorg. Chem.* **29** (1990) 1777.
739 A. A. Eagle, E. R. T. Tiekink and C. G. Young, *Inorg. Chem.* **37** (1997) 6315.
740 W. Adam, J. Putterlik, R. M. Schumann and J. Sundermeyer, *Organometallics* **15** (1996) 4586.
741 A. A. Eagle, G. N. George, E. R. T. Tiekink and C. G. Young, *Inorg. Chem.* **36** (1997) 472.
742 L. L. Blosch, K. Abboud and J. M. Boncella, *J. Am. Chem. Soc.* **113** (1991) 7066.
743 US Patent 5,459,213 (Oct. 17, 1995), D. R. Kelsey to Shell Oil Co.
744 S. G. Feng, P. S. White and J. L. Templeton, *Organometallics* **13** (1994) 1214.
745 S. Ahn and A. Mayr, *J. Am. Chem. Soc.* **118** (1996) 7408.
746 F. G. A. Stone and M. L. Williams, *J. Chem. Soc., Dalton Trans.* (1988) 2467.
747 M. D. Bermudez, E. Delgado, G. P. Elliott, N. H. Tran-Huy, F. Mayor-Real, F. G. A. Stone and M. J. Winter, *J. Chem. Soc., Dalton Trans.* (1987) 1235.
748 S. V. Hoskins, A. P. James, J. C. Jeffery and F. G. A. Stone, *J. Chem. Soc., Dalton Trans.* (1986) 1709.
749 M. D. Bermudez and F. G. A. Stone, *J. Organomet. Chem.* **347** (1988) 115.
750 M. Green, J. A. K. Howard, A. P. James, A. N. de M. Jeffs, C. M. Nunn and F. G. A. Stone, *J. Chem. Soc., Dalton Trans.* (1986) 1697.
751 M. Green, J. A. K. Howard, A. P. James, C. M. Nunn and F. G. A. Stone, *J. Chem. Soc., Dalton Trans.* (1986) 187.
752 J. H. Davis, Jr., C. M. Lukehart and L.-A. Sacksteder, *Organometallics* **6** (1987) 50.
753 R. A. Doyle, R. J. Angelici and F. G. A. Stone, *J. Organomet. Chem.* **378** (1989) 81.
754 S. H. F. Becke, M. D. Bermudez, N. H. Tran-Huy, J. A. K. Howard, O. Johnson and F. G. A. Stone, *J. Chem. Soc., Dalton Trans.* (1987) 1229.

755 J. R. Fernández and F. G. A. Stone, *J. Chem. Soc., Dalton Trans.* (1988) 3035.
756 *Chemistry International* **9** (1987) 216.
757 M. K. Chan and W. H. Armstrong, *Inorg. Chem.* **28** (1989) 3777.
758 J. E. Sheats, R. S. Czernuszewicz, G. C. Dismukes, A. L. Rheingold, V. Petrouleas, J.-A. Stubbe, W. H. Armstrong, R. H. Beer and S. J. Lippard, *J. Am. Chem. Soc.* **109** (1987) 1435.
759 S. L. Dexheimer, J. W. Gohdes, M. K. Chan, K. S. Hagen, W. H. Armstrong and M. P. Klein, *J. Am. Chem. Soc.* **111** (1989) 8923.
760 I. I. Putrenko, *Biochemistry* **35** (1995) 2865.
761 R. W. Thomas, A. Davison, H. S. Trop and E. Deutsch, *Inorg. Chem.* **19** (1980) 2840.
762 R. Alberto, W. A. Herrmann, J. C. Bryan, P. A. Schubiger, F. Baumgärtner and D. Mihalios, *Radiochimica Acta* **63** (1993) 153.
763 R. Alberto, W. A. Herrmann, P. Kiprof and F. Baumgärtner, *Inorg. Chem.* **31** (1992) 895.
764 J. E. Joachim, C. Apostolidis, B. Kanellakopulos, R. Maier and M. Ziegler, *Z. f. Naturforschung* **48b** (1993) 227.
765 J. A. Thomas and A. Davison, *Inorg. Chim. Acta* **190** (1991) 231.
766 J. E. Joachim, C. Apostolidis, B. Kanellakopulos, R. Maier, N. Marques, D. Meyer, J. Müller, A. Pires de Matos, B. Nuber, J. Rebizant and M. Ziegler, *J. Organomet. Chem.* **448** (1993) 119.
767 J. E. Joachim, C. Apostolidis, B. Kanellakopulos, D. Meyer, B. Nuber, K. Raptis, J. Rebizant and M. Ziegler, *J. Organomet. Chem.* **492** (1995) 199.
768 J. E. Joachim, C. Apostolidis, B. Kanellakopulos, R. Maier, D. Meyer, J. Müller, A. J. Rebizant and M. Ziegler, *J. Organomet. Chem.* **455** (1993) 137.
769 A. Duatti, F. Tisato, F. Refosco, U. Mazzi and M. Nicolini, *Inorg. Chem.* **28** (1989) 4564.
770 F. Tisato, C. Bolzati, A. Duatti, G. Bandoli and F. Refosco, *Inorg. Chem.* **32** (1993) 2042.
771 I. A. Degnan, W. A. Herrmann and E. Herdtweck, *Chem. Ber.* **123** (1990) 1347.
772 I. A. Degnan, J. Behm, M. R. Cook and W. A. Herrmann, *Inorg. Chem.* **30** (1991) 2165.
773 C. S. Masui and J. M. Mayer, *Inorg. Chim. Acta* **251** (1996) 325.
774 A. Paulo, A. Domingos and I. Santos, *Inorg. Chem.* **35** (1996) 1798.
775 B. J. Coe, *Polyhedron* **9** (1992) 1085.
776 A. Domingos, J. Marçalo, A. Paulo, A. Pires de Matos and I. Santos, *Inorg. Chem.* **32** (1993) 5114.
777 A. Paulo, A. Domingos, J. Marçalo, A. Pires de Matos and I. Santos, *Inorg. Chem.* **34** (1995) 2113.
778 D. D. DuMez and J. M. Mayer, *Inorg. Chem.* **37** (1998) 445.
779 S. N. Brown and J. M. Mayer, *J. Am. Chem. Soc.* **116** (1994) 2219.

780 S. N. Brown, A. W. Myers, J. R. Fulton and J. M. Mayer, *Organometallics* **17** (1998) 3364.
781 S. N. Brown and J. M. Mayer, *Organometallics* **14** (1995) 2951.
782 S. N. Brown and J. M. Mayer, *Inorg. Chem.* **31** (1992) 4091.
783 P. B. Kettler, Y.-D. Chang, Q. Chen, J. Zubieta, M. J. Abrams and S. K. Larsen, *Inorg. Chim. Acta* **231** (1995) 13.
784 S. N. Brown and J. M. Mayer, *Inorg. Chem.* **34** (1995) 3560.
785 S.-J. Wang and R. J. Angelici, *J. Organomet. Chem.* **352** (1988) 157.
786 D. D. DuMez and J. M. Mayer, *J. Am. Chem. Soc.* **118** (1996) 12416.
787 D. D. DuMez and J. M. Mayer, *Inorg. Chem.* **34** (1995) 6396.
788 D. Nunes, A. Domingos, A. Paulo, L. Patrício, I. Santos, M. F. N. N. Carvalho and A. J. L Pombeiro, *Inorg. Chim. Acta* **271** (1998) 65.
789 K. P. Gable, A. AbuBaker, K. Zientara and A. M. Wainwright, *Organometallics* **18** (1999) 173.
790 S. N. Brown and J. M. Mayer, *J. Am. Chem. Soc.* **118** (1996) 12119.
791 Y. Matano, S. N. Brown, T. O. Northcutt and J. M. Mayer, *Organometallics* **17** (1998) 2939.
792 A. M. LaPointe and R. R. Schrock, *Organometallics* **14** (1995) 1875.
793 G. F. Diaz, M. V. Campos and A. H. O. Klahn, *Vibrational Spectroscopy* **9** (1995) 257.
794 M. Angaroni, G. A. Ardizzoia, G. D'Alfonso, G. La Monica, N. Masciocchi and M. Moret, *J. Chem. Soc., Dalton Trans.* (1990) 1895.
795 T. B. Gunnoe, M. Sabat and W. D. Harman, *J. Am. Chem. Soc.* **120** (1998) 8747.
796 J. A. McCleverty and I. Wolochowicz, *J. Organomet. Chem.* **169** (1979) 289.
797 D. G. Hamilton, X.-L. Luo and R. H. Crabtree, *Inorg. Chem.* **28** (1989) 3198.
798 D. Nunes, A. Domingos, A. Paulo, L. Patrício, I. Santos, M. F. N. N. Carvalho and A. J. L. Pombeiro, *Inorg. Chim. Acta* **271** (1998) 65.
799 A. Paulo, K. R. Reddy, A. Domingos and I. Santos, *Inorg. Chem.* **37** (1998) 6807.
800 J. P. Jesson, S. Trofimenko and D. R. Eaton, *J. Am. Chem. Soc.* **89** (1967) 3158.
801 J. P. Jesson and J. F. Weiher, *J. Chem. Phys.* **46** (1967) 1995.
802 R. B. King and A. Bond, *J. Am. Chem. Soc.* **96** (1974) 1334.
803 S. Anderson, A. F. Hill, A. J. P. White and D. J., Williams, *Organometallics* **17** (1998) 2665.
804 S. Anderson, A. F. Hill, A. M. Z. Slawin, A. J. P. White, and D. J. Williams, *Inorg. Chem.* **37** (1998) 594.
805 F. A. Cotton, B. A. Frenz and A. Shaver, *Inorg. Chim. Acta* **7** (1973) 161.
806 G. Bellachioma, G. Cardaci, V. Gramlich, A. Macchioni, F. Pieroni and L. M. Venanzi, *J. Chem. Soc., Dalton Trans.* (1998) 947.

807 J. D. Oliver, D. F. Mullica, B. Hutchinson and W. O. Milligan, *Inorg. Chem.* **19** (1980) 165.
808 Y. Sohrin, H. Kokusen and M. Matsui, *Inorg. Chem.* **34** (1995) 3928.
809 S. J. Mason, C. M. Hill, V. J. Murphy, D. O'Hare and D. J. Watkin, *J. Organomet. Chem.* **485** (1995) 165.
810 S.-H. Cho, D. Whang and K. Kim, *Bull. Korean Chem. Soc.* **12** (1991) 107.
811 S.-H. Cho, D. Whang, K.-N. Han and K. Kim, *Inorg. Chem.* **31** (1992) 519.
812 H. Fukui, M. Ito, Y. Moro-oka and N. Kitajima, *Inorg. Chem.* **29** (1990) 2868.
813 K.-N. Han, D. Whang, H.-J. Lee, Y. Do and K. Kim, *Inorg. Chem.* **32** (1993) 2597.
814 D. W. Hausler and L. T. Taylor, *Anal. Chem.* **53** (1981) 1223.
815 P. R. Sharp and A. J. Bard, *Inorg. Chem.* **22** (1983) 2689.
816 R. A. Binstead and J. K. Beattie, *Inorg. Chem.* **25** (1986) 1481.
817 H. D. Burrows and S. J. Formosinho, *J. Chem. Soc., Faraday Trans.* **82** (1986) 1563.
818 F. Grandjean, G. J. Long, B. B. Hutchinson, L. N. Ohlhausen, P. Neill and J. D. Holcomb, *Inorg. Chem.* **28** (1989) 4406.
819 G. J. Long and B. B. Hutchinson, *Inorg. Chem.* **26** (1987) 608.
820 B. Hutchinson, L. Daniels, E. Henderson, P. Neill, G. J. Long and L. W. Becker, *J. Chem. Soc., Chem. Commun.* (1979) 1003.
821 S. Calogero, G. Gioia Lobbia, P. Cecchi, G. Valle and J. Friedl, *Polyhedron* **13** (1994) 87.
822 A. Gulino, E. Ciliberto, S. Di Bella and I. Fragalà, *Inorg. Chem.* **32** (1993) 3759.
823 J. K. Beattie, R. A. Binstead and R. J. West, *J. Am. Chem. Soc.* **100** (1978) 3044.
824 S. Zamponi, G. Gambini, P. Conti, G. G. Lobbia, R. Marassi, M. Berettoni and P. Cecchi, *Polyhedron* **14** (1995) 1929.
825 T. Buchen and P. Gütlich, *Inorg. Chim. Acta* **231** (1995) 221.
826 C. Hannay, M.-J. Hubin-Franskin, F. Grandjean, V. Briois, J.-P. Itié, A. Polain, S. Trofimenko and G. J. Long, *Inorg. Chem.* **36** (1997) 5580.
827 S. Anderson, A. F. Hill, A. M. Z. Slawin and D. J. Williams, *J. Chem. Soc. Chem. Commun.* (1993) 266.
828 W. H. Armstrong and S. J. Lippard, *J. Am. Chem. Soc.* **105** (1983) 4837.
829 W. H. Armstrong, A. Spool, G. C. Papaephthymiou, R. B. Frankel and S. J. Lippard, *J. Am. Chem. Soc.* **106** (1984) 3653.
830 W. H. Armstrong and S. J. Lippard, *J. Am. Chem. Soc.* **106** (1984) 4632.
831 W. H. Armstrong and S. J. Lippard, *J. Am. Chem. Soc.* **107** (1985) 3730.

832	B. Hedman, M. S. Co, W. H. Armstrong, K. O. Hodgson and S. J. Lippard, *Inorg. Chem.* **25** (1986) 3708.
833	R. C. Scarrow, M. J. Maroney, S. M. Palmer and L. Que, Jr., *J. Am. Chem. Soc.* **108** (1986) 6832.
834	A. Ericson, B. Hedman, K. O. Hodgson, J. Green, H. Dalton, J. G. Bentsen, R. H. Beer and S. J. Lippard, *J. Am. Chem. Soc.* **110** (1988) 2330.
835	R. S. Czernuszewicz, J. E. Sheats and T. G. Spiro, *Inorg. Chem.* **26** (1987) 2063.
836	S. M. Kauzlarich, B. K. Theo, T. Zirino, S. Burman, J. C. Davis and B. A. Averill, *Inorg. Chem.* **25** (1986) 2781.
837	A. Spool, I. D. Williams and S. J. Lippard, *Inorg. Chem.* **24** (1985) 2156.
838	J. A. Hartmann, R. K. Rardin, P. Chaudhuri, K. Pohl, K. Wieghardt, B. Nuber, J. Weiss, G. C. Papaephthymiou, R. B. Frankel and S. J. Lippard, *J. Am. Chem. Soc.* **109** (1987) 7387.
839	I. M. Arafa, H. M. Goff, S. S. David, B. P. Murch and L. Que, Jr., *Inorg. Chem.* **26** (1987) 2779.
840	S. Yan, D. D. Cox, L. L. Pearce, C. Juarez-Garcia, L. Que., Jr., J. H. Zhang and C. J. O'Connor, *Inorg. Chem.* **28** (1989) 2507.
841	J. Sanders-Loehr, W. D. Wheeler, A. K. Shiemke, B. A. Averill and T. M. Loehr, *J. Am. Chem. Soc.* **111** (1989) 8084.
842	D. Hotzelmann, K. Wieghardt, U. Flörke, H.-J. Haupt, D. C. Weatherburn, J. Bonvoisin, G. Blondin and J.-J. Girerd, *J. Am. Chem. Soc.* **114** (1992) 1681.
843	P. N. Turowski, W. H. Armstrong, M. E. Roth and S. J. Lippard, *J. Am. Chem. Soc.* **112** (1990) 681.
844	P. Turowski, W. H. Armstrong, S. Liu, S. N. Brown and S. J. Lippard, *Inorg. Chem.* **33** (1994) 636.
845	J. P. Kirby, B. T. Weldon and J. K. McCusker, *Inorg. Chem.* **37** (1998) 3658.
846	N. Kitajima, H. Fukui and Y. Moro-oka, *J. Chem. Soc., Chem. Commun.* (1988) 485.
847	E. H. Ha, R. Y. N. Ho, J. F. Kisiel and J. S. Valentine, *Inorg. Chem.* **34** (1995) 2265.
848	S. Ciurli, M. Carrie, J. A. Weigel, M. J. Carey, T. D. P. Stack, G. C. Papaephthymiou and R. H. Holm, *J. Am. Chem. Soc.* **112** (1990) 2654.
849	J. A. Weigel and R. H. Holm, *J. Am. Chem. Soc.* **113** (1991) 4184.
850	S. Ciurli and R. H. Holm, *Inorg. Chem.* **28** (1989) 1685.
851	G. Ferguson and R. J. Restivo, *J. Chem. Soc., Chem. Commun.* (1973) 847.
852	R. J. Restivo and G. Ferguson, D. J. O'Sullivan and F. J. Lalor, *Inorg. Chem.* **14** (1975) 3046.
853	S. Bhambri, D. Tocher, *Polyhedron* **15** (1996) 2763.

854 S. Bhambri and D. A. Tocher, *J. Chem. Soc., Dalton Trans.* (1997) 3367.
855 S. Bhambri and D. A. Tocher, *J. Organomet. Chem.* **507** (1996) 291.
856 S. Bhambri, A. Bishop, N. Kaltsoyannis and D. A. Tocher, *J. Chem. Soc., Dalton Trans.* (1998) 3379.
857 A. M. McNair, D. C. Boyd and K. R. Mann, *Organometallics* **5** (1986) 303.
858 M. O. Albers, H. E. Oosterhuizen, D. J. Robinson, A. Shaver and E. Singleton, *J. Organomet. Chem.* **282** (1985) C49.
859 M. O. Albers, D. J. Robinson, A. Shaver and E. Singleton, *Organometallics* **5** (1987) 2199.
860 N. W. Alcock, I. D. Burns, K. S. Claire and A. F. Hill, *Inorg. Chem.* **31** (1992) 2906.
861 I. D. Burns, A. F. Hill and D. J. Williams, *Inorg. Chem.* **35** (1996) 2685.
862 A. Patel and D. T. Richens, *Inorg. Chem.* **30** (1991) 3789.
863 K. Hiraki, N. Ochi, H. Takaya, Y. Fuchita, Y. Shimokawa and H. Yamashida, *J. Chem. Soc., Dalton Trans.* (1990) 1679.
864 E. Gutiérrez-Puebla, A. Monge, M. Paneque, M. L. Poveda, S. Taboada, M.Trujillo and E. Carmona, *J. Am. Chem. Soc.* **121** (1999) 346.
865 C. Nataro, L. M. Thomas and R. J. Angelici, *Inorg. Chem.* **36** (1997) 6000.
866 B. Moreno, S. Sabo-Etienne, B. Chaudret, A. Rodriguez-Fernandez, F. Jalòn and S. Trofimenko, *J. Am. Chem. Soc.* **116** (1994) 2635.
867 M. H. Halcrow, B. Chaudret and S. Trofimenko, *J. Chem. Soc., Chem. Commun.* (1993) 465.
868 B. Moreno, S. Sabo-Etienne, B. Chaudret, A. Rodriguez, F. Jalòn and S. Trofimenko, *J. Am. Chem. Soc.* **117** (1995) 7441.
869 C. Vicente, G. B. Shul'pin, B. Moreno, S. Sabo-Etienne and B. Chaudret, *J. Mol. Catal.* **98** (1995) L5.
870 W.-C. Chan, C.-P. Lau, Y.-Z. Chen, Y.-Q. Fang and S.-M. Ng, *Organometallics* **16** (1997) 34.
871 M. A. Jiménez Tenorio, M. Jiménez Tenorio, M. C. Puerta and P. Valerga, *J. Chem. Soc., Dalton Trans.* (1998) 3601.
872 M. A. J. Tenorio, M. J. Tenorio, M. C. Puerta and P. Valerga, *Organometallics* **16** (1997) 5528.
873 M. J. Tenorio, M. A. J. Tenorio, M. C. Puerta and P. Valerga, *Inorg. Chim. Acta* **259** (1997) 77.
874 C. Bohanna, M. A. Esteruelas, A. V. Gómez, A. M. López and M.-P. Martínez, *Organometallics* **16** (1997) 4464.
875 S. M. Ng, Y. Q. Fang, C. P. Lau, W. T. Wong and G. Jia, *Organometallics* **17** (1998) 2052.
876 M. M. de V. Steyn, E. Singleton, S. Hietkamp and D. C. Liles, *J. Chem. Soc., Dalton Trans.* (1990) 2991.
877 M. Sørlie and M. Tilset, *Inorg. Chem.* **34** (1995) 5199.

878	M. I. Bruce, D. N. Sharrocks and F. G. A. Stone, *J. Organomet. Chem.* **31** (1971) 269.
879	I. D. Burns, A. F. Hill, A. J. P. White, D. J. Williams and J. D. E. T. Wilton-Ely, *Organometallics* **17** (1998) 1552.
880	Y. Mizobe, M. Hosomizu and M. Hidai, *Inorg. Chim. Acta* **273** (1998) 238.
881	F. Jalòn, A. Otero and A. Rodriguez, *J. Chem. Soc. (Dalton)* (1995) 1629.
882	Y. Maruyama, S. Ikeda and F. Ozawa, *Bull. Chem. Soc. Jpn.* **70** (1997) 689.
883	A. E. Corrochano, F. A. Jalòn, A. Otero, M. M. Kubicki and P. Richard, *Organometallics* **18** (1997) 145.
884	M. Onishi, K. Ikemoto and K. Hiraki, *Inorg. Chim. Acta* **190** (1991) 157.
885	M. Onishi, K. Ikemoto and K. Hiraki, *Inorg. Chim. Acta* **219** (1994) 3.
886	M. Onishi, K. Ikemoto, K. Hiraki and K. Aoki, *Chem. Lett.* (1998) 23.
887	M. Onishi, *Bull. Chem. Soc. Jpn.* **64** (1991) 3039.
888	A. F. Hill, *J. Organomet. Chem.* **395** (1990) C35.
889	N. W. Alcock, A. F. Hill and R. P. Melling, *Organometallics* **10** (1991) 3898.
890	Y.-Z. Chen, W. C. Chan, C. P. Lau, H. S. Chu, H. L. Lee and G. Jia, *Organometallics* **16** (1997) 1241.
891	N.-Y. Sun and S. J. Simpson, *J. Organomet. Chem.* **434** (1992) 341.
892	H. Katayama, T. Yoshida and F. Ozawa, *J. Organomet. Chem.* **562** (1998) 203.
893	M. S. Sanford, L. M. Henling and R. H. Grubbs, *Organometallics* **17** (1998) 5384.
894	C. Gemel, G. Trimmel, C. Slugovc, S. Kremel, K. Mereiter, R. Schmid and K. Kirchner, *Organometallics* **15** (1996) 3998.
895	C. Slugovc, K. Mereiter, R. Schmid and K. Kirchner, *J. Am. Chem. Soc.* **120** (1998) 6175.
896	C. Slugovc, D. Doberer, C. Gemel, R. Schmid, K. Kirchner, B. Winkler and F. Steltzer, *Monatsh. Chem.* **129** (1998) 221.
897	C. Gemel, P. Wiede, K. Mereiter, V. N. Sapunov, R. Schmid and K. Kirchner, *J. Chem. Soc. Dalton Trans.* (1996) 4071.
898	G. Trimmel, C. Slugovc, P. Wiede, K. Mereiter, V. N. Sapunov, R. Schmid and K. Kirchner, *Inorg. Chem.* **36** (1997) 1076.
899	C. Slugovc, P. Wiede, R. Schmid and K. Kirchner, *Organometallics*, **16** (1997) 2768.
900	C. Slugovc, K. Mauthner, M. Kacetl, K. Mereiter, R. Schmid and K. Kirchner, *Chem. Eur. J.* **4** (1998) 2043.
901	C. Slugovc, V. N. Sapunov, P. Wiede, K. Mereiter, R. Schmid and K. Kirchner, *J. Chem. Soc., Dalton Trans.* (1997) 4209.
902	C. Gemel, G. Kickelbick, R. Schmid and K. Kirchner, *J. Chem. Soc. Dalton Trans.* (1997) 2113.

903	C. Slugovc, K. Mereiter, E. Zobetz, R. Schmid and K. Kirchner, *Organometallics* **15** (1996) 5275.
904	C. Gemel, K. Mereiter, R. Schmid and K. Kirchner, *Organometallics* **16** (1997) 2623.
905	C. Slugovc, K. Mereiter, R. Schmid and K. Kirchner, *Organometallics* **17** (1998) 827.
906	K. Hiraki, N. Ochi, T. Kitamura, Y. Sasada and S. Shinoda, *Bull. Chem. Soc. Jpn.* **55** (1982) 2356.
907	T. Tanase, N. Takeshita, S. Yano, I. Kinoshita and A. Ichimura, *New J. Chem.* (1998) 927.
908	D. A. Freedman, T. P. Gill, A. M. Blough, R. S. Koefod and K. R. Mann, *Inorg. Chem.* **36** (1997) 95.
909	A. M. LaPointe and R. R. Schrock, *Organometallics* **12** (1993) 3379.
910	A. M. LaPointe, R. R. Schrock and W. M. Davis, *J. Am. Chem. Soc.* **117** (1995) 4802.
911	W. S. Ng, G. Jia, M. Y. Hung, C. P. Lau, K. Y. Wong and L. Wen, *Organometallics* **17** (1998) 4556.
912	J. L. Koch and P. A. Shapley, *Organometallics* **16** (1997) 4071.
913	T. J. Crevier and J. M. Mayer, *J. Am. Chem. Soc.* **120** (1998) 5595.
914	T. J. Crevier and J. M. Mayer, *Angew. Chem. Int. Ed.* **37** (1998) 1891.
915	K. D. Demadis, E.-S. El-Samanoudy, G. M. Coia and T. J. Meyer, *J. Am. Chem. Soc.* **121** (1999) 535.
916	E.-S. El-Samanoudy, K. D. Demadis, T. J. Meyer and P. S. White, *Inorg. Chem.* submitted
917	K. D. Demadis, E.-S. El-Samanoudy, T. J. Meyer and P. S. White, *Polyhedron,* submitted
918	K. D. Demadis, E.-S. El-Samanoudy, T. J. Meyer and P. S. White, *Inorg. Chem,* submitted
919	T. J. Meyer, personal communication
920	T. J. Crevier, S. Lovell and J. M. Meyer, *Chem.Commun.* (1998) 2371.
921	J. P. Jesson, *J. Chem. Phys.* **45** (1966) 1049.
922	J. P. Jesson, *J. Chem. Phys.* **47** (1967) 582.
923	J. P. Jesson, *J. Chem. Phys.* **47** (1967) 579.
924	D. R. Eaton, *Can. J. Chem.* **47** (1969) 2645.
925	A. N. Kitaigorodskii and A. N. Belayev, *Z. Naturforsch.* **40a** (1985) 1271.
926	A. N. Kitaigorodskii and O. A. Chamayeva, *Izvest. Akad. Nauk S.S.S.R. Ser. Khim.* (1988) 1756.
927	A. N. Kitaigorodskii, *Izv. Akad. Nauk. S.S.S.R. , Ser. Khim.* **36** (1987) 2183.
928	A. Hayashi, K. Nakajima and M. Nonoyama, *Polyhedron* **16** (1997) 4087.
929	O. J. Curnow and B. K. Nicholson, *J. Organomet. Chem.* **267** (1984) 257.

REFERENCES

930 O. J. Curnow and B. K. Nicholson and M. J. Severinsen, *J. Organomet. Chem.* **388** (1984) 379.
931 D. M. Schubert, C. B. Knobler, S. Trofimenko and M. F. Hawthorne, *Inorg. Chem.* **29** (1990) 2364.
932 R. B. King and A. Bond, *J. Am. Chem. Soc.* **96** (1974) 1334.
933 S. Trofimenko, *Inorg. Chem.* **10** (1971) 1372.
934 R. B. King and A. Bond, *J. Organomet. Chem.* **73** (1974) 115.
935 D. J. O'Sullivan and F. J. Lalor, *J. Organomet. Chem.* **65** (1974) C47.
936 M. Cocivera, T. J. Desmond, G. Ferguson, B. Kaitner, F. J. Lalor, D. J. O'Sullivan, M. Parvez and B. Ruhl, *Organometallics* **1** (1982) 1132.
937 M. Cocivera, G. Ferguson, B. Kaitner, F. J. Lalor and D. J. O'Sullivan, *Organometallics* **1** (1982) 1125.
938 R. G. Ball, C. K. Ghosh, J. K. Hoyano, A. D. McMaster and W. A. G. Graham, *J. Chem. Soc., Chem. Commun.* (1989) 341.
939 J.-C. Chambron, E. M. Eichhorn, T. S. Franczyk and D. M. Stearns, *Acta Crystallogr.* **C47** (1991) 1732.
940 S. M. Socol and D. W. Meek, *Inorg. Chim. Acta* **101** (1985) L45.
941 R. Bonnaire, D. Davoust and N. Platzer, *Organic Magnetic Resonance* **22** (1984) 80.
942 M. Cocivera, G. Ferguson, F. J. Lalor and P. Szczecinski, *Organometallics* **1** (1982) 1139.
943 M. Cocivera, G. Ferguson, R. E. Lenkinski and P. Szczecinski, *J. Magn. Reson.* **46** (1982) 168.
944 D. D. Wick and W. D. Jones, *Inorg. Chem.* **36** (1997) 2723.
945 C. K. Ghosh and W. A. G. Graham, *J. Am. Chem. Soc.* **109** (1987) 4726.
946 C. Barrientos, C. K. Ghosh, W. A. G. Graham and M. J. Thomas, *J. Organomet. Chem.* **394** (1990) C31.
947 C. K. Ghosh, D. P. S. Rodgers and W. A. G. Graham, *J. Chem. Soc., Chem. Commun.* (1988) 1511.
948 C. K. Ghosh and W. A. G. Graham, *J. Am. Chem. Soc.* **111** (1989) 375.
949 A. A. Purwoko and A. J. Lees, *Inorg. Chem.* **34** (1995) 424.
950 A. A. Purwoko, D. P. Drolet and A. J. Lees, *J. Organomet. Chem.* **504** (1995) 107.
951 A. A. Purwoko and A. J. Lees, *Inorg. Chem.* **35** (1996) 675.
952 A. A. Purwoko, S. D. Tibensky and A. J. Lees, *Inorg. Chem.* **35** (1996) 7049.
953 P. E. Bloyce, J. Mascetti and A. J. Rest, *J. Organomet. Chem.* **444** (1993) 223.
954 S. E. Bromberg, H. Yang, M. C. Asplund, T. Lian, B. K. McNamara, K. T. Kotz, J. S. Yeston, M. Wilkens, H. Frei, R. G. Bergman and C. B. Harris, *Science* **278** (1997) 260.
955 A. J. Lees, *J. Organomet.Chem.* **554** (1998) 1.

956 M. Paneque, S. Taboada and E. Carmona, *Organometallics* **15** (1996) 2678.
957 W. J. Jones and E. T. Hessell, *J. Am. Chem. Soc.* **115** (1993) 554.
958 W. D. Jones and F. J. Feher, *Organometallics* **2** (1983) 686.
959 W. D. Jones, R. P. Duttweiler, Jr., F. J. Feher and E. T. Hessell, *New J. Chem.* **13** (1989) 725.
960 W. D. Jones and E. T. Hessell, *J. Am. Chem. Soc.* **114** (1992) 6087.
961 W. D. Jones and E. T. Hessell, *Inorg. Chem.* **30** (1991) 778.
962 E. T. Hessell and W. D. Jones, *Organometallics* **11** (1992) 1496.
963 P. J. Pérez, M. L. Poveda and E. Carmona, *Angew. Chem. Int. Ed. Engl.* **34** (1995) 231.
964 R. Jimenéz-Cataño, S. Niu and M. B. Hall, *Organometallics* **16** (1997) 1962.
965 N. G. Connelly, D. J. H. Emslie, B. Metz, A. G. Orpen and M. J. Quayle, *Chem. Commun.* (1996) 2289.
966 V. Chauby, C. S. Le Berre, P. Kalck, J.-C. Daran and G. Commenges, *Inorg. Chem.* **35** (1996) 6354.
967 G. E. Herberich and U. Büschges, *Chem. Ber.* **122** (1989) 615.
968 W. J. Oldham, Jr. and D. M. Heinekey, *Organometallics* **16** (1997) 467.
969 H. Katayama, K. Yamamura, Y. Miyaki and F. Ozawa, *Organometallics* **16** (1997) 4497.
970 A. F. Hill, A. J. P. White, D. J. Williams and J. D. E. T. Wilton-Ely, *Organometallics* **17** (1998) 3152.
971 D. W. Wick, T. O. Northcutt, R. J. Lachicotte and W. D. Jones, *Organometallics* **17** (1998) 4484.
972 U. E. Bucher, T. Lengweiler, D. Nanz, W. Von Philipsborn and L. M. Venanzi, *Angew. Chem., Int. Ed. Engl.* **29** (1990) 548.
973 J. Eckert, A. Albinati, U. E. Bucher and L. M. Venanzi, *Inorg. Chem.* **35** (1996) 1292.
974 R. Gelabert, M. Moreno, J. M. Lluch and A. Lledós, *Organometallics* **16** (1997) 3805.
975 W. J. Oldham, Jr., A. S. Hinkle and D. M. Heinekey, *J. Am. Chem. Soc.* **119** (1997) 11028.
976 S. Ikeda, Y. Maruyama and F. Ozawa, *Oganometallics* **17** (1998) 3770.
977 A. Albinati, U. E. Bucher, V. Gramlich, O. Renn, H. Rüegger and L. M. Venanzi, *Inorg. Chim. Acta* **284** (1999) 191.
978 E. Gutiérrez-Puebla, A. Monge, M. C. Nicasio, P. J. Pérez, M. L. Poveda, L. Rey, C. Ruíz and E. Carmona, *Inorg. Chem.* **37** (1998) 4538.
979 A. Albinati, M. Bovens, H. Rüegger and L. M. Venanzi, *Inorg. Chem.* **36** (1997) 5991.
980 F. M. Alias, M. L. Poveda, M. Sellin and E. Carmona, *J. Am. Chem. Soc.* **120** (1998) 5816.
981 M. J. Fernandez, M. J. Rodriguez and L. A. Oro, *Polyhedron* **14** (1991) 1595.

982	M. Bovens, T. Gerfin, V. Gramlich, W. Petter, L. M. Venanzi, M. T. Haward, S. A. Jackson and O. Eisenstein, *New J. Chem.* **16** (1992) 337.
983	M. J. Fernández, M. J. Rodriguez, L. A. Oro and F. J. Lahoz, *J. Chem. Soc., Dalton Trans.* (1989) 2073.
984	A. Ferrari, E. Polo, H. Rüegger, S. Sostero and L. M. Venanzi, *Inorg. Chem.* **35** (1996) 1602.
985	A. Ferrari, M. Merlin, S. Sostero, O. Traverso, H. Rüegger and L. M. Venanzi, *Helv. Chim. Acta* **81** (1998) 2127.
986	M. A. Ciriano, M. J. Fernández, J. Modrego, M. J. Rodríguez and L. A. Oro, *J. Organomet. Chem.* **443** (1993) 249.
987	R. S. Tanke and R. H. Crabtree, *Inorg. Chem.* **28** (1989) 3444.
988	P. J. Pérez, M. L. Poveda and E. Carmona, *J. Chem. Soc., Chem. Commun.* (1992) 8.
989	P. J. Pérez, M. L. Poveda and E. Carmona, *J. Chem. Soc., Chem. Commun.* (1992) 558.
990	O. Boutry, E. Gutiérrez, A. Monge, M. C. Nicasio, P. J. Pérez and E. Carmona, *J. Am. Chem. Soc.* **114** (1992) 7288.
991	E. Gutiérrez, A. Monge, M. C. Nicasio, M. L. Poveda and E. Carmona, *J. Am. Chem. Soc.* **116** (1994) 791.
992	M. Paneque, M. L. Poveda, L. Rey, S. Taboada, E. Carmona and C. Ruiz, *J. Organomet. Chem.* **504** (1995) 147.
993	M. Paneque, M. L. Poveda, V. Salazar, S. Taboada and E. Carmona, *Organometallics* **18** (1999) 139.
994	Y. Alvarado, P. J. Daff, P. J. Pérez, M. L. Poveda, R. Sánchez-Delgado and E. Carmona, *Organometallics* **15** (1996) 2192.
995	M. J. Fernandez, J. Modrego, M. J. Rodriguez and L. A. Oro, *J. Organomet. Chem.* **438** (1992) 337.
996	M. J. Fernandez, M. J. Rodriguez, M. C. Santamaria and L. A. Oro, *J. Organomet. Chem.* **441** (1992) 155.
997	D. M. Heinekey, W. J. Oldham, Jr. and J. S. Wiley, *J. Am. Chem. Soc.* **118** (1996) 12842.
998	O. Boutry, M. L. Poveda and E. Carmona, *J. Organomet. Chem.* **528** (1997) 143.
999	Y. Alvarado, O. Boutry, E. Gutierrez, A. Monge, M. C. Nicasio, M. L. Poveda, P. J. Pérez, C. Ruiz, C. Bianchini and E. Carmona, *Chem. Eur. J.* **3** (1997) 860.
1000	E. Gutiérrez-Puebla, A. Monge, M. Paneque, M. L. Poveda, V. Salazar and E. Carmona, *J. Am. Chem. Soc.* **121** (1999) 248.
1001	D. M. Heinekey and W. J. Oldham, Jr., *J. Am. Chem. Soc.* **116** (1994) 3137.
1002	M. Paneque, M. L. Poveda and S. Taboada, *J. Am. Chem. Soc.* **116** (1994) 4519.

1003 H. Lehmkuhl, J. Näser, G. Mehler, T. Keil, F. Danowski, R. Benn, R. Mynott, G. Schroth, B. Gabor, C. Krüger and P. Betz, *Chem. Ber.* **124** (1991) 441.
1004 T. R. Belderraín, E. Gutiérrez, A. Monge, M. C. Nicasio, M. Paneque, M. L. Poveda and E. Carmona, *Organometallics* **12** (1993) 4431.
1005 E. Gutiérrez, S. A. Hudson, A. Monge, M. C. Nicasio, M. Paneque and C. Ruiz, *J. Organomet. Chem.* **551** (1998) 215.
1006 A. J. Canty, H. Jin, A. S. Roberts, P. R. Traill, B. W. Skelton and A. H. White, *J. Organomet. Chem.* **489** (1995) 153.
1007 J. Bielawski, T. G. Hodgkins, W. J. Layton, K. Niedenzu, P. M. Niedenzu and S. Trofimenko, *Inorg. Chem.* **25** (1986) 87.
1008 K. Ohkita, H. Kurosawa, T. Hasegawa, T. Shirafuji and I. Ikeda, *Inorg. Chim. Acta* **198-200** (1992) 275.
1009 S. Watanabe and H. Kurosawa, *Organometallics* **17** (1998) 479.
1010 A. J. Canty, N. J. Minchin, L. M. Engelhardt, B. W. Skelton and A. H. White, *J. Chem. Soc., Dalton Trans.* (1986) 645.
1011 M. K. Das, K. Niedenzu and S. Roy, *Inorg. Chim. Acta* **150** (1988) 47.
1012 M. Onishi, K. Hiraki, H. Konda, Y. Ishida, Y. Ohama and Y. Uchibori, *Bull. Chem. Soc. Jpn.* **59** (1986) 201.
1013 M. Onishi and K. Hiraki, *Inorg. Chim. Acta* **224** (1994) 131.
1014 M. Onishi, Y. Ohama, K. Sugimura and K. Hiraki, *Chem. Lett., Chem. Soc. Jpn.* (1976) 955.
1015 O. M. Abu Salah, M. I. Bruce, P. J. Lohmeyer, C. L. Raston, B. W. Skelton, and A. H. White, *J. Chem. Soc., Dalton Trans.* (1981) 962.
1016 K. D. Gallicano and N. L. Paddock, *Can. J. Chem.* **60** (1982) 521.
1017 M. Onishi, K. Hiraki, M. Shironita, Y. Yamaguchi and S. Nakagawa, *Bull. Chem. Soc. Jpn.* **53** (1980) 961.
1018 M. Onishi, T. Ito and K. Hiraki, *J. Organomet. Chem.* **209** (1981) 123.
1019 M. Onishi, H. Yamamoto and K. Hiraki, *Bull. Chem. Soc. Jpn.* **53** (1980) 2540.
1020 J.-M. Valk, F. Maassarani, P. van der Sluis, A. L. Spek, J. Boersma and G. van Koten, *Organometallics* **13** (1994) 2320.
1021 A. J. Canty, J. L. Hoare, B. W. Skelton, A. H. White and G. van Koten, *J. Organomet. Chem.* **552** (1998) 23.
1022 M. Onishi, K. Hiraki, T. Itoh and Y. Ohama, *J. Organomet. Chem.* **254** (1983) 381.
1023 M. Onishi, K. Hiraki, A. Ueno, Y. Yamaguchi and Y. Ohama, *Inorg. Chim. Acta* **82** (1984) 121.
1024 T. Tanase, T. Fukushima, T. Nomura, Y. Yamamoto and K. Kobayashi, *Inorg. Chem.* **33** (1994) 32.
1025 A. J. Canty and H. Jin, *J. Organomet. Chem.* **565** (1998) 135.
1026 A. J. Canty and P. R. Traill, *J. Organomet. Chem.* **435** (1992) C8.
1027 A. J. Canty, R. T. Honeyman, A. S. Roberts, P. R. Traill, R. Colton, B. W. Skelton and A. H. White, *J. Organomet. Chem.* **471** (1994) C8.

1028 A. J. Canty, H. Jin, A. S. Roberts, B. W. Skelton, P. R. Traill, and A. H. White, *Organometallics* **14** (1995) 199.
1029 A. J. Canty, S. D. Fritsche, H. Jin, B.W. Skelton and A. H. White, *J. Organomet. Chem.* **490** (1995) C18.
1030 A. J. Canty, H. Jin, A. S. Roberts, B. W. Skelton and A. H. White, *Organometallics* **15** (1996) 5713.
1031 A. J. Canty, H. Jin, B. W. Skelton and A. H. White, *J. Organomet. Chem.* **503** (1995) C16.
1032 H. C. Clark and L. E. Manzer, *J. Chem. Soc. Chem. Commun.* (1973) 870.
1033 H. C. Clark and L. E. Manzer, *Inorg. Chem.* **13** (1974) 1291.
1034 H. C. Clark and L. E. Manzer, *Inorg. Chem.* **13** (1974) 1996.
1035 P. E. Rush and J. D. Oliver, *J. Chem. Soc., Chem. Commun.* (1974) 996.
1036 J. D. Oliver and P. E. Rush, *J. Organomet. Chem.* **104** (1976) 117.
1037 L. E. Manzer, *Inorg. Chem.* **15** (1976) 2354.
1038 L. E. Manzer and P. Z. Meakin, *Inorg. Chem.* **15** (1976) 3117.
1039 B. W. Davies and N. C. Payne, *Inorg. Chem.* **13** (1974) 1843.
1040 J. D. Oliver and N. C. Rice, *Inorg. Chem.* **15** (1976) 2741.
1041 D. L. Reger, J. C. Baxter and L. Lebioda, *Inorg. Chim. Acta* **165** (1989) 201.
1042 R. B. King and A. Bond, *J. Am. Chem. Soc.* **96** (1974) 1338.
1043 S. Roth, V. Ramamoorthy and P. R. Sharp, *Inorg. Chem.* **29** (1990) 3345.
1044 A. J. Canty, A. Dedieu, H. Jin, A. Milet and M. K. Richmond, *Organometallics* **15** (1996) 2845.
1045 A. J. Canty, S. D. Fritsche, H. Jin, J. Patel, B. W. Skelton and A. H. White, *Organometallics* **16** (1997) 2175.
1046 D. D. Wick and K. I. Goldberg, *J. Am. Chem. Soc.* **119** (1997) 10235.
1047 A. Murphy, B. J. Hathaway and T. J. King, *J. Chem. Soc., Dalton Trans.* (1979) 1646.
1048 N. Kitajima, Y. Moro-oka, A. Uchida, Y. Sasada and Y. Ohashi, *Acta Crystallogr.* **C44** (1988) 1876.
1049 R. E. Marsh, *Acta Crystallogr.* **C45** (1989) 1269.
1050 J. L. Wootton, J. I. Zink, G. D. Fleming and M. C. Vallette, *Inorg. Chem.* **36** (1997) 789.
1051 O. M. Abu Salah , M. I. Bruce and C. Hameister, *Inorg. Syn.* **21** (1982) 107.
1052 M. I. Bruce and A. P. P. Ostazewski, *J. Chem. Soc., Dalton Trans.* (973) 2433.
1053 K. W. Yang, Y. Q. Yin and D. S. Jin, *Chinese Chem. Lett.* **5** (1994) 341.
1054 M. R. Churchill, B. G. DeBoer, F. J. Rotella, O. M. Abu Salah and M. I. Bruce, *Inorg. Chem.* **14** (1975) 2051.
1055 O. M. Abu Salah and M. I. Bruce, *J. Organomet. Chem.* **87** (1975) C15.

1056 C. S. Arcus, J. L Wilkinson, C. Mealli, T. J. Marks and J. A. Ibers, *J. Am. Chem. Soc.* **96** (1974) 7564.

1057 G. G. Lobbia, C. Pettinari, C. Santini, M. Colapietro, P. Cecchi, *Polyhedron* **16** (1997) 207.

1058 G. G. Lobbia, C. Pettinari, F. Marchetti, B. Bovio and P. Cecchi, *Polyhedron* **15** (1996) 881.

1059 C. Mealli, C. S. Arcus, J. L. Wilkinson, T. J. Marks and J. Ibers, *J. Am. Chem. Soc.* **98** (1976) 711.

1060 O. M. Abu Salah, M. I. Bruce, P. J. Lohmeyer, C. L. Raston, B. W. Skelton and A. J. White, *J. Chem. Soc., Dalton Trans.* (1981) 962.

1061 O. M. Abu Salah and M. I. Bruce, *Aust. J. Chem.* **30** (1977) 2293.

1062 R. F. Sterling, and C. Kutal, *Inorg. Chem.* **19** (1980) 1502.

1063 J. S. Thompson, T. J. Marks and J. A. Ibers, *Proc. Natl. Acad. Sci. USA*, **74** (1977) 3114 .

1064 J. S. Thompson, T. J. Marks and J. A. Ibers, *J. Am. Chem. Soc.* **101** (1979) 4180.

1065 J. S. Thompson, R. L. Harlow and J. F. Whitney, *J. Am. Chem. Soc.* **105** (1983) 3522.

1066 J. S. Thompson and J. F. Whitney, *Acta Crystallogr.* **C40** (1984) 756.

1066a J. S. Thompson, *Biol. Inorg. Copper Chem., Proc. Conf. Copper Coord. Chem., 2nd* (Pub. 1986) **2** (1984) 1.

1067 P. J. Pérez, M. Brookhart and J. L. Templeton, *Organometallics* **12** (1993) 261.

1068 J. S. Thompson, *J. Am. Chem. Soc.* **106** (1984) 4057.

1069 N. Kitajima, T. Koda and Y. Moro-oka, *Chem. Letters* (1988) 347.

1070 N. Kitajima, T. Koda, S. Hashimoto, T. Kitagawa and Y. Moro-oka, *J. Chem. Soc., Chem. Commun.* (1988) 151.

1071 N. Kitajima, T. Koda, Y. Iwata and Y. Moro-oka, *J. Am. Chem. Soc.* **112** (1990) 8833.

1072 N. Kitajima, T. Koda, S. Hashimoto, T. Kitagawa and Y. Moro-oka, *J. Am. Chem. Soc.* **113** (1991) 5664.

1073 A. Bérces, *Inorg. Chem.* **36** (1997) 4831.

1074 K. Fujisawa, Y. Moro-oka and N. Kitajima, *J. Chem. Soc., Chem. Commun.* (1994) 623.

1075 M. H. W. Lam, Y.-Y. Tang, K.-M. Fung, X.-Z. You and W.-T. Wong, *Chem. Commun.* (1997) 957.

1076 O. M. Abu Salah and M. I. Bruce, *J. Organomet. Chem.* **87** (1974) C15.

1077 O. M. Abu Salah, G. S. Ashby, M. I. Bruce, E. A. Pederzolli and J. D. Walsh, *Aust. J. Chem.* **32** (1979) 1613.

1078 M. I. Bruce and J. D. Walsh, *Aust. J. Chem.* **32** (1979) 2753.

1079 C. Santini, G. G. Lobbia, C. Pettinari, M. Pellei, G. Valle and S. Calogero, *Inorg. Chem.* **37** (1998) 890.

1080 C. Santini, C. Pettinari, G. G. Lobbia, D. Leonesi, G. Valle and S. Calogero, *Polyhedron* **17** (1998) 3201.

1081 Effendy, G. G. Lobbia, C. Pettinari, C. Santini, B. W. Skelton and A. H.White, *J. Chem. Soc., Dalton Trans.* (1998) 2739.
1082 N. F. Borkett and M. I. Bruce, *Inorg. Chim. Acta* **12** (1975) L33.
1083 N. F. Borkett, M. I. Bruce and J. D. Welsh, *Aust. J. Chem.* **33** (1980) 949.
1084 K. Nakata, S. Kawabata and K. Ichikawa, *Acta Crystallogr.* **C51** (1995) 1092.
1085 K.-W. Yang, Y.-Z. Wang, Z.-X. Huang and J. Sun, *Polyhedron* **16** (1997) 1297.
1086 A. Looney and G. Parkin, *Inorg. Chem.*, **33** (1994) 1234.
1087 P. N. V. P. Kumar and D. S. Marynick. *Inorg. Chem.* **32** (1993) 1857.
1088 A. Looney, R. Han, I. B. Gorrell, M. Cornebise, K. Yoon, G. Parkin and A. L. Rheingold, *Organometallics,* **13** (1995) 274.
1089 U. Brand, M. Rombach and H. Vahrenkamp, *Chem. Commun.* (1998) 2717.
1090 W. R. McWhinnie, Z. Monsef-Mirzai, M. C. Perry, N. Shaikh and T. A. Hamor, *Polyhedron* **12** (1993) 1193.
1091 D. L. Reger, S. S. Mason, A. L. Rheingold and R. L. Ostrander, *Inorg. Chem.* **32** (1993) 5216.
1092 D. L. Reger, S. M. Myers, S. S. Mason, A. L. Rheingold, B. S. Haggerty and P. D. Ellis, *Inorg. Chem.* **34** (1995) 4996.
1093 D. L. Reger, S. M. Myers, S. S. Mason, D. J. Darensbourg, M. W. Holtcamp, J. H. Reibenspies, A. S. Lipton and P. D. Ellis, *J. Am. Chem. Soc.* **117** (1995) 10998.
1094 D. L. Reger and S. S. Mason, *Organometallics* **12** (1993) 2600.
1095 A. J. Canty, B. W. Skelton and A. H. White, *Aust. J. Chem.* **40** (1987) 1609.
1096 G. G. Lobbia, F. Bonati, P. Cecchi and C. Pettinari, *Gazz. Chim. Ital.* **121** (1991) 355.
1097 G. G. Lobbia, P. Cecchi, F. Bonati and G. Rafaiani, *Synth. React. Inorg. Met.-Org. Chem.* **22** (1992) 775.
1098 S. Aime, G. Digilio, R. Gobetto, P. Cecchi, G. G. Lobbia and M. Camalli, *Polyhedron,* **18** (1994) 2695.
1099 G. G. Lobbia, P. Cecchi, F. Giordano and C. Santini, *J. Organomet. Chem.* **515** (1996) 213.
1100 G. G. Lobbia, P. Cecchi, R. Gobetto, G. Digilio, R. Spagna and M. Camalli, *J. Organomet. Chem.* **539** (1997) 9.
1101 J. Bielawski and K. Niedenzu, *Inorg. Chem.* **25** (1986) 85.
1102 A. H. Cowley, C. J. Carrano, R. L. Geerts, R. A. Jones and C. M. Nunn, *Angew. Chem. Int. Ed. Engl.* **27** (1988) 277.
1103 R. Han, A. Looney and G. Parkin, *J. Am. Chem. Soc.* **111** (1989) 7276.
1104 A. Looney and G. Parkin, *Polyhedron* **9** (1990) 265.
1105 D. L. Reger and Y. Ding, *Organometallics* **12** (1993) 4485.
1106 C. H. Dungan, W. Maringgele, A. Meller, K. Niedenzu, H. Nöth, J. Serwatowska and J. Serwatowski, *Inorg. Chem.* **30** (1991) 4799.

1107 D. L. Reger, D. G. Garza, A. L. Rheingold and G. P. A. Yap, *Organometallics* **17** (1998) 3624.
1108 D. L. Reger, S. S. Mason, A. L. Rheingold and R. Ostrander, *Inorg. Chem.* **33** (1994) 1803.
1109 D. L. Reger, S. S. Mason, L. B. Reger, A. L. Rheingold and R. Ostrander, *Inorg. Chem.* **33** (1994) 1811.
1110 D. L. Reger, S. S. Mason, A. L. Rheingold, B. S. Haggerty and F. P Arnold, *Organometallics* **13** (1994) 5049.
1111 D. L. Reger and P. S. Coan, *Inorg. Chem.* **34** (1995) 6226.
1112 A. Frazer, B. Piggott, M. Harman, M. Mazid and M. B. Hursthouse, *Polyhedron* **11** (1992) 3013.
1113 C. Janiak, S. Temizdemir and T. G. Scharmann, *Z. anorg. allg. Chem.* **624** (1998) 755.
1114 M. Onishi, K. Hiraki and S. Nakagawa, *Bull. Chem. Soc. Jpn.* **53** (1980) 1459.
1115 D. L. Reger and P. S. Coan, *Inorg. Chem.* **35** (1996) 258.
1116 A. C. Filippou, P. Portius and G. Kociok-Köhn, *Chem.Commun.* (1998) 2327.
1117 A. H. Cowley, R. L. Geerts, C. M. Nunn and C. J. Carrano, *J. Organomet. Chem.* **341** (1988) C27.
1118 D. L. Reger and Y. Ding, *Polyhedron* **13** (1994) 869.
1119 D. L. Reger, S. J. Knox, M. F. Huff, A. L. Rheingold and B. S. Haggerty, *Inorg. Chem.* **30** (1991) 1754.
1120 M. N. Hansen, K. Niedenzu, J. Serwatowska, J. Serwatowski and K. R. Woodrum, *Inorg. Chem.* **30** (1991) 866.
1121 D. L. Reger, M. F. Huff, S. J. Knox, R. J. Adams, D. C. Appleby and R. K. Harris, *Inorg. Chem.* **32** (1993) 4472.
1122 D. L. Reger, Y. Ding, A. L. Rheingold and R. L. Ostrander, *Inorg. Chem.* **33** (1994) 4226.
1123 K. Niedenzu, H. Nöth, J. Serwatowska and J. Serwatowski, *Inorg. Chem.* **30** (1991) 3249.
1124 S. K. Lee and B. K. Nicholson, *J. Organomet. Chem.* **309** (1986) 257.
1125 G. G. Lobbia, F. Bonati, P. Cecchi, A. Cingolani and A. Lorenzotti, *J. Organomet. Chem.* **378** (1989) 139.
1126 G. G. Lobbia, F. Bonati, P. Cecchi and D. Leonesi, *J. Organomet. Chem.* **391** (1990) 155.
1127 G. G. Lobbia, F. Bonati, P. Cecchi, A. Lorenzotti and C. Pettinari, *J. Organomet. Chem.* **403** (1991) 317.
1128 O.-S. Jung, J. H. Jeong and Y. S. Sohn, *J. Organomet. Chem.* **439** (1992) 23.
1129 O.-S. Jung, J. H. Jeong and Y. S. Sohn, *J. Organomet. Chem.* **399** (1990) 235.
1130 B. K. Nicholson, *J. Organomet. Chem.* **265** (1984) 153.

REFERENCES

1131 S. Calogero, L. Stievano, G. G. Lobbia, A. Cingolani, P. Cecchi and G. Valle, *Polyhedron,* **14** (1995) 1731.
1132 G. G. Lobbia, S. Zamponi, R. Marassi, M. Berrettoni, S. Stizza and P. Cecchi, *Gazz. Chim. Ital.* **123** (1993) 589.
1133 S. A. A. Zaidi, A. A. Hashmi and K. S. Siddiqi, *J. Chem. Research (S)* (1988) 410.
1134 D. L. Reger, M. F. Huff, A. L. Rheingold and B. S. Haggerty, *J. Am. Chem. Soc.* **114** (1992) 579.
1135 D. L. Reger, J. E. Collins, A. L. Rheingold, L. M. Liable-Sands and G. P. A. Yap, *Inorg. Chem.* **36** (1997) 345.
1136 R. L. Reger, *Synlett.* (1992) 469.
1137 A. H. Cowley, R. L. Geerts and C. M. Nunn, *J. Am. Chem. Soc.* **109** (1987) 6523.
1138 K. W. Bagnall, A. C. Tempest, J. Takats and A. P. Masino, *Inorg. Nucl. Chem. Lett.* **12** (1976) 555.
1139 M. V. R. Stainer and J. Takats, *Inorg. Chem.* **21** (1982) 4050.
1140 M. V. R. Stainer and J. Takats, *J. Am. Chem. Soc.* **105** (1983) 410.
1141 R. A. Faltynek, *J. Coord. Chem.* **20** (1989) 73.
1142 A. Seminara and A. Mesumeci, *Inorg. Chim. Acta* **95** (1984) 291.
1143 W. D. Moffat, M. V. R. Stainer and J. Takats, *Inorg. Chim. Acta* **139** (1987) 75.
1144 D. L. Reger, S. J. Knox, J. A. Lindeman and L. Lebioda, *Inorg. Chem.* **29** (1990) 416.
1145 H. Kokusen, Y. Sohrin, H. Hasegawa, S. Kihara and M Matsui, *Bull. Chem. Soc. Jpn.* **68** (1995) 172.
1146 C. Apostolidis, A. Carvalho, A. Domingos, B. Kanellakopulos, R. Maier, N. Marques, A. Pires de Matos and J. Rebizant, *Polyhedron* **18** (1998) 263.
1147 M. Onishi, K. Itoh, K. Hiraki, Y. Ohira, N. Nagaoka and K. Nishimura, *Kidorui* **26** (1995) 264.
1148 M. Onishi, N. Nagaoka, K. Hiraki and K. Itoh, *J. Alloys and Compounds* **236** (1996) 6.
1149 M. Onishi, K. Itoh and K. Hiraki, *Reports of the Faculty of Engineering, Nagasaki University,* **27** (1997) 167.
1150 M. Onishi, K. Itoh, K. Hiraki, R. Oda and K. Aoki, *Inorg. Chim. Acta* **277** (1998) 8.
1151 M. Onishi, N. Nagaoka, K. Hiraki, R. Oda and K. Itoh, *Kidorui* **28** (1996) 286.
1152 M. A. J. Moss, R. A. Kresinski, C. J. Jones and W. J. Evans, *Polyhedron* **12** (1993) 1953.
1153 G. H. Maunder, A. Sella and D. A. Tocher, *J. Chem. Soc., Chem. Commun.* (1994) 885.
1154 X. Zhang, G. R. Loppnow, R. McDonald and J. Takats, *J. Am. Chem. Soc.* **117** (1995) 7828.

1155 J. Takats, X. W. Zhang, V. W. Day and T. A. Eberspacher, *Organometallics* **12** (1993) 4286.
1156 U. Kilimann and F. T. Edelmann, *J. Organomet. Chem.* **444** (1993) C15.
1157 R. J. H. Clark, S. Y. Liu, G. M. Maunder, A. Sella and M. R. J. Elsegood, *Dalton Trans.* (1997) 2241.
1158 A. Domingos, J. Marçalo, N. Marques, A. Pires de Matos, A. Galvão, P. C. Isolani, G. Vicentini and K. Zinner, *Polyhedron* **14** (1995) 3067.
1159 C.-D. Sun and W. T. Wong, *Inorg. Chim. Acta* **255** (1997) 355.
1160 T. Sanada, T. Suzuki and S. Kasaki, *J. Chem. Soc., Dalton Trans.* (1998) 959.
1161 J. Takats, *J. Alloys and Compounds* **249** (1997) 52.
1162 K. W. Bagnall and J. Edwards, *J. Organomet. Chem.* **80** (1974) C14.
1163 K. W. Bagnall and J. Edwards, *J. Less-Common Met.* **48** (1976) 159.
1164 K. W. Bagnall, J. Edwards and F. Heatley, *Transplutonium Elements*, W. Müller and R. Lindner, Eds., North-Holland, Amsterdam (1976) 119.
1165 K. W. Bagnall, J. Edwards, J. G. H. du Preez and R. F. Warren, *J. Chem. Soc., Dalton Trans.* (1975) 140.
1166 K. W. Bagnall, A. Beheshti and F. Heatley, *J. Less-Common Met.* **61** (1978) 171.
1167 K. W. Bagnall, A. Beheshti, F. Heatley, and A. C. Tempest, *J. Less-Common Met.* **64** (1979) 267.
1168 K. W. Bagnall, A. Beheshti, J. Edwards, F. Heatley, and A. C. Tempest, *J. Chem. Soc., Dalton Trans.* (1979) 1241.
1169 I. Ahmed, K. W. Bagnall, L. Xing-Fu, P. Po-Jung, *J. Chem. Soc., Dalton Trans.* (1984) 19.
1170 D. G. Karraker, *Inorg. Chem.* **22** (1983) 503.
1171 I. Santos, J. Marçalo, N. Marques and A. Pires de Matos, *Inorg. Chim. Acta* **134** (1987) 315.
1172 I. Santos, N. Marques and A. Pires de Matos, *Inorg. Chim. Acta* **139** (1987) 89.
1173 A. Domingos, J. Marçalo, I. Santos and A. Pires de Matos, *Polyhedron*, **9** (1990) 1645.
1174 A. Domingos, A. Pires de Matos and I. Santos, *J. Less-Common Met.* **149** (1989) 279.
1175 A. Domingos, J. Marçalo and A. Pires de Matos, *Polyhedron* **11** (1992) 909.
1176 A. Domingos, A. Pires de Matos and I. Santos, *Polyhedron* **11** (1992) 1601.
1177 M. P. C. Campello, A. Domingos and I. Santos, *J. Organomet. Chem.* **484** (1994) 37.
1178 N. Marques, A. Pires de Matos, and K. W. Bagnall, *Inorg. Chim. Acta* **95** (1984) 75.
1179 N. Marques, A. Pires de Matos, and K. W. Bagnall, *Inorg. Chim. Acta* **94** (1984) 90.

REFERENCES

1180 R. G. Ball, F. Edelman, J. G. Matisons, J. Takats, N. Marques, J. Marçalo, A. Pires de Matos and K. W. Bagnall, *Inorg. Chim. Acta* **132** (1987) 137.
1181 R. Maier, J. Müller, B. Kanellakopulos, C. Apostolidis, A. Domingos, N. Marques and A. Pires de Matos, *Polyhedron* **12** (1993) 2801.
1182 N. Marques, J. Marçalo, A. Pires de Matos, K. W. Bagnall and J. Takats, *Inorg. Chim. Acta* **139** (1987) 79.
1183 N. Marques, J. Marçalo, T. Almeida, J. M. Carretas, A. Pires de Matos, K. W. Bagnall and J. Takats, *Inorg. Chim. Acta* **139** (1987) 83.
1184 N. Marques, J. Marçalo, A. Pires de Matos, I. Santos and K. W. Bagnall, *Inorg. Chim. Acta* **134** (1987) 309.
1185 A. Domingos, J. Marçalo, N. Marques and A. Pires de Matos, J. Takats and K. W. Bagnall, *J. Less-Common Met.* **149** (1989) 271.
1186 A. Domingos, J. Marçalo, N. Marques and A. Pires de Matos, *Polyhedron* **11** (1992) 501.
1187 M. Silva, N. Marques and A. Pires de Matos, *J. Organomet. Chem.* **493** (1995) 129.
1188 J. Marçalo, N. Marques, A. Pires de Matos and K. W. Bagnall, *J. Less-Common Met.* **122** (1986) 219.
1189 I. Santos, N. Marques and A. Pires de Matos, *J. Less-Common Metals* **122** (1986) 215.
1190 A. Domingos, N. Marques and A. Pires de Matos, *Polyhedron*, **9** (1990) 69.
1191 A. Domingos, J. P. Leal, J. Marçalo, N. Marques, A. Pires de Matos, I. Santos, M. Silva, B. Kanellakopulos, R. Maier, C. Apostolidis and J. A. Martinho Simões, *Eur. J. Solid State Inorg. Chem.* **28** (1991) 413.
1192 Z. Liang, A. G. Marshall, J. Marçalo, N. Marques, A. Pires de Matos, I. Santos and D. A. Well, *Organometallics* **10** (1991) 2794.
1193 A. Domingos, N. Marques, A. Pires de Mastos, I. Santos and M. Silva, *Organometallics* **13** (1994) 654.
1194 J. Collin, A. Pires de Matos and I. Santos, *J. Organomet. Chem.* **463** (1993) 103.
1195 J. P. Leal, N. Marques, A. Pires de Matos, M. J. Calhorda, A. M. Galvão and J. A. Marinho Simões, *Organometallics* **11** (1992) 1632.
1196 J. P. Leal and J. A. M. Simoes, *J. Chem. Soc., Dalton Trans.* (1994) 2687.
1197 I. Santos, N. Maques, and A. Pires de Matos, *Inorg. Chim. Acta* **139** (1987) 87.
1198 Y. Sun, R. McDonald, J. Takats, V. W. Day and T. A. Eberspacher, *Inorg. Chem.* **33** (1994) 4433.
1199 R. Krentz, Ph.D. Dissertation, University of Alberta at Edmonton (1989).
1200 A. Kremer-Aach, W. Kläui, R. Bell, A. Strerath and H. Wunderlich and D. Mootz, *Inorg. Chem.* **36** (1997) 1552.

1201 M. H. Chisholm, N. W. Eilerts and J. C. Huffman, *Inorg. Chem.* **35** (1996) 445.
1202 C. W. Eigenbrot and K. N. Raymond, *Inorg. Chem.* **20** (1981) 1553.
1203 C. W. Eigenbrot and K. N. Raymond, *Inorg. Chem.* **21** (1982) 2653.
1204 H. Schumann, P. R. Lee and J. Loebel, *Angew. Chem. Int. Ed. Engl.* **28** (1989) 1033.
1205 G. B. Deacon, B. M. Gatehouse, S. Nickel and S. N. Platts, *Aust. J. Chem.* **44** (1991) 613.
1206 J. E. Cosgriff, G. B. Deacon and B. M. Gatehouse, *Aust. J. Chem.* **46** (1993) 1881.
1207 J. E. Cosgriff, G. B. Deacon, B. M. Gatehouse, M. Hemling and H. Schumann, *Angew. Chem. Int. Ed. Engl.* **32** (1993) 874.
1208 J. E. Cosgriff, G. B. Deacon, B. M. Gatehouse, M. Hemling and H. Schumann, *Aust. J. Chem.* **47** (1994) 874.
1209 J. E. Cosgriff, G. B. Deacon, B. M. Gatehouse, P. R. Lee and H. Schumann, *Z. anorg. allg. Chem.* **622** (1996) 1399.
1210 I. A. Guzei, A. G. Baboul, G. P. A. Yap, A. L. Rheingold, H. B. Schlegel and C. H. Winter, *J. Am. Chem. Soc.,* **119** (1997) 3387.
1211 G. B. Deacon, E. E. Delbridge, B. W. Skelton and A. H. White, *Angew. Chem. Int. Ed.* **37** (1998) 2251.
1212 E. M. Holt, S. L. Holt, K. J. Watson and B. Olsen, *Cryst. Struct. Commun.* **7** (1978) 613.
1213 F. A. Cotton, C. A. Murillo and B. R. Stults, *Inorg. Chim. Acta* **22** (1977) 75.
1214 G. Parkin, *Chem. Rev.* **93** (1993) 887.
1215 J. L. Kisko, T. Hascall and G. Parkin, *J. Am. Chem. Soc.* **120** (1998) 10561.
1216 V. W. Miner and J. H. Prestegard, *J. Magn. Res.* **50** (1982) 168.
1217 E. Hertweck, F. Peters, W. Scherer and M. Wagner, *Polyhedron* **17** (1998) 1149.
1218 G. Linti, H. Nöth and R. T. Paine, *Chem. Ber.* **126** (1993) 875.
1219 N. C. Harden, J. C. Jeffery, J. A. McCleverty, L. H. Rees and M. D. Ward, *New J. Chem.* (1998) 661.
1220 P. Cecchi, G. G. Lobbia, F. Marchetti, G. Valle and S. Calogero, *Polyhedron* **13** (1994) 2173.
1221 A. S. Lipton, S. S. Mason, D. L. Reger and P. D. Ellis, *J. Am. Chem. Soc.* **116** (1994) 10182.
1222 T. J. Desmond, F. J. Lalor, G. Ferguson and M. Parvez, *J. Organomet. Chem.* **277** (1984) 91.
1223 G. G. Lobbia, P. Cecchi, S. Bartolini, C. Pettinari and A. Cingolani, *Gazz. Chim. Ital.* **123** (1993) 641.
1224 G. G. Lobbia, S. Calogero, B. Bovio and P. Cecchi, *J. Organomet. Chem.* **440** (1992) 27.

1225 C. Hannay, R. Thissen. V. Briois, M.-J. Huben-Franskin, F. Grandjean, G. J. Long and S. Trofimenko, *Inorg. Chem.* **33** (1994) 5983.
1226 G. J. Long, F. Grandjean and S. Trofimenko, *Inorg. Chem.* **32** (1993) 1055.
1227 C. G. Young, L. J. Laughlin, S. Colmanet and S. D. B. Scrofani, *Inorg. Chem.* **35** (1996) 5368.
1228 A. H. Cowley, R. L. Geerts, C. M. Nunn and S. Trofimenko, *J. Organomet. Chem.* **365** (1989) 19.
1229 I. B. Gorrell and G. Parkin, *Inorg. Chem.* **29** (1990) 2452.
1230 R. Han and G. Parkin, *Inorg. Chem.* **31** (1992) 983.
1231 R. Han, M. Bachrach and G. Parkin, *Polyhedron* **9** (1990) 1775.
1232 R. Han and G. Parkin, *J. Am. Chem. Soc.* **112** (1990) 3662.
1233 R. Han and G. Parkin, *Polyhedron* **9** (1990) 2655.
1234 M. H. Chisholm and N. W. Eilerts, *Chem. Commun.* (1996) 853.
1235 M. Madan, M. R. Bond, T. Oteno and C. J. Carrano, *Inorg. Chem.* **34** (1995) 1233.
1236 M. Cano, J. A. Campo, J. V. Heras, E. Pinilla, C. Rivas and A. Monge, *Polyhedron* **13** (1994) 2463.
1237 I. A. Koval, K. B. Yatsimirsky, S. Trofimenko and V. V. Pavlishchuk, *Teor. Eksp. Khim.* **34** (1998) 351.
1238 S. M. Carrier, C. E. Ruggiero, R. P. Houser and W. B. Tolman, *Inorg. Chem.* **32** (1993) 4889.
1239 C. E. Ruggiero, S. M. Carrier, W. E. Antholine, J. W. Whittaker, C. J. Cramer and W. B. Tolman, *J. Am. Chem. Soc.* **115** (1993) 11285.
1240 R. Alsfasser, A. K. Powell and H. Vahrenkamp, *Angew. Chem. Int. Ed. Engl.* **29** (1990) 898.
1241 R. Han, I. B. Gorrell, A. G. Looney and G. Parkin, *J. Chem. Soc., Chem. Commun.* (1991) 717.
1242 K. Yoon and G. Parkin, *J. Am. Chem. Soc.* **113** (1991) 8414.
1243 K. Yoon and G. Parkin, *Inorg. Chem.* **31** (1992) 1656.
1244 R. Alsfasser, A. K. Powell and H. Vahrenkamp, *Angew. Chem. Int. Ed. Engl.* **29** (1990) 898.
1245 R. Alsfasser and H. Vahrenkamp, *Chem. Ber.* **126** (1993) 695.
1246 T. W. Hambley, M. J. Lynch and E. S. Zvargulis, *J. Chem. Soc., Dalton Trans.* (1996) 4283.
1247 A. Looney, A. Saleh, Y. Zhang and G. Parkin, *Inorg. Chem.* **33** (1994) 1158.
1248 D. L. Reger, S. S. Mason, J. Takats, X. W. Zhang, A. L. Rheingold and B. S. Haggerty, *Inorg. Chem.* **32** (1993) 4345.
1249 H. V. R. Dias, L. Huai, W. Jin and S. G. Bott, *Inorg. Chem.* **34** (1995) 1973.
1250 A. Frazer, P. Hodge and B. Piggott, *J. Chem. Soc., Chem. Commun.* (1996) 1727.
1251 P. Hodge and B. Piggott, *Chem. Commun.* (1998) 1933.

1252 E. Gutierrez, S. A. Hudson, A. Monge, M. C. Nicasio, M. Paneque and E. Carmona, *J. Chem. Soc., Dalton Trans.* (1992) 2651.
1253 J. L. Detrich, R. Konecny, W. M. Vetter, D. Doren, A. L. Rheingold and K. H. Theopold, *J. Am. Chem. Soc.* **118** (1996) 1703.
1254 J. A. Campo, M. Cano, J. V. Heras, E. Pinilla, A. Monge and J. A. McCleverty, *J. Chem. Soc., Dalton Trans.* (1998) 3065.
1255 D. D. LeCloux and W. B. Tolman, *J. Am. Chem. Soc.* **115** (1993) 1153.
1256 D. D. LeCloux, C. J. Tokar, M. Osawa, R. P. Houser, M. C. Keyes and W. B. Tolman, *Organometallics* **13** (1994) 2855.
1257 M. C. Keyes, B. M. Chamberlain, S. A. Caltagirone, J. A. Halfen and W. B. Tolman, *Organometallics* **17** (1998) 1984.
1258 H. Brunner, U. P. Singh, T. Boeck, S. Altmann, T. Scheck and B. Wrackmeyer, *J. Organomet. Chem.* **443** (1993) C16.
1259 D. M. Eichhorn and W. H. Armstrong, *Inorg. Chem.* **29** (1990) 3607.
1260 F. Harmann, W. Kläui, A. Kremer-Aach, R. Bell, A. Strerath and H. Wunderlich and D. Mootz, *Z. Anorg. Allg. Chem.* **619** (1993) 2071.
1261 D. Sanz, M. D. Santa Maria, R. M. Claramunt, M. Cano, J. V. Heras, J. A. Campo, F. A. Ruiz, E. Pinilla and A. Monge, *J. Organomet. Chem.* **526** (1996) 341.
1262 M. A. Halcrow, J. E. Davies and P. R. Raithby, *Polyhedron* **16** (1997) 1535.
1263 J. Perkinson, S. Brodie, K. Yoon, K. Mosny, P. J. Carroll, T. V. Morgan and S. J. N. Burgmayer, *Inorg. Chem.* **30** (1991) 719.
1264 B. Domhöver, W. Kläui, A. Kremer-Aach, R. Bell and D. Mootz, *Angew. Chem. Int. Ed.* **37** (1998) 3050.
1265 R. Alsfasser, A. K. Powell, S. Trofimenko and H. Vahrenkamp, *Chem. Ber.* **126** (1993) 685.
1266 K.-W. Yang, Y.-Q. Yin, Y.-Z. Wang and Z.-X. Huang, *Huaxue Xuebao* **55** (1997) 209.
1267 K. Yang, Q. Yin and D. Jin, *Polyhedron* **14** (1995) 1021.
1268 D. J. Darensbourg, M. W. Holtcamp, M. E. Longridge, B. Khandelwal, K. K. Klausmeyer and J. H. Reibenspies, *J. Am. Chem. Soc.* **117** (1995) 318.
1269 D. J. Darensbourg, M. W. Holtcamp, B. Khandelwal, K. K. Klausmeyer and J. H. Reibenspies, *Inorg. Chem.* **34** (1995) 2389.
1270 A. S. Lipton, S. S. Mason, S. M. Myers, D. L. Reger and P. D. Ellis, *Inorg. Chem.* **35** (1996) 7111.
1271 D. J. Darensbourg, M. W. Holtcamp, B. Khandelwal, K. K. Klausmeyer and J. H. Reibenspies, *J. Am. Chem. Soc.* **117** (1995) 538.
1272 D. J. Darensbourg, S. A. Niezgoda, M. W. Holtcamp, J. D. Draper and J. H. Reibenspies, *Inorg. Chem.* **36** (1997) 2426.
1273 A. Frazer, B. Piggott, M. B. Hursthouse and M. Mazid, *J. Am. Chem. Soc.* **116** (1994) 4127.

REFERENCES

1274 G. Ferguson, M. C. Jennings, F. J. Lalor and C. Shanahan, *Acta Crystallogr.* **C47** (1991) 2079.
1275 P. Ghosh and G. Parkin, *Chem. Commun.* (1996) 1239.
1276 P. Ghosh and G. Parkin, *Polyhedron* **16**, (1997) 1255.
1277 M. Cano, J. V. Heras, A. Monge, E. Gutierrez, C. J. Jones, S. L. W. McWhinnie and J. A. McCleverty, *J. Chem. Soc., Dalton Trans.* (1992) 2435.
1278 M. Cano, J. V. Heras, A. Monge, E. Pinilla, C. J. Jones and J. A. McCleverty, *J. Chem. Soc., Dalton Trans.* (1994) 1555.
1279 J. L. Schneider, S. M. Carrier, C. E. Ruggiero, V. G. Young, Jr., and W. B. Tolman, *J. Am. Chem. Soc.* **120** (1998) 11408.
1280 A. J. Amoroso, A. M. C. Thompson, J. C. Jeffery, P. L. Jones, J. A. McCleverty and M. D. Ward, *J. Chem. Soc., Chem. Commun.* (1994) 2751.
1281 A. J. Amoroso, J. C. Jeffery, P. L. Jones, J. A. McCleverty, L. Rees, A. L. Rheingold, Y. Sun, J. Takats, S. Trofimenko, M. D. Ward and G. P. A. Yap, *J. Chem. Soc., Chem. Commun.* (1995) 1881.
1282 A. J. Amoroso, J. C. Jeffery, P. L. Jones, J. A. McCleverty and M. D. Ward, *Polyhedron* **12** (1996) 2023.
1283 A. J. Amoroso, J. C. Jeffery, P. L. Jones, J. A. McCleverty, E. Psillakis and M. D. Ward, *J. Chem. Soc., Chem. Commun.* (1995) 1175.
1284 P. J. Jones, A. J. Amoroso, J. C. Jeffery, J. A. McCleverty, E. Psillakis, L. H. Rees and M. D. Ward, *Inorg. Chem.* **36** (1997) 10.
1285 A. J. Amoroso, J. C. Jeffery, P. L. Jones, J. A. McCleverty, P. Thornton, and M. D. Ward, *Angew. Chem. Int. Ed. Engl.* **34** (1995) 1443.
1286 P. L. Jones, J. C. Jeffery, J. P. Maher, J. A. McCleverty, P. H. Rieger and M. D. Ward, *Inorg. Chem.* **36** (1997) 3088.
1287 D. A. Bardwell, J. C. Jeffery, P. L. Jones. J. A. McCleverty and M. D. Ward, *J. Chem. Soc. Dalton Trans.* (1995) 2921.
1288 G. G. Lobbia, P. Cecchi, C. Santini, S. Calogero, G. Valle and F. E. Wagner, *J. Organomet. Chem.* **513** (1996) 139.
1289 G. G. Lobbia, B. Bovio, C. Santini, C. Pettinari and F. Marchetti, *Polyhedron* **16** (1997) 671.
1290 G. G. Lobbia, C. Santini, F. Giordano, P. Cecchi and K. Coacci, *J. Organomet. Chem.* **552** (1998) 31.
1291 M. D. Olson, S. J. Rettig, A. Storr, J. Trotter and S. Trofimenko, *Acta Crystallogr.* **C47** (1991) 1543.
1292 J. C. Calabrese, P. J. Domaille, J. S. Thompson and S. Trofimenko, *Inorg. Chem.* **29** (1990) 4429.
1293 M. D. Olson, S. J. Rettig, A. Storr and S. Trofimenko, *Acta Crystallogr.* **C47** (1991) 1544.
1294 S. J. Nieter Burgmayer, personal communication
1295 N. Kitajima and Y. Moro-oka, *J. Chem. Soc., Dalton Trans.* (1993) 2665.

1296　N. Kitajima, S. Hikichi, M. Tanaka and Y. Moro-oka, *J. Am. Chem. Soc.* **115** (1993) 5496.
1297　N. Kitajima, U. P. Singh, H. Amagai, M. Osawa and Y. Moro-oka, *J. Am. Chem. Soc.* **113** (1991) 7757.
1298　N. Kitajima, H. Komatsuzaki, S. Hikichi, M. Osawa and Y. Moro-oka, *J. Am. Chem. Soc.* **116** (1994) 11596.
1299　N. Kitajima, M. Osawa, N. Tamura, Y. Moro-oka, T. Hirano, M. Hirobe and T. Nagano, *Inorg. Chem.* **32** (1993) 1879.
1300　H. Komatsuzaki, S. Ichikawa, S. Hikichi, M. Akita and Y. Moro-oka, *Inorg. Chem.* **37** (1998) 3652.
1301　M. Osawa, K. Fujisawa, N. Kitajima and Y. Moro-oka, *Chem. Lett.* (1997) 919.
1302　N. Kitajima, M. Osawa, S. Imai, K. Fujisawa, Y. Moro-Oka, K. Heerwegh, C. A. Reed and P. D. W. Boyd, *Inorg. Chem.* **33** (1994) 4613.
1303　N. Kitajima, M. Osawa, M. Tanaka and Y. Moro-oka, *J. Am. Chem. Soc.* **113** (1991) 8952.
1304　M. Osawa, M. Tanaka, K. Fujisawa, N. Kitajima and Y. Moro-oka, *Chem. Lett.* (1996) 397.
1305　H. Komatsuzaki, Y. Nagasu, K. Suzuki, T. Shibasaki, M. Satoh, F. Ebina, S. Hikichi, M. Akita and Y. Moro-oka, *J. Chem. Soc (Dalton)* (1998) 511.
1306　M. Osawa, U. P. Singh, M. Tanaka, Y. Moro-oka and N. Kitajima, *J. Chem. Soc., Chem. Commun.* (1993) 310.
1307　M. Ito, H. Amagai, H. Fukui, N. Kitajima and Y. Moro-oka, *Bull. Chem. Soc. Jpn.* **69** (1996) 1937.
1308　N. Kitajima, M. Ito, H. Fukui and Y. Moro-oka, *J. Am. Chem. Soc.* **115** (1993) 9335.
1309　T. Ogihara, S. Hikichi, M. Akita and Y. Moro-oka, *Inorg. Chem.* **37** (1998) 2614.
1310　N. Kitajima, H. Fukui and Y. Moro-oka, *J. Am. Chem. Soc.* **112** (1990) 6402.
1311　N. Kitajima, N. Tamura, M. Tanaka and Y. Moro-oka, *Inorg. Chem.* **31** (1992) 3342.
1312　K. Kim and S. J. Lippard, *J. Am. Chem. Soc.* **118** (1996) 4914.
1313　Y. Takahashi, M. Akita, S. Hikichi and Y. Moro-oka, *Inorg. Chem.* **37** (1998) 3186.
1314　S. Hikichi, H. Komatsuzaki, N. Kitajima, M. Akita, M. Mukai, T. Kitagawa and Y. Moro-oka, *Inorg. Chem.* **36** (1997) 266.
1315　S. Hikichi, H. Komatsuzaki, M. Akita and Y. Moro-oka, *J. Am. Chem. Soc.* **120** (1998) 4699.
1316　M. Akita, N. Shirasawa, S. Hikichi and Y. Moro-Oka, *Chem. Commun.* (1998) 973.
1317　M. Akita, T. Miyaji, S. Hikichi and Y. Moro-Oka, *Chem. Commun.* (1998) 1005.

REFERENCES

1318 M. Akita, K. Ohta, Y. Takahashi, S. Hikichi and Y. Moro-oka, *Organometallics* **16** (1997) 4121.
1319 N. Kitajima, K. Fujisawa and Y. Moro-oka, *J. Am. Chem. Soc.* **111** (1989) 8975.
1320 M. J. Baldwin, D. E. Root, J. E. Pate, K. Fujisawa, N. Kitajima and E. I. Solomon, *J. Am. Chem. Soc.* **114** (1992) 10421.
1321 S. Hikichi, M. Tanaka, Y. Moro-oka and N. Kitajima, *J. Chem. Soc., Chem. Commun.* (1994) 1737.
1322 N. Kitajima, K. Fujisawa and Y. Moro-oka, *J. Am. Chem. Soc.* **112** (1990) 3210.
1323 N. Kitajima, K. Fujisawa, M. Tanaka and Y. Moro-oka, *J. Am. Chem. Soc.* **114** (1992) 9232.
1324 D. Qiu, L. T. Kilpatrick, N. Kitajima and T. S. Spiro, *J. Am. Chem. Soc.* **116** (1994) 2585.
1325 N. Kitajima, K. Fujisawa, C. Fujimoto and Y. Moro-oka, *Chem. Lett.* (1989) 421.
1326 N. Kitajima, K. Fujisawa, S. Hikichi and Y. Moro-oka, *J. Am. Chem. Soc.* **115** (1993) 7874.
1327 N. Kitajima, T. Katayama, K. Fujisawa, Y. Iwata and Y. Moro-oka, *J. Am. Chem. Soc.* **115** (1993) 7872.
1328 S. Hikichi, M. Tanaka, Y. Moro-oka and N. Kitajima, *J. Chem. Soc., Chem. Commun.* (1992) 814.
1329 A. Looney, R. Han, K. McNeill and G. Parkin, *J. Am. Chem. Soc.* **115** (1993) 4690.
1330 M. C. Kuchta, J. B. Bonnano and G. Parkin, *J. Am. Chem. Soc.* **118** (1996) 10914.
1331 M. C. Kuchta and G. Parkin, *Inorg. Chem.* **36** (1997) 2492.
1332 M. C. Kuchta and G. Parkin, *J. Chem. Soc., Dalton Trans.* (1998) 2279.
1333 M. C. Kuchta and G. Parkin, *Main Group Chem.* **1** (1996) 291.
1334 M. C. Kuchta, H. V. R. Dias, S. G. Bott and G. Parkin, *Inorg. Chem.* **35** (1996) 943.
1335 M. C. Kuchta and G. Parkin, *J. Am. Chem. Soc.* **117** (1995) 12651.
1336 H. V. R. Dias and H.-L. Lu, *Inorg. Chem.* **34** (1995) 5380.
1337 H. V. R. Dias, H.-L. Lu, J. D. Gorden and W. Jin, *Inorg. Chem.* **35** (1996) 2149.
1338 K. Weis, M. Rombach and H. Vahrenkamp, *Eur. J. Inorg. Chem.* (1998) 263.
1339 H. V. R. Dias and W. Jin, *J. Am. Chem. Soc.* **117** (1995) 11381.
1340 H. V. R. Dias and W. Jin, *Inorg. Chem.* **35** (1996) 267.
1341 H. V. R. Dias, Z. Wang and W. Jin, *Inorg. Chem.* **36** (1997) 6205.
1342 H. V. R. Dias and W. Jin, *Inorg. Chem.* **35** (1996) 3687.
1343 C. Santini, G. G. Lobbia, M. Pellei, C. Pettinari, G. Valle and S. Calogero, *Inorg. Chim. Acta* **282**, (1998) 1.

1344 M. Cano, J. V. Heras, C. J. Jones, J. A. McCleverty and S. Trofimenko, *Polyhedron* **9** (1990) 619.
1345 M. Cano, J. V. Heras, A. Monge, E. Pinilla, E. Santamaria, H. A. Hinton, C. J. Jones and J. A. McCleverty, *J. Chem. Soc., Dalton Trans.* (1995) 2281.
1346 O. M. Reinaud, A. L. Rheingold and K. H. Theopold, *Inorg. Chem.* **33** (1994) 2306.
1347 O. M. Reinaud and K. H. Theopold, *J. Am. Chem. Soc.* **116** (1994) 6979.
1348 J. L. Detrich, O. M. Reinaud, A. L. Rheingold and K. H. Theopold, *J. Am. Chem. Soc.* **117** (1995) 11745.
1349 K. Yoon and G. Parkin, *Polyhedron* **14** (1995) 811.
1350 J. L. Kersten, R. R. Kucharczyk, G. P. A. Yap, A. L. Rheingold and K. H. Theopold, *Chem. Eur. J.* **3** (1997) 1668.
1351 A. Hess, M. R. Hörz, L. M. Liable-Sands, D. C. Lindner, A. L Rheingold and K. H. Theopold, *Angew.Chem. Int. Ed.* **38** (1999) 166.
1352 R. Alsfasser, S. Trofimenko, A. Looney, G. Parkin and H. Vahrenkamp, *Inorg. Chem.* **30** (1991) 4098.
1353 R. Alsfasser, M. Ruf, S. Trofimenko and H. Vahrenkamp, *Chem. Ber.* **126** (1993) 703.
1354 A. Looney, G. Parkin, R. Alsfasser, M. Ruf and H. Vahrenkamp, *Angew. Chem. Int. Ed. Engl.* **31** (1992) 92.
1355 M. Ruf, F. A. Schell, R. Walz and H. Vahrenkamp, *Chem. Ber. /Recueil* **130** (1997) 101.
1356 R. Walz, K. Weis, M. Ruf and H. Vahrenkamp, *Chem. Ber./Recueil* **130** (1997) 975.
1357 Brandsch, F.-A. Schell, K. Weiss, M. Ruf, B. Müller and H. Vahrenkamp, *Chem. Ber./ Recueil* **130** (1997) 283.
1358 K. Weis and H. Vahrenkamp, *Eur. J. Inorg. Chem.* (1998) 271.
1359 S. Kiani, J. R. Long and P. Stavropoulos, *Inorg. Chim. Acta* **263** (1997) 357.
1360 G. H. Maunder, A. Sella and D. A. Tocher, *J. Chem. Soc., Chem. Commun.* (1994) 2689.
1361 A. F. Hill and J. D. E. T. Wilton-Ely, *J. Chem. Soc., Dalton Trans.* (1998) 3501.
1362 A. C. Hiller, A. Sella and M. R. J. Elsegood, *J. Chem. Soc., Dalton Trans.* (1998) 3871.
1363 K. Fujisawa, M. Tanaka, Y. Moro-oka and N. Kitajima, *J. Am. Chem. Soc.* **116** (1994) 12079.
1364 S. Hikichi, T. Ogihara, K. Fujisawa, N. Kitajima, M. Akita and Y. Moro-oka, *Inorg. Chem.* **36** (1997) 4539.
1365 H. Komatsuzaki, N. Sakamoto, M. Satoh, S. Hikichi, M. Akita and Y. Moro-oka, *Inorg. Chem.* **37** (1998) 6554.
1366 C. K. Ghosh, J. K. Hoyano, R. Krentz and W. A. G. Graham, *J. Am. Chem. Soc.* **111** (1989) 5480.

REFERENCES

1367   E. Del Ministro, O. Renn, H. Rüegger, L. M. Venanzi, U. Burckhardt and V. Gramlich, *Inorg. Chim. Acta* **240** (1995) 631.
1368   M. Ruf and H. Vahrenkamp, *Chem. Ber.* **129** (1996) 1025.
1369   M. Ruf and H. Vahrenkamp, *Inorg. Chem.* **35** (1996) 6571.
1370   R. Burth and H. Vahrenkamp, *Z. Anor. Allg. Chem.* **624** (1998) 381.
1371   M. Ruf, K. Weis and H. Vahrenkamp, *Inorg. Chem.* **36** (1997) 2130.
1372   M. Ruf, K. Weis and H. Vahrenkamp, *J. Am. Chem. Soc.* **118** (1996) 9288.
1373   M. Ruf, B. C. Noll, M. D. Groner, G. T. Lee and C. G. Pierpont, *Inorg. Chem.* **36** (1997) 4860.
1374   M. Ruf, A. M. Lawrence, B. C. Noll and C. G. Pierpont, *Inorg. Chem.* **37** (1998) 1992.
1375   M. Ruf and C. G. Pierpont, *Angew. Chem. Int. Ed.* **37** (1998) 1736.
1376   K. Weis and H. Vahrenkamp, *Inorg. Chem.* **36** (1997) 5592.
1377   K. S. Siddiqi, M. A. Neyazi, Z. A. Siddiqi, S. J. Majid and S. A. A. Zaidi, *Indian J. Chem.* **21A** (1982) 932.
1378   K. S. Siddiqi, M. A. Neyazi and S. A. A. Zaidi, *Synth. React. Inorg. Met.-Org. Chem.* **12** (1981) 253.
1379   A. J. Canty, A. Dedieu, H. Jin, A. Milet, B. W. Skelton, S. Trofimenko and A. H. White, *Inorg. Chim. Acta* **287** (1999) 27.
1380   K. S. Siddiqi, S. Khan and S. A. A. Zaidi, *Synth. React. Inorg. Met.-Org. Chem.* **13** (1983) 425.
1381   S. A. A. Zaidi, S. A. Shaheer, S. R. A. Zaidi and M. Shakir, *Indian J. Chem.* **25A** (1986) 863.
1382   K. S. Siddiqi, M. A. Neyazi, S. Tabassum, Luftullah and S. A. A. Zaidi, *Indian J. Chem.* **30A** (1991) 724.
1383   S. Hikichi, M. Yoshizawa, Y. Sasakura, M. Akita and Y. Moro-oka, *J. Am. Chem. Soc.* **120** (1998) 10567.
1384   A. Wlodarczyk, R. M. Richardson, M. D. Ward, J. A. McCleverty, M. H. B. Hursthouse and S. J. Coles, *Polyhedron* **15** (1996) 27.
1385   G. G. Lobbia, P. Cecchi, R. Spagna, M. Colapietro, A. Pifferi and C. Pettinari, *J. Organomet.Chem.* **485** (1995) 45.
1386   S. Calogero, P. Valle, P. Cecchi and G. G. Lobbia, *Polyhedron* **9** (1996) 1465.
1387   E. Bernarducci, W. F. Schwindinger, J. L. Hughey IV, K. Krogh-Jespersen and H. J. Schugar, *J. Am. Chem. Soc.* **103** (1981) 1686.
1388   O.-S. Jung, J. H. Jeong and Y. S. Sohn, *Organometallics* **10** (1991) 2217.
1389   N. Yasuda, H. Kokusen, Y. Sohrin, S. Kihara and M. Matsui, *Bull. Chem. Soc. Jpn.* **65** (1992) 781.
1390   H. Kokusen, Y. Sohrin, M. Matsui, Y. Hata and H. Hasegawa, *J. Chem. Soc., Dalton Trans.* (1996) 195.
1391   G. Paolucci, S. Cacchi and L. Caglioti, *J. Chem. Soc., Perkin Trans.* (1979) 1129.
1392   J. R. Jezorek and W. H. McCurdy, Jr., *Inorg. Chem.* **14** (1975) 1939.

1393 F. Mani, *Inorg. Chim. Acta* **117** (1986) L1.
1394 P. Dapporto, F. Mani and C. Mealli, *Inorg. Chem.* **17** (1978) 1323.
1395 C. Santini-Scampucci and G. Wilkinson, *J. Chem. Soc., Dalton Trans.* (1976) 807.
1396 D. H. Williamson, C. Santini-Scampucci and G. Wilkinson, *J. Organomet. Chem.* **77** (1974) C25.
1397 M. DiVaira and F. Mani, *J. Chem. Soc., Dalton Trans.* (1990) 191.
1398 A. Bencini, M. DiVaira and F. Mani, *J. Chem. Soc., Dalton Trans.* (1991) 41.
1399 A. Pizzano, L. Sanchez, M. Altmann, A. Monge, C. Ruiz and E. Carmona, *J. Am. Chem. Soc.* **117** (1995) 1759.
1400 M. del Mar Conejo, A. Pizzano, L. J. Sánchez and E. Carmona, *J. Chem. Soc., Dalton Trans.* (1996) 3687.
1401 L. Contreras, A. Pizzano, L. Sánchez and E. Carmona, *J. Organomet. Chem.* **500** (1995) 61.
1402 A. F. Hill, J. A. K. Howard, T. P. Spaniol, F. G. A. Stone and J. Szameitat, *Angew. Chem. Int. Ed. Engl.* **28** (1989) 210.
1403 M. D. Bermudez, F. P. E. Brown and F. G. A. Stone, *J. Chem. Soc., Dalton Trans.* (1989) 1139.
1404 A. Owunwanne, H. Abdel-Dayem and T. Yacoub, *NucCompact* **18** (1987) 268.
1405 W. H. Armstrong, M. E. Roth and S. J. Lippard, *J. Am. Chem. Soc.* **109** (1987) 6318.
1406 M. M. T. Khan, P. S. Roy, K. Venkatasubramanian and N. H. Khan, *Inorg. Chim. Acta* **176** (1990) 49.
1407 M. M. T. Khan, R. K. Veera, R. C. Bhardwaj and H. C. Bajaj, *Proc. - Indian Acad. Sci., Chem. Sci.* **97** (1986) 9.
1408 A. F. Hill, A. J. P. White, D. J. Williams and J. D. E. T. Wilton-Ely, *Organometallics* **17** (1998) 4249.
1409 J. Guggenberger, C. T. Prewitt, P. Meakin, S. Trofimenko and J. P. Jesson, *Inorg. Chem.* **12** (1973) 508.
1410 H. M. Echols and D. Dennis, *Acta Crystallogr.* **B32** (1976) 1627.
1411 D. A. Clemente and M. Cingi-Biagini, *Inorg. Chem.* **26** (1987) 2350.
1412 E. Gutiérrez, M. C. Nicasio, M. Paneque, C. Ruyiz and V. Salazar, *J. Organomet. Chem.* **549** (1997) 167.
1413 O. M. Abu Salah, M. I. Bruce and J. D. Walsh, *Aust. J. Chem.* **32** (1979) 1209.
1414 M. I. Bruce, J. D. Walsh, B. W. Skelton and A. H. White, *J. Chem. Soc., Dalton Trans.* (1981) 956.
1415 D. Ajò, A. Bencini and F. Mani, *Inorg. Chem.* **27** (1988) 2437.
1416 M. M. Díaz-Requejo, M. C. Nicasio and P. P. Pérez, *Organometallics* **17** (1998) 3051.
1417 S. Trofimenko, *Inorg. Chem.* **8** (1969) 1714.
1418 D. L. Reger, S. J. Knox and L. Lebioda, *Organometallics* **9** (1990) 2218.

1419	D. L. Reger, S. J. Knox and L. Lebioda, *Inorg. Chem.* **28** (1989) 3092.
1420	R. E. Marsh, *Inorg. Chem.* **29** (1990) 1449.
1421	B. K. Nicholson, R. A. Thompson and F. D. Watts, *Inorg. Chim. Acta* **48** (1988) 101.
1422	D. K. Dey, M. K. Das and R. K. Bansal, *J. Organomet. Chem.* 535 (1997) 7.
1423	D. L. Reger, S. J. Knox, A. L. Rheingold and B. S. Haggerty, *Organometallics* **9** (1990) 2581.
1424	D. L. Reger, P.-T. Chou, S. L. Studer, S. J. Knox and M. L. Martinez, *Inorg. Chem.* **30** (1991) 2397.
1425	Y. Sun, J. Takats, T. Eberspacher and V. Day, *Inorg. Chim. Acta* **229** (1995) 315.
1426	U. B. Ceipidor, M. Tomasetti, F. Bonati, R. Curini and G. D'Ascenzo, *Thermochim. Acta* **63** (1983) 59.
1427	F. A. Cotton, M. Jeremic and A. Shaver, *Inorg. Chim. Acta* **6** (1972) 543.
1428	J. L. Calderon, F. A. Cotton and A. Shaver, *J. Organomet. Chem.* **42** (1972) 419.
1429	A. Pizzano, L. Sánchez, E. Gutiérrez, A. Monge and E. Carmona, *Organometallics* **14** (1995) 14.
1430	L. J. Laughlin, J. M. Gulbis, E. R. T. Tiekink and C. G. Young, *Aust. J. Chem.* **47** (1994) 471.
1431	M. Rombach, C. Maurer, K. Weis, E. Keller and H. Vahrenkamp, *Chem. Eur. J.* **5** (1999) 1013.
1432	F. Bonati, F. Minghetti and G. Banditelli, *J. Organomet. Chem.* **87** (1975) 365.
1433	Z. R. Reeves, K. L. V. Mann, J. C. Jeffery, J. A. McCleverty, M. D. Ward, F. Barigelletti and N. Armaroli, *J. Chem. Soc., Dalton Trans.* (1999) 349.
1434	M. O. Albers, S. F. A. Cosby, D. C. Liles, D. J. Robinson, A. Shaver and E. Singleton, *Organometallics* **6** (1987) 2014.
1435	D. L. Reger, C. A. Swift and L. Lebioda, *J. Am. Chem. Soc.* **105** (1983) 5343.
1436	R. P. Houser and W. B. Tolman, *Inorg. Chem.* **34** (1995) 1632.
1437	D. L. Reger, M. J. Pender, D. L. Caulder, L. B. Reger, A. L. Rheingold and L. M. Liable-Sands, *J. Organomet. Chem.* **512** (1996) 91.
1438	A. Carvalho, A. Domingos, P. Gaspar, N. Marques, A. Pires de Matos and I. Santos, *Polyhedron* **12** (1992) 1481.
1439	D. L. Reger, J. A. Lindeman and L. Lebioda, *Inorg. Chem.* **27** (1988) 1890.
1440	E. Frauendorfer and G. Agrifoglio, *Inorg. Chem.* **21** (1982) 4122.
1441	V. Rodriguez, J. Full, B. Donnadieu, S. Sabo-Etienne and B. Chaudret, *New. J. Chem.* **21** (1997) 847.
1442	P. Ghosh and G. Parkin, *Chem. Commun.* (1998) 413.

1443	P. Ghosh, J. B. Bonnano and G. Parkin, *J. Chem. Soc., Dalton Trans.* (1998) 2779.
1444	A. Bardwell, J. C. Jeffery, J. A. McCleverty and M. D. Ward, *Inorg. Chim. Acta* **267** (1998) 323.
1445	E. Psillakis, J. C. Jeffery, J. A. McCleverty and M. D. Ward, *Chem. Commun.* (1997) 479.
1446	J. S. Fleming, E. Psillakis, S. M. Couchman, J. C. Jeffery, J. A. McCleverty and M. D. Ward, *J. Chem. Soc., Dalton Trans.* (1998) 537.
1447	F. G. Herring, D. J. Patmore and A. Storr, *J. Chem. Soc., Dalton Trans.* (1975) 711.
1448	S. J. Rettig, M. Sandercock, A. Storr and J. Trotter, *Can. J. Chem.* **68** (1990) 59.
1449	H. M. Echols and D. Dennis, *Acta Crystallogr.* **B30** (1974) 2173.
1450	F. A. Cotton and G. N. Mott, *Inorg. Chem.* **22** (1983) 1136.
1451	F. A. Cotton and R. L. Luck, *Inorg. Chem.* **28** (1989) 3210.
1452	F. A. Cotton, B. A. Frenz and A. G. Stanislowski, *Inorg. Chim. Acta* **7** (1973) 503.
1453	F. A. Cotton and V. W. Day, *J. Chem. Soc., Chem. Commun.* (1974) 415.
1454	F. A. Cotton, B. W. S. Kolthammer and G. N. Mott, *Inorg. Chem.* **20** (1981) 3890.
1455	H. C. Clark and S. Goel, *J. Organomet. Chem.* **165** (1979) 383.
1456	L. Komorowski, W. Maringgele, A. Meller, K. Niedenzu and J. Serwatowski, *Inorg. Chem.* **29** (1990) 3845.
1457	H. C. Clark, C. R. Jablonski and K. von Werner, *J. Organomet. Chem.* **82** (1974) C51.
1458	H. C. Clark and K. von Werner, *J. Organomet. Chem.* **101** (1975) 347.
1459	B. W. Davies and N. C. Payne, *J. Organomet. Chem.* **102** (1975) 245.
1460	S. Trofimenko, J. C. Calabrese and J. S. Thompson, *Angew. Chem. Int. Ed. Engl.* **28** (1989) 205.
1461	S. Trofimenko, F. B. Hulsbergen and J. Reedijk, *Inorg. Chim. Acta* **183** (1991) 203.
1462	C. W. Heitsch, *Abstr. of Papers, 153rd Natl. ACS Meeting*, Miami Beech, Florida (1967) L109.
1463	M. Yalpani, R. Boese and K. Köster, *Chem. Ber.* **123** (1990) 1285.
1464	F. A. Cotton and C. A. Murillo, *Inorg. Chim. Acta* **17** (1976) 121.
1465	S.-F. Lush, S.-H. Wang, G.-H. Lee, S.-M. Peng, S.-L. Wang and R.-S. Liu, *Organometallics* **9** (1990) 1862.
1466	V. O. Atwood, D. A. Atwood, A. H. Cowley and S. Trofimenko, *Polyhedron* **6** (1992) 711.
1467	W. Weber and K. Niedenzu, *J. Organomet. Chem.* **205** (1981) 147.
1468	M. Cano, J. V. Heras, E. Santamaria, E. Pinilla, A. Monge, C. J. Jones and J. A. McCleverty, *Polyhedron* **12** (1993) 1711.
1469	P. Ghosh and G. Parkin, *J. Chem. Soc., Dalton Trans.* (1998) 2281.

1470 N. Marques, personal communication
1471 A. L. Companion, F. Liu and K. Niedenzu, *Inorg. Chem.* **24** (1985) 1738.
1472 P. v. R. Schleyer and M. Bühl, *Angew. Chem. Int. Ed. Engl.* **29** (1990) 304.
1473 M. Dabrowski, T. Klis, S. Lulinski, I. Madura, J. Serwatowski and J. Zachara, *J. Organomet. Chem.* **570** (1998) 31.
1474 T. G. Hodgkins, K. Niedenzu, K. S. Niedenzu and S. S. Seelig, *Inorg. Chem.* **20** (1981) 2097.
1475 M. K. Das, A. L. DeGraffenreid, K. D. Edwards, L. Komorowski, J. F. Mariategui, B.W. Miller, M. T. Mojesky and K. Niedenzu, *Inorg. Chem.* **27** (1988) 3085.
1476 K. Niedenzu and H. Nöth, *Chem. Ber.* **116** (1983) 1132.
1477 E. Hanecker, T. G. Hodgkins, K. Niedenzu and H. Nöth, *Inorg. Chem.* **24** (1985) 459.
1478 C. M. Clarke, M. K. Das, E. Hanecker, J. F. Mariategui, K. Niedenzu, P. M. Niedenzu, H. Nöth and K. R. Warner, *Inorg. Chem.* **26** (1987) 2310.
1479 K. Niedenzu and P. M. Niedenzu, *Inorg. Chem.* **23** (1984) 3713.
1480 M. Yalpani, R. Boese and R. Köster, *Chem. Ber.* **123** (1990) 713.
1481 M. K. Das, K. Niedenzu and H. Nöth, *Inorg. Chem.* **27** (1988) 1112.
1482 J. Bielawski and K. Niedenzu, *Inorg. Chem.* **25** (1986) 1771.
1483 C. Habben, L. Komorowski, W. Maringgele, A. Meller and K. Niedenzu, *Inorg. Chem.* **28** (1989) 2659.
1484 F. Jäkle, T. Priermeier and M. Wagner, *J. Chem. Soc., Chem. Commun.* (1995) 1765.
1485 F. Jäkle, T. Priermeier and M. Wagner, *Organometallics* **15** (1996) 2033.
1486 E. Herdtweck, F. Jäkle, G. Opromolla, M. Spiegler, M Wagner and P. Zanello, *Organometallics* **15** (1996) 5524.
1487 E. M. Holt, S. L. Tebben, S. L. Holt and K. J. Watson, *Acta Crystallogr.* **B33** (1977) 1986.
1488 B.C. Brock, K. Niedenzu, E. Hanecker and H. Nöth, *Acta Crystallogr.* **C41** (1985) 1458.
1489 L.-Y. Hsu, J. F. Mariategui, K. Niedenzu and S. G. Shore, *Inorg. Chem.* **26** (1987) 143.
1490 J. Bielawski, M. K. Das, E. Hanecker, K. Niedenzu and H. Nöth, *Inorg. Chem.* **25** (1986) 4623.
1491 M. Yalpani, R. Boese and R. Köster, *Chem. Ber.* **123** (1990) 1275.
1492 M. Yalpani, R. Köster, and R. Boese *Chem. Ber.* **124** (1991) 1699.
1493 J. Takats, personal communication
1494 C. M. Clarke, K. Niedenzu, P. M. Niedenzu and S. Trofimenko, *Inorg. Chem.* **24** (1985) 2648.
1495 J. Bielawski, T. G. Hodgkins, W. J. Layton, K. Niedenzu, P. M. Niedenzu and S. Trofimenko, *Inorg. Chem* **25** (1986) 87.
1496 C. E. May, K. Niedenzu and S. Trofimenko, *Z. Naturforsch.* **31b** (1976) 1662.

1497 C. E. May, K. Niedenzu and S. Trofimenko, *Z. Naturforsch.* **33b** (1978) 220.
1498 D. J. Crowther, R. A. Fisher, J. A. M. Canich, G. G. Hlatky and H. W. Turner, for EXXON Co. *US Patent Appl.* 907,098, 01 July 1992; *Chem. Abstr.* **123** (1995) 33861t.
1499 T. Obara, S. Ueki, for Toa Nenryo Kogyo K. K., *Jpn. Kokai Tokkyo Koho JP* 01 95,110 [89 95,110], 08 Oct 1987; *Chem. Abstr.* **111** (1989) 233883r.
1500 K. J. Jens, M. Tilset, A. Heuman, and J. Cockbain, for Borealis A.S., *PCT Int. Appl.* WO 97 17,379 15 May 1997; *Chem. Abstr.* **127** (1997) 51112m.
1501 W. Kläui and B. Domhover, to BASF A.-G., *PCT Int. Appl.* WO 97 23,492, 3 July 1997; *Chem. Abstr.* **127** (1997) 136149d.
1502 L. K. Johnson, A. M. Bennett;, S. D. Ittel, L. Wang, A. Parthasarathy, E. Hauptman, R. D. Simpson, J. Feldman and E. B. Coughlin, to Du Pont, *WO 98/30609*, 16 July 1998.
1503 Y. Suzuki and Y. Kiso to Mitsui Petrochemical Ind. Ltd., *EP 0 795 542 A1* 17.09.1997 Bulletin 1997/38.
1504 M. Tachibana, K. Sasaki, A. Ueda, M. Sakai, Y. Sakakibara, A. Ohno and T. Okamoto, *Chem. Lett.* (1991) 993.
1505 T. L. F. Favre, R. Hage, K. van der Heim-Rademacher, H. Koek, R. J. Martens, T. Swarthoff and M. R. P. van Vliet, to Unilever Co., *European Patent Applic.* 0 458 397 A2, 15.05.1991
1506 T. L. F. Favre, R. Hage, K. van der Heim-Rademacher, H. Koek, R. J. Martens, T. Swarthoff and M. R. P. van Vliet, to Unilever Co., *European Patent Applic.* 0 458 398 A2, 15.05.1991
1507 E. C. Plappert, K.-H. Dahmen, H. van den Bergh, T. Stumm and R. Hauert, *Proc. Mat. Res. Soc. Symp.* **337** (1994) 697.
1508 E. C. Plappert, T. Stumm, H. van den Bergh, R. Hauert, and K.-H. Dahmen, *Chem. Vap. Deposition* **3** (1997) 37.
1509 R.-R. Lang, E. C. Plappert, K.-H. Dahmen, R. Hauert, P. Nebiker and M. Döbeli, *J. Non-Crystalline Solids* **187** (1995) 430.
1510 N. Yasuda, N. Nakao, Y. Sohrin, S. Kihara and M. Matsui, *Proc. Symp. Solvent Extr.* (1987) 203.
1511 H. Kokusen, T. Ishido, Y. Sohrin, S. Kihara and M. Matsui, *Proc. Symp. Solvent Extr.* (1989) 61.
1512 A. G. Orpen, L. Brammer, F. H. Allen, O. Kennard, D. G. Watson and R. Taylor, *J. Chem. Soc., Dalton Trans.* (1989) S1.
1513 R. Colton, A. D'Agostino, J. C. Traeger and W. Kläui, *Inorg. Chim. Acta* **233** (1995) 51.
1514 J.-L. Aubagnac, R.-M. Claramunt, C. Lopez and J. Elguero, *Rapid Commun. Mass Spectr.* **5** (1991) 113.
1515 G. Bruno, G. Centineo, E. Ciliberto, S. di Bella and I. Fragalà, *Inorg. Chem.* **23** (1984) 1832.

REFERENCES

1516 P. Ghosh, P. J. Desrosiers and G. Parkin, *J. Am. Chem. Soc.* **120** (1998) 10416.
1517 V. Skagestad and M. Tilset, *J. Am. Chem. Soc.* **115** (1993) 5077.
1518 M. K. Das and A. Bhaumik, *Indian J. Chem.* **36B** (1997) 1020.
1519 T. Cecchi, F. Pucciarelli and P. Passamonti, *J. Liq. Chrom. and Re. Technol.* **20** (1997) 3329.
1520 C. M. Bastos and T. D. Ocain, *US Pat.* 5,708,022 to Procept, Inc., 28 Oct 1994.
1521 J. A. Larrabee, C. M. Alessi, E. T. Asiedu, J. O. Cook, K. R. Hoerning, L. J. Klingler, G. S. Okin, S. G. Santee and T. L. Volkert, *J. Am. Chem. Soc.* **119** (1997) 4182.
1522 S. J. Nieter Burgmayer, *Structure and Bonding* **92** (1998) 101-103.
1523 C. G. Young and A. G. Wedd, in *Encyclopedia of Inorganic Chemistry*, ed. R. B. King, Wiley, Chichester, UK (1994) 2330.
1524 N. Kitajima, *Advances in Inorganic Chemistry* **39** (1992) 1.
1525 B. Hutchinson, M. Hoffbauer and J. Takemoto, *Spectrochim. Acta* **A32** (1976) 1785.
1526 C. Bergquist and G. Parkin, *Inorg. Chem.* **38** (1999) 422.
1527 K. H. Theopold, personal communication
1528 D. D. Wick and W. D. Jones, *Organometallics* **18** (1999) 495.
1529 J. D. Jewson, L. M. Liable-Sands, G. P. A. Yap, A. L. Rheingold and K. H. Theopold, *Organometallics* **18** (1999) 300.
1530 R. G. Lawrence, C. J. Jones and R. A. Kresinski, *Inorg. Chim. Acta* **285** (1999) 283.
1531 H. A. Hinton, H. Chen, T. A. Hamor, F. S. McQuillan and C. J. Jones, *Inorg. Chim. Acta* **285** (1999) 55.
1532 S. Trofimenko, *J. Am. Chem. Soc.* **91** (1969) 2139.
1533 D. K. Dey, M. K. Das, K. Chinnakali, H.-K. Fun and I. A. Razak, *Acta Crystallogr.* **C55** (1999) 20.
1534 P. J. Desrochers, personal communication.
1535 C. J. Jones, *Chemical Society Reviews* **27** (1998) 289.
1536 W. J. Jones, personal communication.
1537 D. M. Heinekey, personal communication.
1538 R. R. Conry, G. Ji and A. A. Tipton, *Inorg. Chem.* **38** (1999) 906.
1539 R. A. Kresinski, *J. Chem. Soc., Dalton Trans.* (1999) 401.
1540 F. Malbosc, P. Kalck, J.-C. Daran and M. Etienne, *J. Chem. Soc., Dalton Trans.* (1999) 271.
1541 Y.-H. Lo, Y.-C. Lin, G.-H. Lee and Y. Wang, *Organometallics* **18** (1999) 982.
1542 L. M. Liable-Sands, personal communication.
1543 E. L. Hegg, R. Y. N. Ho and L. Que, Jr., *J. Am. Chem.Soc.* **121** (1999) 1072.
1544 N. Shirasawa, M. Akita, S. Hikichi and Y. Moro-oka, *Chem. Commun.* (1999) 417.

1545 S. Anderson, D. J. Cook and A. F. Hill, *Organometallics* **16** (1997) 5595.
1546 J. S. Thompson, T. Sorrell, T. J. Marks and J. A. Ibers, *J. Am. Chem. Soc.* **101** (1979) 4193.
1547 J. S. Thompson, *Biol. Inorg. Copper Chem., Proc. Conf. Copper Coord. Chem., 2nd* (Pub. 1986) **2** (1984) 1.
1548 R. T. Stibrany, S. Knapp, J. A. Potenza and H. J. Schugar, *Inorg. Chem.* **38** (1999) 132.
1549 D. L. Reger, J. E. Collins, R. Layland and R. D. Adams, *Inorg. Chem.* **35** (1996) 1372.
1550 G. Bellachioma, G. Cardacci, A. Macchioni, G. Reichenbach and S. Terenzi, *Organometallics* **15** (1996) 4349.
1551 D. L. Reger, J. E. Collins, M. A. Matthews, A. L. Rheingold, L. M. Liable-Sands and I. A. Guzei, *Inorg. Chem.* **36** (1997) 6266.
1552 M. Yamaguchi, T. Iida and T. Yamagishi, *Inorg. Chem. Commun.* **1** (1998) 299.
1553 J. A. McCleverty, personal communication.
1554 H. Adams, R. J. Cubbon, M. J. Sarsfield and M. J. Winter, *Chem. Commun.* (1999) 491.
1555 F. C. McQuillan, H. Chen, T. A. Hamor, C. J. Jones, H. A. Jones and R. P. Sidebotham, *Inorg. Chem.* **38** (1999) 1555.
1556 D. D. Wick, K. A. Reynolds and W. D. Jones, *J. Am. Chem. Soc.* **121** (1999) 3974.
1557 E. S. Domnina, V. N. Voropayev, G. G. Skvortsova, M. V. Sigalov, T. K. Voropayeva and G. S. Muraveiskaya, *Ko-ord. Khim.* **9** (1983) 1101.
1558 B. Buriez, I. D. Burns, A. F. Hill, A. J. P. White, D. J. Williams and J. D. E. T. Wilton-Ely, *Organometallics* **18** (1999) 1504.
1559 A. Antiñolo, F. Carillo-Hermosilla, A. E. Corrochano, F. Fernández-Baeza, M. Lanfranchi, A. Otero and M. A. Pellinghelli, *J. Organomet. Chem.* **577** (1999) 174.
1560 E. R. Humphrey, K. L. V. Williams, Z. R. Reeves, A. Behrendt, J. C. Jeffery, J. P. Maher, J. A. McCleverty and M. D. Ward, *New J. Chem.* **23** (1999) 417.
1561 D. J. Cook and A. F. Hill, *Chem. Commun.* (1997) 955.
1562 K. J. Harlow, A. F. Hill and J. D. E. T. Wilton-Ely, *J. Chem. Soc., Dalton Trans.* (1999) 285.
1563 Z. A. Siddiqi, S. Khan, K. S. Siddiqi and S. A. A. Zaidi, *Synth. React. Inorg. Met.-Org. Chem* **14** (1984) 303.
1564 G. Shengli, L. Shengming, Y. Yuanqi and Y. Kaibei, *J. Coord. Chem.* **46** (1998) 145.
1565 A. Paulo, J. Ascenso, A. Domingos, A. Galvão and I. Santos, *J. Chem. Soc., Dalton Trans.* (1999) 1293.
1566 Y. Alvarado, M. Busolo and F. López-Linares, *J. Mol. Cat. A: Chemical* **142** (1999) 163.
1567 M. Costas and A. Llobet, *J. Mol. Cat. A: Chemical* **142** (1999) 113.

REFERENCES

1568   E. R. Humphrey, Z. Reeves, J. C. Jeffery, J. A. McCleverty and M. D. Ward, *Polyhedron* **18** (1999) 1335.

Index

actinides, 95-97
activation
 of halogen 63
 of C—H bonds 71, 75, 76, 80, 81, 98-104
 of dioxygen, 11, 142
agostic bonds, 4, 5, 11, 35, 36, 40, 55, 56, 70, 95, 106, 123, 134, 136, 144, 151, 152, 155, 157-165, 167-171
aldehyde oxidoreductase, 190
aluminum,
 homoscorpionates of, Tp, Tp* 89, Tp^{tBu} 107, 119; Tp^{tBu*} 107; 119; Tp^{Ph*} 107, 124
 heteroscorpionates of, Bp^{tBu} 163; Bp^{tBu*} 163
Angelici, 11, 54
Arsenic,
 homoscorpionates of pzTp 93

Bagnall, 95
Barium,
 homoscoprionates of Tp, Tp*, pzTp 31
 heteroscorpionates of, Bp 157
Bergman, 75, 200

beryllium,
 homoscorpionates of Tp, Tp*, pzTp, 30, 205; Tp^{tBu}, 11, Tp, Tp*, pzTp, 30, 205; Tp^{tBu}, 11, 116, 117
 heteroscorpionates of, Bp 157
bite size, 107, 160
blue copper proteins, 74, 195
boron ,
 homoscoprionates of, Tp 89, EtTp 109, Tp^{Me} 114
bromoperoxidase, 189
Bruce, 85

cadmium
 homoscorpionates of, Tp, Tp*, pzTp 87-88, Tp^{Me} 113, Tp^{tBu} 11, 119; Tp^{Ph} 124, 125, 187; Tp^{Ms}, Tp^{Ms*}127; Tp^{4Cl} 132; Tp^{4Br} 132; Tp^{iPr2} 140; Tp^{tBu2} 140; Tp^{tBu,Me} 144
calcium
 homoscorpionates of, Tp, Tp*, pzTp 31
 heteroscorpionates of, Bp 157
Canty, 12
carbon
 homoscorpionates of, RTp 91

heteroscorpionates of, Et$_2$Bp 169
carbonic anhydrase, 143, 196
Carmona, 80
catalysis
    acyclic diene metathesis, 52
    addition to acetylenes, 71, 188
    carbene/nitrene transfer, 57, 86, 159, 172, 187
    decarboxylation, 124, 188
    carbene/nitrene transfer, 57, 86, 159, 172, 187
    decarboxylation, 124, 188
    dimerization
        of acetylenes, 72, 184
        of acrylonitrile, 186
    epoxidation, 68
    oligomerization
        of 1,9-decadiene, 183
        of propylene, 186
    oxidation, 68, 188
    polymerization
        of acetylenes, 71, 77
        of cyclooctene, 59, 183, 184
        of L,L-dilactide, 117, 185
        of epoxides and CO$_2$, 125, 187
        of ethylene, 185, 186
        of norbornene, 70, 184, 185
        of olefins, 60, 70, 186
        of olefins and CO, 124, 186
        of phenylacetylenes, 184
        ROMP, 60, 70
    reduction of ketones, 185
catechol dehydrogenase, 193
cerium,
    homoscorpionates of, Tp, Tp* 93-95
    heteroscorpionates of, Bp* 162; Bp^{Bipy} 166
cesium
    homoscorpionates of, Tp, Tp*, pzTp 30
chromium
    homoscorpionates of, Tp, Tp*, pzTp 36, 37, 206; Tp^{tBu,Me} 142, 143
    heteroscorpionates of, Bp 157, 158, Et$_2$Bp 167
Churchill, 10
cobalt
    homoscorpionates of, Tp, Tp*, pzTp 10, 28, 74, 75, 205-207; BuTp 109; MeS(CH$_2$)$_2$Tp 110; PhTp 110; TolTp 111; (p-BrPh)Tp 111; (C$_6$D$_5$)Tp 111; Tp^{iPr*} 114; pz^{o}Tp^{iPr} 114, 207; Tp^{tBu} 116, 117, 207; Tp^{Cbu}120; Tp^{Np} 120; Tp^{Cpe} 121; Tp^{Menth} 122; Tp^{Ph} 123, 207; Tp^{An}126; Tp$^{(4ClPh)}$ 126; Tp^{Tn} 127; Tp^{Ms}, Tp^{Ms*} 127; Tp^{Ant} 128; Tp$^{\alpha Nt}$ 128; Tp$^{\beta Nt}$ 128; Tp^{CHPh_2} 130, 131; pz^{o}Tp^{4Me} 132; Tp^{4iPr} 132; Tp^{4Cl} 132; Tp^{4Br} 132; Tp^{iPr,4Br} 134, 207; Tp^{Cy,4Br} 134; Tp^{iPr,4tBu}135; Tp^a 136; Tp^b 136; Tp^{a*} 136; Tp^{Bo,3Me} 136; Tp^{Bo,3tBu}136; Tp$^{(Bo,3tBu)*}$137; Tp^{Bn,4Ph} 137; Tp^{iPr2} 138, 139; Tp^{iPr,Me} 11, 142; Tp^{tBu,Me} 11, 142, 143, 187; Tp^{tBu,iPr} 145; Tp^{tBu,Tn} 145; Tp^{Cum,Me} 146; Tp^{Bn,Me} 148; Tp^{4Bo} 148; Tp^{4Bo,5Me} 149; Tp^{4Bo,5Et} 149; Tp^{4Bo,5tBu}149; Tp^{4Bo,5Ph}149; Tp^{4Bo,4,6Me_2} 150; Tp^{Me3} 150, 207; Tp*^{Bu}151; Tp^{Br3} 152; Tp^{Ph,Me,Ph} 152; Tp$^{(Ph,Me,Ph)*}$153; Tp^{4Bo,3Me} 153; Tp^{a*,3Me} 153;
    heteroscorpionates of, Bp 157, 158; Bp* 160; Bp^{iPr} 163; Bp^{Ph} 164; Bp^{Et2} 164; Bp^{tBu,iPr} 165; Et$_2$Bp 167; (BBN)Bp 169; F$_2$Bp* 172; (TolS)Bp* 174
Cone angles, 105
copper
    homoscorpionates of, Tp, Tp*, pzTp 28, 85, 86, 187, 194-196, 204, 205, 207; BuTp 109; PhTp 110; Tp^{Me} 113, 204; Tp^{CF3} 114; Tp^{C_2F_5} 114; Tp^{C_3F_7} 114; Tp^{tBu} 11,

118, 196, 207; pz^{o}Tp^{iPr} 5, 114;
Tp^{CyH} 121; Tp^{Menth} 122; Tp^{Ph} 124,
207; Tp^{Tol} 125; Tp$^{(4FPh)}$ 126; Tp^{Fn}
127; Tp^{Ms} 127; Tp^{Py} 130; pz^{o}Tp^{4Me}
132; Tp^{4Cl} 132; Tp^{4Br} 132; Tp^{iPr2}
11, 138, 139, 195; Tp$^{(CF_3)2}$ 141;
Tp^{Ph2} 141, 196; Tp^{tBu,Me} 142;
Tp^{iPr,4Br} 207; Tp^{Me3} 207; Tp^{Ph2}
141, 196; Tp^{tBu,Me} 142;
heteroscorpionates of, Bp 157, 159,
187; Bp^{iPr} 163; Bp^{tBu} 163, 164;
Bp$^{(CF3)2}$ 165; Bp^{Bipy} 166; Me$_2$Bp
167; Et$_2$Bp 167; Ph$_2$Bp 171;
Ph$_2$Bp^{pm} 172; (TolS)Bp* 174,
195; Tp^{Tn,Me} 148

copper proteins, 194
Cotton, 11
Curtis, 5
cytochrome c oxidase, 161, 196

deoxyhemerythrin, 192
diformazan peroxidase, 191
diformazan superoxidase, 191
dimethylsulfoxide reductase, 189
dysprosium
    homoscorpionates of, Tp, Tp* 93,
94

enzymes,
    modelling of 188-196
erbium
    homoscorpionates of, Tp, Tp* 93,
94
    heteroscorpionates of, Bp^{Bipy} 166
Etienne, 11, 12
europium
    homoscorpionates of, Tp, Tp* 93-
95; Tp^{Py} 129; Tp^{Py6Me} 130;
Tp^{tBu,Me} 144; Tp^{tBu,iPr} 144;
Tp^{CF3,Me} 145; Tp^{Cum,Me} 147;
Tp^{tBuPh,Me} 147; Tp^{*Cl} 151
    heteroscorpionates of, Bp^{Py} 165

formate dehydrogenase, 190

formylmethanofuran dehydrogenase, 190
free acids,
    3, 10, 28-30; (BBN)BpH 170;
(BBN)Bp^{Ph} 171; (BBN)Bp*H 171;
(BBN)Bp^{Ph,Me}H 171; (BBN)Bp^{Fc}H
171; (MeNCH$_2$CH$_2$NMe)BpH 173

gadolinium
    homoscorpionates of, Tp, Tp* 93,
94; Tp^{Py6Me} 130
    heteroscorpionates of, Bp^{Bipy} 166
galactose oxidase, 195
gallium
    homoscorpionates of, Tp, Tp*,
pzTp 89, 90; Tp^{tBu2} 140
    heteroscorpionates of, Bp 159; Bp*
162; Me$_2$Bp 167
germanium
    homoscorpionates of, Tp*, pzTp
91, 92;
gold
    homoscorpionates of, Tp, Tp*,
pzTp 28, 29, 87; Tp$^{(CF_3)2}$ 141
galactose oxidase, 195
Gmelin, 12
Graham, 75, 198

hafnium
    homoscorpionates of, Tp 33
hemerythrin, 67, 192
heteroscorpionates, 155-182
holmium
    homoscorpionates of, Tp, Tp* 93,
94
homoscoprionates,
    first generation 27-95
    second generation 99-154
        B-substituted 109-113
        3-monosubstituted 113-131
        4-monosubstituted 131, 132
        5-monosubstituted 132
        3,4-disubstituted 134-137
        3,5-disubstituted 137-148
        4,5-disubstituted 148-150

trisubstituted 150-153
hydrogenase, 194

indium
    homoscorpionates of, Tp, Tp*, pzTp 90, 91; Tp^{tBu} 119; Tp^{Ph} 125; Tp^{tBu2} 140
    heteroscorpionates of, Bp 159, Bp* 162

iridium
    homoscorpionates of, Tp, Tp*, pzTp 78-82, 201-203, 206; Tp^{Me} 101, 108, 206; Tp^{Me*} 101, 108; Tp^{iPr} 114, 206; Tp^{iPr*} 114, 206; Tp^{CF_3,Me} 145, 206; Tp^{Ph,Me} 146, 206; Tp^{Me3} 150, 206; Tp^{*Cl} 151, 206; Tp^{*Br} 152; Tp^{iPr2} 206
    heteroscorpionates of, Bp* 161

iron
    homoscorpionates of, Tp, Tp*, pzTp 66-68, 187, 192, 205; BuTp 109; PhTp 110; PhTp^{tBu} 110; FcTp 111; FcTp^{4SiMe_3} 111; FcTp^{Me} 111; Tp^{Me} 113; pz^{o}Tp^{iPr} 5,114; Tp^{Menth} 122; pz^{o}Tp^{4Me} 132; Tp^{4Cl} 132; Tp^{4Br} 132; Tp^a 136; Tp^{iPr2} 137-139, 192, 193; Tp^{tBu,iPr}145; Tp^{4Bo,5Me} 149; Tp^{4Bo,5Et} 149; Tp^{4Bo,4,6Me_2} 150; Tp^{Me3} 150; Tp^{*Bn}151
    heteroscorpionates of, Bp 157, 158; Et$_2$Bp 167; (pz*)Bp 176; (pz)Bp* 176

Janiak, 12

α-ketoglutarate, 68, 188
ketones,
    reduction of 69
Kirchner, 13, 71
Kitajima, 11, 12

Lalor, 11
lanthanides, 93-95, 125, 144

lanthanum
    homoscorpionates of, Tp, Tp*, pzTp 93-95; Tp^{Ph} 125; Tp^{Tn} 127;
    heteroscorpionates, Bp^{Py} 165

lead
    homoscorpionates, Tp, Tp*, pzTp 92,93, Tp^{Me} 113; Tp^{Py} 129; Tp^{4Cl} 132; Tp^{4Br} 132;
    heteroscorpionates, Bp 160, Bp^{Py} 165

ligand profile, 107
lipoxygenase, 188
lithium, 28
    homoscorpionates of, Tp, Tp*, pzTp 30

lutetium
    homoscorpionates of, Tp, Tp* 94, 95;

magnesium
    homoscorpionates of, Tp, Tp*, pzTp 30, 31, 205; BuTp 109; Tp^{tBu} 11, 117, 185; Tp^{Tol} 125; Tp^{tBuPh,Me} 147
    heteroscorpionates, Bp 157

manganese
    homoscorpionates of, Tp, Tp*, pzTp 61, 62, 191; BuTp 109; PhTp 110; Tp^{Me} 113; pz^{o}Tp^{iPr} 114; Tp^{Menth}122; Tp^{Py} 129; pz^{o}Tp^{4Me} 132; Tp^{4Cl} 132; Tp^{4Br} 132; Tp^{iPr2} 138, 177, 191; Tp$^{(CF3)2}$ 141; Tp^{tBu,iPr} 145; Tp^{Me3} 150
    heteroscorpionates, Bp 157, Bp^{Bipy} 166, Et$_2$Bp 167

manganese superoxide dismutase, 191
Marques, 12
Mayer, 62, 72
McCleverty, 11, 12, 13
metals
    deposition of, 204
    extraction of, 94, 160, 205
methane monooxygenase, 192
methanol monooxygenase, 193

methemerythrin, 192
methionine synthase, 146
mercury
    homoscorpionates of, Tp, Tp*, pzTp 88, 89; Tp^{Me} 113; Tp^{4Br} 132; Tp^{Ph2} 141; Tp^{Me3} 150; $Tp*^{Cl}$ 151;
molybdenum
    homoscorpionates of, Tp, Tp*, pzTp 10-12, 82, 184, 189, 190, 206, 207; $Tp^{iPr,4Br}$104; BuTp 109; PhTp 110; Tp^{Me}-Tp^{Me} 112, Tp^{Me} 113; pz^{oTpiPr} 114; Tp^{tBu}119; Tp^{Cy} 121; Tp^{Ph} 123; $pz^{o}Tp^{Ph}$ 125; Tp^{Tol} 125; Tp^{An} 126; $Tp^{(4ClPh)}$ 126; Tp^{Ms}, 128; Tp^{4Cl} 132; Tp^{4Br} 132; Tp^{a*} 136; Tp^{Et2} 137; $Tp^{iPr,Me}$ 142; Tp^{4Bo} 148; $Tp^{4Bo,5Me}$ 149; $Tp^{4Bo,5tBu}$149; $Tp^{4Bo,4,6Me2}$ 150; Tp^{Me3} 150; $Tp*^{Et}$ 150; $Tp*^{Bun}$151; $Tp*^{Am}$ 151; $Tp*^{Bn}$151; $Tp*^{Cl}$ 151 heteroscorpionates, Bp 157, Bp* 160, 161; Bp^{Ph} 164; Bp^{Et2} 164; Bp^{Ph2} 165, Bp^{Br3} 165, Et_2Bp 168; Ph_2Bp 171; $(i$-$PrO)Bp*$ 175
molybdopterins, 49
Moro-oka, 12

neodymium,
    homoscorpionates, Tp, Tp* 93-95
neptunium
    homoscorpionates, Tp 96
    heteroscorpionates of, Bp 160
nickel
    homoscorpionates, Tp, Tp*, pzTp 82, 194, 205, 207; $Tp^{iPr,4Br}$ 104; BuTp 109; Tp^{Me}-Tp^{Me} 112; Tp^{Me} 113; $pz^{o}Tp^{iPr}$ 5, 114; Tp^{Menth} 122, 123; Tp^{Ph} 123; Tp^{Ph} 124, 186; Tp^{Tol} 125, 186; Tp^{An} 126; $pz^{o}Tp^{4Me}$ 132; Tp^{4iPr} 132; Tp^{4Cl} 132; Tp^{4Br} 132; $Tp^{iPr,4Br}$ 134; Tp^{iPr2} 138, 139; $Tp^{tBu,Me}$ 142; $Tp^{tBu,Tn}$ 145; $Tp^{Ph,Me}$ 146; $Tp^{Cum,Me}$ 147; Tp^{Me3} 150, 207; $Tp^{Ph,Tn}$ 146
heteroscorpionates of, Bp 157, 159; Bp* 160; $H_2B(pz)(pz*)$ 162; Bp^{iPr} 163; Bp^{tBu} 163; Bp^{Ph} 164; Bp^{Et2} 164; Bu_2Bp 169; $Bp^{[2,4(OMe)2Ph]}$ 166; Et_2Bp 167; Me_2Bp 167; Bu_2Bp 169; $(BBN)Bp$ 170; Ph_2Bp 171; F_2Bp* 172; $(pz^{4CN})Bp$ 176
Niedenzu, 12
niobium
    homoscorpionates, Tp, Tp* 34-36, 185; Tp^{Cpr*} 116; $Tp*^{Cl}$ 151, 185
    heteroscorpionates of, Bp 158
nitrate reductase, 189
nitrite reductase, 49, 118, 141
nitrous oxide reductase, 161, 196
non-linear optics, 47, 53, 57

Osmium
    homoscorpionates, Tp, Tp* 72, 73
oxyhemocyanin, 86

palladium
    homoscorpionates, Tp, Tp*, pzTp 5, 28, 82, 83; Tp-Tp 112; Tp^{Ph} 123; $pz^{o}Tp^{Ph}$ 125; Tp^{Ms}, Tp^{Ms*} 128; Tp^{iPr2} 139; Tp^{4Bo} 149; $(BBN)Bp^{Fc}$ 171
heteroscorpionates of, Bp 159; Bp* 161; Et_2Bp 168; Et_2Bp^{Fc} 169
Parkin, 11, 12
phenylalanine hydroxylase, 124, 194
phosphorus
    homoscorpionates, Tp* 93
plastocyanin, 174, 195
platinum
    homoscorpionates, Tp, Tp*, pzTp 84, 85, 204; Tp^{4Bo} 148
    heteroscorpionates of, Bp 159; Bp* 161; Et_2Bp 168; Ph_2Bp 171
plutonium
    homoscorpionates, Tp, Tp* 97
polypyrazolylborates,
    definition of, 1, 3

also see "scorpionates"
potassium
    homoscorpionates of, Tp, Tp*, pzTp 30, 205, 207; Tp^{Me} 100; Tp^{Tol} 125; Tp^{An} 126; Tp$^{(4ClPh)}$ 126; Tp^{Py} 129; Tp$^{(CF_3)2}$ 141; Tp^{Ph,Me,Ph} 152
    heteroscorpionates of, Pr$_2$Bp 169; (BBN)Bp^{Me} 170
praseodymium
    homoscorpionates of, Tp, Tp* 93-95;
purple acid phosphatase, 67, 192
pyrazaboles, 29, 30 159, 177-182
    unsymmetrical, 181
    symmetrical, 178
    synthesis of, 177-179

rearrangement
    of Tp^x ligands, 3, 6, 101, 107,108
    of Bp^x ligands, 163
Reger, 11, 12
regiochemistry in ligand synthesis 99-102
rhenium
    homoscorpionates of, Tp, Tp*, pzTp 28, 62-66, 203
rhodium
    homoscorpionates of, Tp, Tp*, pzTp 75-78, 184, 198, 199, 207; MeTp^{Me} 109, 113, 206; Tp^{tBu} 117, 200; Tp^{Menth} 122; Tp^{Ph} 124, 199, 200; Tp^{Ms}, Tp^{Ms*} 128; Tp^{iPr,4Br} 134; Tp^a 136; Tp^{a*} 136; Tp^{CF_3,Me} 145, 206; Tp^{CF_3,Tn} 145; Tp^{Ph,Me} 146; Tp^{Me3} 150, 206; Tp*^{Cl} 151, 206; Tp^{4Bo,3Me} 153; Tp^{a*,3Me} 153; Tp^{iPr2} 184; Tp^{Ph2} 184; Tp$^{(CF_3)2}$ 185; Tp^{iPr} 200, 206; Tp^{iPr,4Br} 206
    heteroscorpionates of, Bp 158; Bp* 161, 200; Et$_2$Bp 168; (BBN)Bp^{Me} 170; Ph$_2$Bp 171; Ph(Me)Bp 173
ribonucleotide reductase, 67
rubidium
    homoscorpionates of, Tp, Tp*, pzTp 30
ruthenium
    homoscorpionates of, Tp, Tp*, pzTp 68-72, 184, 185, 203, 204; MeTp^{Me} 109; pz^{o}Tp^{4Me} 132; Tp^{iPr,4Br} 134;Tp^{iPr2,Br} 152
    heteroscorpionates of, Bp 158; Bp* 161; Bp$^{(CF_3)2}$ 165

samarium
    homoscorpionates of, Tp, Tp* 93-95; Tp^{Py} 129; Tp^{tBu,Me} 144
    heteroscorpionates of, Bp 160; Bp* 162; CpBp* 176
Santos, 12
scandium
    homoscorpionates of, Tp 31
scorpionates
    abbreviation system for, 5-8
    analogs of,
        via pyrazole replacement, 23,24
        via boron replacement, 24
            with Al or In, 24
            with Ga, 24
            with C, 24
            with Si, 25
            with other elements, 25
    comparison to Cp ligands, 9
    historical development, 10
    reviews of, 12, 13
    synthesis of, 13-18
Shaver, 12
silicon
    homoscorpionates of, pzTp 91
    heteroscorpionates of, Ph$_2$Bp 172
silver
    homoscorpionates of, Tp, Tp*, pzTp 86; Tp^{Me} 113, 114; pz^{o}Tp^{Me} 114; Tp^{CF_3} 114; Tp^{oAn} 126; Tp^{Py} 129; Tp^{4Br} 132; Tp$^{(CF_3)2}$ 141; Tp^{Ph2} 141

heteroscorpionates of, Bp 159;
Ph$_2$Bp 171
sodium
  homoscorpionates of,Tp, Tp*, pzTp
  30, 205, 207
  heteroscorpionates of, H$_2$B(pz)(pz*)
  162
steric effects
  quantification of, 102-107
strontium
  homoscorpionates of, Tp, Tp*,
  pzTp 31
sulfite oxidase, 49, 189

Takats, 13
tantalum
  homoscorpionates of, Tp, Tp*,
  pzTp 36
  heteroscorpionates of, Bp 158; Bp*
  161
technetium
  homoscorpionates of, Tp, Tp* 62
  heteroscorpionates of, Bp 158
terbium
  homoscorpionates of, Tp, Tp* 93,
  94; Tp^{Py6Me} 130
  heteroscorpionates of, Bp 160;
  Bp^{Py} 165
thallium
  homoscorpionates of, Tp, Tp*,
  pzTp 91, 105, 106; Tp^{Cpr} 105, 106;
  Tp^{tBu} 105, 106; Tp^{tBu,Me} 105, 106;
  Tp^{Cbu} 105, 106; Tp^{Cpe} 105, 106,
  121; Tp^{Cy} 105, 106; Tp^{Cy,4Br} 105,
  106; Tp^{iPr,4Br} 105, 106; Tp^{Trip} 122;
  pz^{o}Tp^{Ph} 125; Tp^{Tol} 125; Tp^{oAn}
  126; Tp^{Ms}, Tp^{Ms*} 107, 127; Tp^{Ant}
  128; Tp$^{(2,4(OMe)2Ph)}$ 128; Tp$^{(4ClPh)}$
  126; Tp^{CHPh_2} 130; Tp$^{CO(NC_4H_8)}$
  131; Tp^{CF_3,Tn} 145; Tp^{Br3} 105, 106,
  152; Tp^{4Bo,3Me} 105, 106; Tp$^{(CF_3)2}$
  105, 106; Tp^{3Bo,7tBu} 105, 106;
  Tp^{tBu,R} 172; Tp^{tBu,4Br} 135; Tp^{Cy,4Br}
  135; Tp^{3Bo,7Me} 136; Tp^{3Bo,7tBu} 136

  heteroscorpionates of, Bp^{tBu} 163;
  BpTrip 164; H$_2$B(pz)(pz^{tBu2}) 162;
  Bp^{Fc} 164; Bp^{Py} 165; Bp^{Bipy} 166;
  (BBN)Bp 170; Ph$_2$Bp^{pm} 172;
Templeton, 12, 55
Theopold, 11
thorium
  homoscorpionates of, Tp, Tp*,
  pzTp 95, 96; Tp^{Py} 129;
  heteroscorpionates of, Bp 160
thullium
  homoscorpionates of, Tp, 94;
tin
  homoscorpionates of, Tp, Tp*,
  pzTp 92; Tp^{Me} 114; pz^{o}Tp^{Me} 114;
  Tp^{4Me} 131, 132; pz^{o}Tp^{4Me} 132;
  Tp^{tBu,Me} 144; Tp^{4Bo,NO_2} 149;
  pz^{o}Tp^{4Bo,NO_2} 149; Tp^{Me3} 150
  heteroscorpionates of, Bp 159, 160;
  Bp^{Me} 163; Et$_2$Bp 169; Ph$_2$Bp 172;
  Cl$_2$Bp 173; Br$_2$Bp 173; I$_2$Bp 173
titanium
  homoscorpionates of, Tp, Tp* 32,
  205; Tp^{iPr} 115; Tp^{Menth*} 122, 123;
  Tp^{iPr,4Br} 134
  heteroscorpionates of, Bp 158
Tolman, 11
Trofimenko, 12
tungsten
  homoscorpionates of, Tp, Tp*,
  pzTp 11, 52-61, 183, 184, 187,
  190, 191, 206; Tp^{iPr} 115; Tp^{tBu}
  117; Tp^{Ph} 124; Tp^{*Bn}151
  heteroscorpionates of, Bp 158;
  Et$_2$Bp 168; (MeBnS)Bp* 174
tyrosine hydroxylase, 193

uranium
  homoscorpionates of, Tp, Tp* ,
  pzTp 95-97, 108; Tp^{Py} 129
  heteroscorpionates of, Bp 160; Bp*
  162; Ph$_2$Bp 172

vanadium
  homoscorpionates of, Tp, Tp* 33, 34, 186, 189
  heteroscorpionates of, Bp 158
Vahrenkamp, 11
Venanzi, 78

Ward, 13
Wedd, 12
wedge angles, 106

xanthine oxidase, 49, 50, 189, 190, 191

ytterbium
  homoscorpionates of, Tp, Tp* 93-95; Tp^{tBu,Me} 144
  heteroscorpionates of, Bp* 162
yttrium,
  homoscorpionates of, Tp, Tp* 31, 186
  heteroscorpionaters of, Bp* 162
Young, 12

zinc
  homoscorpionates of, Tp, Tp*, pzTp 87, 188, 198, 205, 207; BuTp 109; PhTp 110; Tp^{Me} 113; pz^{o}Tp^{iPr} 5, 114; Tp^{tBu} 11, 118, 197; Tp^{Menth} 122, Tp^{Menth*} 123; Tp^{Ph} 123, 124, 188; Tp^{Tol} 125; Tp^{An} 126; Tp$^{(4ClPh)}$ 126; Tp^{Ms}, Tp^{Ms*} 127; pz^{o}Tp^{4Me} 132; Tp^{4Cl} 132; Tp^a 136; Tp^{iPr2} 138-140; Tp^{tBu2} 140; Tp^{Ph2} 141; Tp^{tBu,Me} 14Tp^{tBu,Me} 11, 142, 197; Tp^{tBu,Tn} 145; Tp^{Ph,Me} 145, 146, 197; Tp^{Tol,Me} 146; Tp^{Cum,Me} 11, 146, 147, 197; Tp^{Cum,Me*} 146; Tp^{3Py,Me} 148; Tp^{3Pic,Me} 148, 197; Tp^{4Bo,5Me} 149; Tp^{Me3} 150, 207; Tp^{*Bu}151; Tp$^{Tn, Me}$ 146
  heteroscorpionates of, Bp 157, H$_2$B(pz)(pz^{tBu2}) 162; H$_2$B(pz*)(pz^{tBu2}) 163; H$_2$B(pz^{Trip})(pz^{tBu2}) 163; Bp^{iPr} 163; Bp^{tBu2} 163; Bp^{Ph} 164; Bp$^{(CF3)2}$ 165; Bp^{tBu,iPr} 165; Bp^{Bipy} 166; Me$_2$Bp 167; Et$_2$Bp 167; (BBN)Bp 170; (Ph$_2$CHO)Bp^{tBu,iPr} 174; (Ph$_2$CHS)Bp^{tBu,iPr}175; (MeO)Bp^{tBu} 175; (EtO)Bp^{tBu}175; (i-PrO)Bp^{tBu} 175; (MeO)Bp^{tBu,iPr} 175; (HCOO)Bp^{tBu,iPr} 175; (3-(CMe$_2$OH)-5-iPrpz)Bp^{iPr2} 177

Zirconium
  homoscorpionates of, Tp, Tp*, 32; iPrTp 109; BuTp 109; FcTp 111
  heteroscorpionates of, Bp 158